DATA-PROCESSING IN PHYTOSOCIOLOGY

Advances in vegetation science 1

Edited by

EDDY VAN DER MAAREL

Dr. W. JUNK bv PUBLISHERS THE HAGUE – BOSTON – LONDON 1980

Data-processing in phytosociology

Report on the activities of the Working-group for data-processing in phytosociology of the International society for vegetation science, 1969-1978

Edited by

EDDY VAN DER MAAREL, LÁSZLÓ ORLÓCI & SANDRO PIGNATTI

Reprinted from *Vegetatio*, 1976–1979, except for the last two articles, which are published here for the first time

Dr. W. JUNK bv PUBLISHERS THE HAGUE – BOSTON – LONDON 1980

Distributors:

for the United States and Canada

Kluwer Boston, Inc.
190 Old Derby Street
Hingham, MA 02043
USA

for all other countries

Kluwer Academic Publishers Group
Distribution Center
P.O. Box 322
3300 AH Dordrecht
The Netherlands

Library of Congress Cataloging in Publication Data CIP

Main entry under title:
Data-processing in phytosociology
(Advances in Vegetation Science v. 1)
1. Plant communities – data-processing – addresses, essays, lectures.
2. Vegetation classification – data-processing – addresses, essays, lectures.
3. Botany – ecology – data-processing – addresses, essays, lectures.
I. Maarel, Eddy van der. II. Orlóci, László. III. Pignatti, Sandro. IV. Working-group for data-processing in phytosociology. V. Vegetation. VI. Series.

QK911.D29 581.5'247'02854 80-21182

ISBN-13: 978-94-009-9196-5 e-ISBN-13: 978-94-009-9194-1
DOI: 10.1007/978-94-009-9194-1

CONTENTS

INTRODUCTION*

Eddy VAN DER MAAREL

This book is a collection of papers published in Vegetatio by members of the Working-Group for Data-Processing of the International Society for Vegetation Science. The Working-Group was formed in 1969 during the 13th International Symposium of the Society in Rinteln, and it finished its activities in 1977 at a meeting of the Group in Nijmegen. As both the name of the Group and the title of this book suggest, the Working-Group has aimed at introducing data-processing and multivariate analysis into the world of 'classical' phytosociology. From the start both Braun-Blanquet phytosociologists with an interest in numerical methods and quantitative plant ecologists with an interest in formal vegetation typology have been among the members of the Group. Therefore it has been able to act as an effective link between different types of vegetation scientists active in, or connected with, the International Society.

Although the number of active members of the Group has never been more than 30 and the numbers at its meetings have been low, the influence of the Group has reached many ecologists and plant sociologists. This is mainly because the results of research on data-sets and methods of data-processing were published soon after they had been discussed at one of the Group's meetings instead of being kept back until the Group had finished its activities. Secondly various members of the Group have been invited to give courses or guest lectures and have thus been able to encourage many young scientists to develop data-processing methods for their own material.

Over this period the development of multivariate methods of classification and ordination progressed. I'll return to this later. This development stimulated the Working-Group to finish its activities. During 1978 the last manuscripts were prepared for publication and then a plan could be realized which had already been put forward some years ago; namely to collect all contributions, arrange them and publish them in a book together with some additional information about the meetings and the data sets we have been working with.

The order of the contributions was not difficult to decide; the paper by the editors on 'Data-processing, retrospect and anticipation' provides a logical introduction. Since systems of data storage and coding form a very important element of our work and many papers deal with selections from our main phytosociological data set, i.e. European salt marshes, Pignatti's paper on data storage and coding was placed next. The papers which follow were written by representatives of groups with different approaches to the methods for the treatment of data. Romane and co-workers describe the 'écothèque' development in Montpellier; Moore and O'Sullivan give an explanation of Moore's PHYTO-approach, which originated in Dublin, the Nijmegen group of TABORD, a pragmatic program for the rearrangement of vegetation tables, and computationally effective; and Holzner and Stockinger explain their approach based on the rapid detection of species groups. These methods have been exchanged and tried out by other groups.

The following paper by Orlóci and Feoli on concentration analysis shows how useful it can be to have experts on multivariate methods amongst phytosociologists and forms a logical transition to the next group of papers. In the first of these van der Maarel summarizes the problems of transformation. The other papers dealing with classification and ordination, particularly by Orlóci and Feoli, are mainly a result of our joint efforts towards a multivariate phytosociology. These papers do not fully cover the recent developments to which the authors themselves too have contributed. Orlóci's successful textbook, with the second edition published soon after the first, is a case in point!

We must admit that the Working-Group did not pay enough attention to the numerical treatment of environmental data, nor to the elaboration of the analysis of vegetation structure. The Montpellier group concentrated on such work from the start, but the amount of time which the active members of the Group could give to data-processing was insufficient. The paper on the concept of 'optimal limits' by three colleagues from Montpellier is the only example of numerical synecological work. Rejmánek's paper on vegetation structure is the only one of its kind too. We hope that much more effort will be given to numerical synecology and a numerical synmorphology will be established in the near future!

The contributions by Neuhäusl and Moravec introduce the field of numerical syntaxonomy. This approach could possibly have raised the interest of the group of classical syntaxonomists also formed at the 1969 Rinteln symposium, which had the 'Prodrome of European plant communities' as its aim. Unfortunately they have not proceeded in this way and with the speed we had all hoped for and it is only possible to give one example of close cooperation between the supporters of the classical and the numerical approach, that of *Spartinetea*, reported by Kortekaas, van der Maarel and Beeftink.

However the road to a numerical syntaxonomy, and beyond that, to a numerical synsystematics, is still open.

The last two papers deal with very large data sets of salt marsh relevés. The paper on CLUSLA by Louppen and van der Maarel shows the phytosociological treatment of such sets. The 'climax' paper by Lausi and Feoli describing a hierarchical classification of European salt marsh vegetation based on numerical methods, demonstrates the wide perspective such methods give for the greater ecological and geographical understanding of large and complex systems of plant communities.

As previously mentioned many multivariate methods have been developed by those outside our Working Group. For example, recent volumes of the Journal of Ecology and Vegetatio contain a great number of new methods of classification and ordination and comparisons between them. Naturally, the Working Group wished to extend its scope and exchange the views it had developed within the views developed by its members with those colleagues responsible for the new developments. Thus, at its final meeting at Nijmegen in 1977 the idea was conceived of organizing a larger symposium to recruit members for an extended Working Group. This symposium was held in May of 1979, again in Nijmegen, and it was then generally agreed, much to the satisfaction of former members of the Group, that this extended Working Group should concern itself with the theoretical aspects of vegetation science (we had already outlined the possible aims of such a group in 1976 when introducing the series of data processing papers). Thus although the name of the Group will not be changed for the time being, the subject of the forthcoming series of annual symposia will be the theoretical aspects of the approaches used in the various branches of vegetation science.

The Proceedings of the Nijmegen symposium on Advances in Vegetation Science will be published in two volumes, one dealing with classification and ordination and the other with succession. It is good that the present book will appear just before these Proceedings! It may now serve a double purpose: firstly to confirm the introduction of data processing into phytosociology and secondly to point to the way for future development in phytosociology, as I and many others envisage it. We want to work towards a general science of vegetation, with a solid framework of phytosociological concepts using reliable numerical methods for the analysis of data and the detection of trends and the testing of hypotheses, alongside the erection of systems of vegetation types which can be well reproduced and interpreted.

DATA-PROCESSING IN PHYTOSOCIOLOGY: RETROSPECT AND ANTICIPATION*

Eddy VAN DER MAAREL[1], László ORLÓCI[2] & Sandro PIGNATTI[3]

[1] Division of Geobotany, University of Nijmegen, Toernooiveld, Nijmegen, The Netherlands
[2] Department of Plant Sciences, University of Western Ontario, London, Canada N6A 5B7
[3] Istituto ed Orto Botanico, I 34100, Trieste, Italy

Keywords:
Classification, Data-Processing, Multivariate analysis, Numerical phytosociology, Ordination, Salt marsh vegetation, Theoretical phytosociology.

Introduction

The Working Group for Data-Processing in Phytosociology of the International Society for Vegetation Science is now to finish its activities. Through four conferences of the Working Group at Trieste (1969), the Netherlands (1971), Prague (1973) and Montpellier (1975), during the 14th Symposium of the Society which was devoted to Basic problems and methods in phytosociology and included 12 contributions from members of the Group (Van der Maarel & Tüxen 1972; see Van der Maarel 1971), during small meetings at the Society's Symposia in Rinteln, as well as through numerous personal contacts, the Working Group largely achieved its original aim, viz. the assimilation of computer-based data-processing in Braun-Blanquet phytosociology. Furthermore, theories and new methods were developed which are of general interest for quantitative and theoretical plant ecology.

The Working Group produced a number of circulars and discussion papers which have been distributed among the ca. 60 members of the Group. Some contributions to the symposia have already been published (see Van der Maarel 1974 for a survey). Two general reports on the progress of the Group have also been published in Vegetatio (Van der Maarel 1971a, 1974).

Since the Prague conference in 1973 it has been planned to publish the proceedings including papers, data treatments and discussions, under its provisional title Data-Processing in Phytosociology (Van der Maarel 1974). For various reasons the Group decided to offer all contributions for publication in Vegetatio. The offprints could be bound and reproduced afterwards in a separate volume. The papers will now appear roughly in their chronological order of presentation. The discussion, which followed many contributions after their introduction during one of the conferences, will be summarized and presented either immediately after the relevant contribution, or in one of the final retrospective papers. The three authors of this paper will edit the proceedings and organise the series for Vegetatio. The Working Group is grateful to Dr. W. Junk Publishers for taking up the papers in Vegetatio.

This introduction will look back on the Group's activities, recount publications which are relevant to these activities and outline contributions from some 20 colleagues.

History of the Working Group

The establishment of the Working Group in 1969 was preceded by presentation of a discussion paper by Pignatti, Cristofolini & Lausi (1968) at the 12th Symposium of the International Society of Vegetation Science. This paper was the outcome of discussions amongst the authors, and with Van der Maarel since 1964. On the suggestion of Pignatti Cristofolini studied the storage system with visual punched cards as developed by Ellenberg (cf. Ellenberg & Cristofolini 1964), whilst Van der Maarel (in Van der Maarel, Westhoff & Van Leeuwen 1964) presented ideas concerning numerical phytosociology. The 1968 discussion paper considered the use of electronic computers for storage and elaboration of phytosociological relevés. Collections of

*Contribution from the Working Group for Data-Processing in Phytosociology, International Society for Vegetation Science.

2000-5000 relevés, found in regional prodromes of plant communities (e.g. Tüxen 1937, Oberdorfer 1957) were considered as a maximum that could be handled by the conventional methods. The total number of available relevés of the European plant communities was then estimated at over 70000. It was proposed to collect as many of the relevés as possible in a data bank to allow simple selections of desired sets.

The idea of a data bank received much interest among colleagues, many of whom were already engaged in the numerical treatment of relevés. With these colleagues the Working Group for Data Processing in Phytosociology was formed at the 13th Symposium of the International Society for Vegetation Science at Rinteln in 1969. The Working Group is also known as the Arbeitsgruppe für Datenverarbeitung in der Phytosoziologie and Groupe d'Etude pour l'élaboration des données en phytosociologie. S. Pignatti was elected president and E. van der Maarel secretary. The Working Group received formal recognition of the International Society for Vegetation Science at one of its several official working groups. The Working Group for the Prodrome of European plant communities is one of them with which closer cooperation was envisaged.

It was hoped by the founders that the Working Group for Data-Processing could carry out the data collection and storage, and could also play a role in the construction of a comprehensive system of vegetation units on the basis of numerical relationships. The basic units were seen to be either single relevés or regional relevé tables documenting local plant communities in the major syntaxa already accepted as valid groups by phytosociologists in Europe. These syntaxa were at the level of classes or groups of related classes.

In a parallel effort, groups of syntaxonomical and synecological specialists were to complete classical treatment of available data in each class of the syntaxonomical system. Tüxen (1971) provided a preliminary list of classes to be treated and the names of specialists to be involved in the prodrome, and he also summarised the procedure (cf. Westhoff & van der Maarel 1973 for full description and explanations):

1. All regional (local) tables are stripped of incomplete or atypical relevés in associations or subassociations.

2. The class editors compose for each association a synoptic table of relevés with presence indicated in absolute figures for each species in each column representing a local type, the local types arranged in geographical order.

3. In each association, the association table is converted into a constancy-table with roman figures used to indicate constancy classes.
Then, a structured alliance, order or class table is composed, with one column for each association.

4. An area distribution map of associations is drawn. Ideally the numerical treatments should be applied in step I in the Prodrome work, so that the identity and homogeneity of local types could be established before the larger units are distinguished.

To promote interaction with the Prodrome work, and also for practical reasons, the Working Group decided to start the analysis with relevés from European coastal and continental salt marshes. This choice was motivated as follows (Van der Maarel 1971a):

1. Salt marsh communities are realitvely well-known and elaborate syntaxonomical systems are available so that comparison between the results of numerical and classical treatments would be rewarding.
2. There are many relevés easily available for the project.
3. The communities occur in almost every European country.
4. All character-species are recognised halophytes which were known as such a priori; consequently it was relatively easy to define salt marsh vegetation and to select relevé sets.

Storage of salt marsh relevés

Coding system

Colleagues from Trieste, who originally suggested the idea, started work on a species code for the recording of floristic and other relevé data.

By the time of the 1st conference at Trieste in November 1969, they could present a draft for discussion (Cristofolini, Lausi & Pignatti 1969).

The final version is to be published shortly after this introduction in a comprehensive paper by Pignatti (see also Pignatti 1973). For names of plant taxa (primarily for vascular plants), we may mention four essentials of their coding system:

1) The system is consistent in using a seven digit code, four digits for the genus and three for the species. Because of the primitive state of knowledge of infraspecific taxa, no extension to them has so far been attempted.

2) Relevés are stored as part of the phytosociological table in which they have been placed originally for the documentation of some local vegetation type, whether published or not. These tables are punched by row, a row

containing cover-abundance data for a species in all relevés of a table.
A row may be punched on one or several cards to accomodate tables consisting of any number of relevés. The arguments for this option of storage are both theoretical and practical: practical because the majority of relevés is available in table form and table punching is relatively fast; theoretical because tables tend to reflect a certain type identity and homogeneity which is more or less maintained in subsequent treatments. The system allows single relevés to be stored as well, either collected in ad hoc tables with a deviating code or as single relevé tables. So far both types of tables were stored.

3) For cover-abundance data one column is reserved. This requires the scale values to be transformed on a 0-9 scale. The Braun-Blanquet, Hult-Sernander and Domin scales give no difficulties in this respect. Van der Maarel will discuss problems in transformation in a subsequent paper.

4) Additional cards are used for bibliographic data of the table and for data concerning the stand and date of analysis. So far these data-sets have not been processed. For detailed environmental data such as soil texture, profile, physical and chemical properties and climatological characteristics no coding system has been developed. In consultation with F. Romane, Centre d'Etudes Phytosociologiques et Ecologiques Louis Emberger, Montpellier, the Working Group decided to consider adopting the code developed at this Institute (Godron et al. 1968, 1969, Romane et al., 1972). We expect to receive a contribution on this subject from colleagues at Montpellier (see also Long, 1974 for a survey of the CEPE-approach, and Westhoff & Van der Maarel 1973 for a summary).

Salt marsh relevé selections

The Trieste group prepared a first selection of 1507 relevés in which the major monographs on salt marsh vegetation and additional works (mainly from Southern Europe) are included. This selection (to be referred to as *selection a*) was considered sufficiently large to fall beyond the possibilities of manual treatment and still not too large to become unmanageable in most of the computers available to the Working Group. This selection is stored both on punch cards and magnetic tape. It was distributed to six centres involved in the Working Group.

As a partial project, all available relevés of *Spartina*-communities were collected by W.M. Kortekaas to be processed numerically in order to arrive at a comparison between numerical and classical syntaxonomy. The 576 relevés containing *Spartina* species derived for a small part from *selection a*, and for the greater part from a set of tables collected for this project. This selection is referred to as *selection b*.

A second selection of 1383 relevés was made largely through the efforts of W.G. Beeftink who, in 1972, presented a systematical and geographical survey of relevé sets from the various salt marsh alliances. This selection was an addition to the first selection, so that the ensemble gives a geographically representative but regionally still incomplete collection of 2890 relevés (to be referred to as *selection c*). This selection was stored on punch cards and tape at Trieste and distributed to four centres.

In order to facilitate testing of methods three small selections of relevés were made at Nijmegen which, in accordance with the wishes of members, included:

Selection d: 60 relevés taken at random from *selection a*.

Selection e: 58 relevés from *selection a*, representing tables assigned to communities from the order *Glauco-Puccinellietalia*; this selection covers relatively species-rich communities.

Selection f: 50 relevés of *Spartina* communities chosen at random from an extreme type of community in *selections a* and *b* with the overall dominance of one species and only very few companion species.

Following the outcome of discussion during the Prague conference, W.M. Kortekaas and W.G. Beeftink enlarged *selection c* to 5888 relevés. In this *selection g* (thus including *selections a, b* and *c*) the regional gaps in *selection c* have largely been filled. *Selection g* may thus be considered a reasonably good representation of the total number of potentially available relevés for salt marshes. The potential number is estimated at 15000-18000 (Beeftink 1972, who already registered 11390 relevés, Van der Maarel 1974).

Survey of treatments on salt marsh selections

Various treatments have been applied to different selections:
R.B. Ivimey-Cook tested associations analysis on *selection a;*
J.J. Moore improved his PHYTO-table rearrangement program and applied it to *selection a*;
L. Fresco developed a compound analysis method on the basis of a set of ordinations of overlapping relevé groups and subjected *selection a* to this analysis;
J. Janssen devised a simple clustering procedure for the classification of large data-sets which he applied to *selection c*;
J. Louppen & E. van der Maarel further elaborated on J.

Janssen's procedure and applied it so *selection g;*
D. Lausi & E. Feoli made an attempt to treat *selection c*, starting with the original tables as synoptic tables.

Further treatments have been made by F. Stockinger & W. Holzner and by A. Češka & H. Roemer.

The small selections have been used by A. Česka, E. Feoli, W.M. Kortekaas & E. van der Maarel, L. Orlóci, F. Romane and G. Spatz for testing table rearrangement programs and for developing new classification and ordination methods.

Selection b was used for comparison of numerical and longhand methods in syntaxonomy.

Main lines of development

We may distinguish five lines of development in the treatments mentioned above and in the other contributions presented at the various conferences. We may structure the forthcoming series of papers along these lines. However, the papers will appear roughly in order of submittance in Vegetation, which may not coincide with our categorisation.

Systematical data-processing

A major achievement is the coding system for plant names and the storage system as developed in Trieste. The system was tested by the Working Group. It may well serve as a starting point for the establishment of a computerbased storage of phytosociological data, as was originally aimed at by the Trieste Group. There are several data banks that code and store phytosociological data concerning European vegetation. A second group, with K. Zukrigl, Vienna, as present secretary is concerned with *Fagus*-woodlands. Holzner and Stockinger stored a large collection of relevés of weed communities in Vienna. The Trieste group developed an archive for alpine grassland data and the Nijmegen group for coastal dune grassland data using the Trieste system, which now comprises 2450 relevés.

We think that phytosociology may profit from further efforts in this direction and the paper by Pignatti should be considered as a good starting point. The items discussed by Pignatti include criteria for an efficient species code, the most efficient storage system, the relation between data collection and data processing and types of output from the data bank (cf. also Cutbill 1971).

Additional developments may be expected from the inclusion of environmental data in the archives. The Working Group has concentrated on the floristic-sociological aspects so far, mainly for practical reasons. However, after some enquiries, the Nijmegen group discovered that hardly any reasonably large set of consistent environmental data pertaining to a phytosociological collection of any geographical range is available.
Combined floristic/environmental analyses, as are carried out by some centres, notably the CEPE at Montpellier (cf. Romane 1972, Bottliková et al. 1975) treat local situations.

Rearrangement of phytosociological tables

The structuring of a raw phytosociological table in order to arrive at blocks of relevés, characterised by blocks of species, is generally considered the crux of the Braun-Blanquet method of synthesis. After Ellenberg (1956) wrote a first general account on the procedure, various others presented similar surveys (e.g. Shimwell 1972, Westhoff & Van der Maarel 1973, Mueller-Dombois & Ellenberg 1974).

It has been recognised in the early days of computer-based plant ecology that the procedure in manual structuring of phytosociological tables could be formalised and programmed for automated processing by a digital computer. Benninghoff & Southworth (1964) were probably the first to develop a computer program for this.Many colleagues followed Benninghoff and produced their own computer programs. Within the Working Group, J.J. Moore (1971, 1972), G. Spatz (1972, Spatz & Siegmund 1973, see also Mueller-Dombois & Ellenberg 1974), A. Češka & H. Roemer (1971), F. Stockinger & W. Holzner (1972, 1973) and J. Janssen & E. van der Maarel (Janssen 1972, see also Westhoff & Van der Maarel 1973) adopted, tested and further refined programs. Some of these programs will be described in more detail and most of them will be demonstrated in subsequent papers.

Westhoff & Van der Maarel (1973) give a summary of the literature using table rearrangement programs, and distinguishe two types. One approach is through grouping of species with similar distribution patterns in the table; this approach is recommended for use with large sets of relevés. The other approach is through grouping of similar relevés; this is effective with small sets of relevés containing many species. Most of the programs require interference by the user to give the finishing touch by hand. Further computerisation, including the use of display terminals, have been adopted (Dale & Quadraccia 1973), in tabular sorting will be considered in a paper from the Nijmegen group. All in all, we can foresee a rapidly increasing interest in the use of table rearrangement programs as well as in their automation and improvement.

Development of classification and ordination techniques

During the first years of the Working Group's existence, multivariate methods in plant ecology have evolved rapidly. Co-authors in Whittaker (1973) give a thorough review of achievements up to about 1972, whilst recent volumes of Vegetatio reflect the latest developments. Members of the Working Group contributed to these developments (e.g. Orlóci 1972, 1973, 1974, 1975, Jancey 1974, see also Orlóci 1975a). Several of the new techniques have been successfully applied to salt marsh relevé selections or have lead to the development of new approaches in numerical phytosociology. Reports on these in subsequent papers may be expected from E. Feoli (cf. 1973, 1975), E. van der Maarel (cf. 1969, 1972), F. Romane (cf. 1972, Bottliková et al 1975), as well as from L. Orlóci.

Among the promising developments we may mention ordination of community types, ordination on the basis of structural characters and higherlevel taxa, the development of polythetic divisive classification methods (e.g. Hill, Bunce & Shaw 1975), application of reciprocal ordering (correspondence analysis: Lacoste 1976, Bouxin 1976, Hill 1973).

Equally interesting is the comparison of the results from numerical treatments with results achieved by use of other phytosociological techniques. Working Group members have evaluated association analysis (Ivimey-Cook 1972, Coetzee & Werger 1975) in this respect. It appears in general from this and other accounts that the method of Braun-Blanquet and numerical classification may lead to different results in some cases and to identical results in others (cf. Stanek 1973, Dale & Webb 1975, Bouxin 1975). This problem will be discussed in a paper at the end of the series. Another point that is elaborated in different papers is the elucidation of relationships between community types in ordinations (cf. van der Maarel 1969, Feoli 1973, Whittaker, 1972).

Numerical syntaxonomy

Numerical syntaxonomy, or the application of numerial methods in the solution of syntaxonomical problems, offers the kind of methodology from which the Braun-Blanquet approach would benefit most through comparison of the results of classical syntaxonomical treatments with results from numerical systems, and through tests of homogeneity, definition of similarity levels etc. So far only one direct comparison is available: the numerical syntaxonomy of the *Spartinea* communities has been compared in discussion papers by Kortekaas & Van der Maarel (1972, 1973) with the *Spartinetea* volume of the Prodrome of European plant communities (Beeftink & Géhu 1973) - the first volume that has already appeared in this series. Since Beeftink also cooperated in the numerical treatments, his account of comparisons could be interesting indeed. One of the first papers to be published in the series will report on results of this study. We hope to present one or two other papers on the same subject later on.

Further in this connection, a comprehensive treatment has been performed by D. Lausi & E. Feoli on 189 tables. In their report they will outline the resulting classification system

Further research in the Working Group included homotoneity (table homogeneity) and similarity tests of types. J. Moravec (1975, see also 1971 and Westhoff & van der Maarel 1973) already published results on this topic, and a paper by R. Neuhäusl will appear in an early issue of Vegetatio. There is a point of general interest in comparing the intuitive defined homogeneity in community types and the numerically assessed homogeneity and average similarity. This, of course, brings up the crucial point of weighing species in syntaxonomy which hopefully will be discussed in some of the papers.

Theoretical phytosociology

During discussions in the Working Group it was stressed over and over again that numerical phytosociology and numerical syntaxonomy, should not in themselves be the ultimate goal. They should rather be seen as a powerful approach to a more general type of phytosociology. Various contributions were already published on this topic (Lausi 1972, Van der Maarel 1972, Whittaker 1972) at the Society's Symposium on basic problems (Van der Maarel & Tüxen 1972). W.G. Beeftink, D. Lausi, E. van der Maarel and M. Rejmánek have contributions in preparation for the forthcoming series.

Perspective

After the various announced papers will have been published, members may be interested to present general discussions on the perspectives on theoretical points. The recent jubilee contributions to J. Braun-Blanquet by Pignatti (1975) and Van der Maarel (1975) are along this line. (see also Pignatti & Pignatti 1975). The questions: 'is phytosociology at the junction of dividing roads' as Pignatti asks, or 'can we integrate syntaxonomy with numerical phytosociology, or both with synecology and syndynamics' as others put the problem, will have to be addressed and answered.

The members of the Working Group believe in an integrated approach, and accordingly, decided to enlarge the group's activities. It will now operate under the provisional name Working Group for Theoretical Phytosociology and its first activity is concerned with organisation of a symposium in 1978. By that time, the series of papers on data-processing in phytosociology will have been completed.

Summary

The existence of the Working Group for Data-Processing in Phytosociology spans about six years, 1969-1975. Its activities are summarized and a series of papers to appear in Vegetatio, presenting results of various numerical treatments, mainly of salt marsh vegetation, is introduced. A survey is given of seven relevé selections to which treatments were applied. Five main fields of activities are discussed: systematical data-processing, rearrangement of phytosociological tables, development of classification and ordination techniques, numerical syntaxonomy, and theoretical phytosociology.

The establishment of the Working Group for Theoretical Phytosociology and its first symposium on theoretical aspects is announced.

Zusammenfassung

Die Aktivitäten der Arbeitsgruppe für Datenverarbeitung in der Phytosoziologie 1969-1975 werden zusammengefasst und eine Serie von Arbeiten in Vegetatio über die Ergebnisse der verschiedenen numerischen Bearbeitungen, hauptsächlich von Salzwiesen, wird eingeleitet. Es folgt eine Übersicht über sieben · Auswahle von Aufnahmen mit welchen gearbeitet wurde. Das Kode-System, das dem Data-Bank zugrunde liegt wird kurz diskutiert. Fünf Hauptentwicklungslinien in den Aktivitäten der Arbeitsgruppe können unterschieden werden: systematische Datenverarbeitung, Umordnung von pflanzensoziologischen Tabellen, Entwicklung von Klassifikations- und Ordinationsverfahren, numerische Syntaxonomie, und theoretische Phytosoziologie.

Der allmähliche Übergang der Gruppe zu einer Arbeitsgruppe für theoretische Phytosoziologie wird angedeutet und ein Symposium über theoretische Pflanzensoziologie angekündigt.

References

Beeftink, W.G. 1972. Übersicht über die Anzahl der Aufnahmen europäischer und nordafrikanischer Salzpflanzengesellschaften für das Projekt der Arbeitsgruppe für Datenverarbeitung. In E. van der Maarel & R. Tüxen (eds.), Grundfragen und Methoden der Pflanzensoziologie. Ber. Int. Symp. Rinteln 1970. p. 371-396. Junk, Den Haag.

Beeftink, W.G. & J.-M. Géhu. 1973. Spartinetea maritimae. Prodrome des groupements végétaux d'Europe. Vol. 1. Cramer, Lehre. 43 pp.

Benninghoff, W.S. & W.C. Southworth. 1964. Ordering of tabular arrays of phytosociological data by digital computer. Abstr. 10. Int. Bot. Congr. Edinburgh: 331-332.

Bottliková, A., Ph. Dagelet, J. Drdoš, J.R. Guillerm, F. Romane & H. Ružičková. 1975. Quelques résultats obtenus par l'analyse factorielle et les profils écologiques sur des observations phyto-écologiques recueillies dans la vallée de Liptov (Tchéchoslovaquie). Vegetatio 31: 79-91.

Bouxin, G. 1975. Ordination and classification in the savanna vegetation of the Akagera Park (Rwanda, Central Africa). Vegetatio 29: 155-167.

Bouxin, G. 1976. Ordination and classification in the upland Rugege forest. Vegetatio 32: 97–115.

Češka, A. & H. Roemer. 1971. A computer program for identifying species relevé groups in vegetation studies. Vegetatio 23: 255-276.

Coetzee, B.J. & M.J.A. Werger. 1975. On association-analysis and · the classification of plant communities. Vegetatio 30: 201-206.

Cristofolini, G., D. Lausi & S. Pignatti. 1969. Survey of the system for coding plant sociological records used by the Trieste Group. Paper presented at the Trieste Conference of the Working Group for Data-Processing in Phytosociology. 13 pp.

Cristofolini, G., D. Lausi & S. Pignatti. 1970. Über statistische Eigenschaften der Charakterarten und deren Verwertung zur Aufstellung einer empirischen Systematik der Pflanzengesellschaften. In R. Tüxen (ed.), Gesellschaftsmorphologie, Ber. Int. Symp. Rinteln 1966, p. 8-15, Junk, Den Haag.

Cutbill, J.L. (ed.) 1971. Data processing in biology and geology. Academic Press London, New York, XV + 346 pp.

Dale, M.B. & L. Quadraccia. 1973. Computer assisted tabular sorting of phytosociological data. Vegetatio 28: 57-73.

Dale, M.B. & L.J. Webb, 1975. Numerical methods for the establishment of associations. Vegetatio 30: 77-87.

Ellenberg, H. 1956. Grundlagen der Vegetationsgliederung. 1. Teil: Aufgaben und Methoden der Vegetationskunde. In: H. Walter. Einführung in die Phytologie 4 (1). 136 pp. Stuttgart.

Ellenberg, H. & G. Cristofolini. 1964. Sichtlochkarten als Hilfsmittel zur Ordnung und Auswertung von Vegetationsaufnahmen. Ber. Geobot. Inst. Rübel 35: 124–134.

Feoli, E. 1973. Un esempio di ordinamento di tipi fitosociologici mediante l'analisi delle componenti principali. Not. Fitosoc. 7: 21-27.

Feoli, E. 1975. Sull'uso della combinazione specifica caratteristica nei confronti tra tipi di vegetazione. Giorn. Bot. Ital. 109: 87-96.

Fresco, L.F.M. 1971. Compound analysis: a preliminary report on a new numerical approach in phytosociology. Acta Bot. Neerl. 20: 589-599.

Godron, M. et al. 1968. Code pour le relevé méthodique de la végétation et du milieu (principes et transcription sur cartes perforées). CNRS, Paris, 292 pp.

Godron, M. et al. 1969. Vade-Mecum pour le relevé méthodique de la végétation et du milieu. CNRS, Paris, 169 pp.

Hill, M.O. 1973. Reciprocal averaging: an eigenvector method of ordination. J. Ecol. 61: 237-249.

Hill, M.O., R.G.H. Bunce & M.W. Shaw. 1975. Indicator species analysis, a divisive polythetic method of classification, and its application to a survey of native pinewoods in Scotland. J. Ecol. 63: 597-613.

Ivimey-Cook, R.B. 1972. Association analysis – some comments on its use. In E. van der Maarel & R. Tüxen (eds.), Grundfragen und Methoden der Pflanzensoziologie. Ber. Int. Symp. Rinteln 1970, p. 89-97. Junk, Den Haag.

Jancey, R.C. 1974. Algorithm for detection of discontinuities in data sets. Vegetatio 29: 131-133.

Janssen, J.G.M. 1972. Detection of some micropatterns of winter annuals in pioneer communities of dry sandy soils. Acta Bot. Neerl. 21: 603-610.

Janssen, J.G.M. 1975. A simple clustering procedure for preliminary classification of very large sets of phytosociological relevés. Vegetatio 30: 67-71.

Kortekaas, W. & E. van der Maarel. 1972. A numerical classification of Spartinetum vegetations. Preliminary paper Int. Colloquium Rinteln 1972, Report Div. Geobotany Nijmegen, 9. PP. + tables.

Kortekaas, W.M. & E. van der Maarel. 1973. A numerical classification of Spartinetum vegetations. II. Comparison of the computer-based numerical system with the system published in the Prodrome des Groupements Végetaux. Paper Prague Conference Working Group for Data-Processing in Phytosociology.

Lacoste, A. 1976. Relations floristiques entre les groupements prairiaux du Triseto-Polygonion et les Megaphorbiaies (Adenostylion) dans les Alpes occidentales. Vegetatio 31:

Lausi, D. 1972. Die Logik der pflanzensoziologischen Vegetationsanalyse. Ein Deutungsversuch. In E. van der Maarel & R. Tüxen (eds.), Grundfragen und Methoden der Pflanzensoziologie, Ber. Int. Symp. Rinteln. 1970, pp. 17-24. Junk, Den Haag.

Lausi, D. 1973. Contributo delle elaborazioni quantitative per analisi e sintesa in fitosociologica. Not. Fitosoc. 7: 1-15.

Long, G. 1974. Diagnostic phyto-écologique et aménagement du territoire. I. Principes généraux et méthodes. Masson et Cie, Paris. 252 pp.

Long, G. 1975. Diagnostic phyto-écologique et aménagement du territoire. II. Application du diagnostic phyto-écologique. Masson et Cie, Paris. VIII + 222 pp.

Maarel, E. van der. 1969. On the use of ordination models in phytosociology. Vegetatio 19: 21-46.

Maarel, E. van der. 1971. Basic problems and methods in phytosociology. Rep. 14th Int. Symp. Rinteln 1970. Vegetatio 22: 275-283.

Maarel, E. van der. 1971. Bericht über die Arbeitsgruppe für Datenverarbeitung der Internationalen Vereinigung für Vegetationskunde. Vegetatio 22: 269-273.

Maarel, E. van der. 1972. Ordination of plant communities on the basis of their plant genus, family and order relationships. In E. van der Maarel & R. Tüxen (eds.), Grundfragen und Methoden der Pflanzensoziologie, Ber. Int. Symp. Rinteln 1970, pp. 183-192. Junk, Den Haag.

Maarel, E. van der. 1974. The Working Group for Data-Processing of the International Society for Plant Geography and Ecology in 1972-1973. Vegetatio 29: 63-67.

Maarel, E. van der. 1975. The Braun-Blanquet approach in perspective. Vegetatio 30: 213-219.

Maarel, E. van der & R. Tüxen (eds.). 1972. Grundfragen und Methoden in der Pflanzensoziologie. (Basic problems and methods in phytosociology). Ber. Int. Symp. Rinteln 1970. Junk, Den Haag. XIX+523 pp.

Maarel, E. van der, V. Westhoff & C.G. van Leeuwen. 1964. European approaches to the variation in vegetation. Paper 10th Int. Bot. Congress Edinburgh.

Moore, J.J. 1971. Phyto. A suite of programs in Fortran IV

for the manipulation of phytosociological tables according to the principles of Braun-Blanquet. Mscr. Dept. Botany Univ. Coll. Dublin. 11 pp.
Moore, J.J. 1972. An outline of computer-based methods for the analysis of phytosociological data. In E. van der Maarel & R. Tüxen (eds.) Grundfragen und Methoden in der Pflanzensoziologie. Ber. Int. Symp. Rinteln 1970, pp. 29-38. Junk, Den Haag.
Moravec, J. 1971. A simple method for estimating homotoneity of sets of phytosociological relevés. Folia Geobot. Phytotax., Praha 6: 147-170.
Moravec, J. 1975. Die Anwendung von Stetigkeitsartengruppen zur numerischen Ordnung von pflanzensoziologischen Tabellen. Vegetatio 30: 41-47.
Mueller-Dombois, D. & H. Ellenberg, 1974. Aims and methods of vegetation ecology. John Wiley & Sons, New York XX + 547 pp.
Oberdorfer, E. 1957. Süddeutsche Pflanzengesellschaften. G. Fischer Verlag, Jena, XXVIII + 564 pp.
Orlóci, L. 1972. On information analysis in phytosociology. In E. van der Maarel & R. Tüxen (eds.), Grundfragen und Methoden in der Pflanzensoziologie. Ber. Int. Symp. Rinteln 1970, pp. 75-88. Junk, Den Haag.
Orlóci, L. 1973. An algorithm for cluster seeking in ecological collections. Vegetatio 27: 339-345.
Orlóci, L. 1974. On in formation flow in ordination. Vegetatio 29: 11-16.
Orlóci, L. 1975. Measurement of redundancy in species collections. Vegetatio 31: 65-67.
Orlóci, L. 1975a. Multivariate analysis in vegetation research. Junk, The Hague. VIII + 276 pp.
Pignatti, S. 1973. Problemi di codifica dei dati floristici in fitosociologia. Not. Fitosoc. 7: 17-20.
Pignatti, S. 1975. Pflanzensoziologie am Scheideweg. Vegetatio 30: 149-152.
Pignatti, S., G. Cristofolini & D. Lausi. 1968. Verwendungsmöglichkeiten einer elektronischen Datenverarbeitungsanlage für die Pflanzensoziologische Dokumentation. Discussion paper 12th Symposium Int. Soc. Vegetation Sc. Rinteln. 6 pp.
Pignatti, E. & S. Pignatti. 1975. Syntaxonomy of the Sesleria varia grasslands of the calcareous Alps. Vegetatio 30: 5-14.
Romane, F., 1972. Un exemple d'analyse factorielle des correspondances en écologie végétale. In E. van der Maarel & R. Tüxen (eds.), Grundfragen und Methoden in der Pflanzensoziologie. Ber. Symp. Rinteln, 1970, pp. 151-162. Junk, Den Haag.
Romane, F. et al. 1972. Un exemple d'organisation du traitement des observations phyto-écologiques. Programme de phyto-écologie fondamentale et générale: Projet écothèque. CNRS-CEPE Louis Emberger, Montpellier. 27 pp.
Shimwell, D.W. 1971. Description and classification of vegetation. London Sidgwick & Jackson. XIV + 322 pp.
Spatz, G. 1972. Eine Möglichkeit zum Einsatz der elektronischen Datenverarbeitung bei der pflanzensoziologischen Tabellenarbeit. In E. van der Maarel & R. Tüxen (eds.), Grundfragen und Methoden in der Pflanzensoziologie, Ber. Int. Symp. Rinteln 1970, pp. 251-258. Junk, Den Haag.
Spatz, G. & J. Siegmund. 1973. Eine Methode zur tabellarischen Ordination, Klassifikation und ökologischen Auswertung von pflanzensoziologischen Bestandsaufnahmen durch den Computer. Vegetatio 28: 1-17.
Stanek, W. 1973. A comparison of Braun-Blanquet's method with sum-of-squares agglomeration for vegetation classification. Vegetatio 27: 323-338.
Stockinger, F. & W. Holzner. 1972. Rationelle Methode zur Auswertung pflanzensoziologischer Aufnahmen mittels Elektronenrechner. In E. van der Maarel & R. Tüxen (eds.), Grundfragen und Methoden in der Pflanzensoziologie, Ber. Int. Symp. Rinteln 1970, pp. 239-248. Junk, Den Haag.
Stockinger, F. & W. Holzner. 1973. Prinzipien einer Implementierung des pflanzensoziologischen Zerlegungsproblems. Mitt. Bot. Linz 4: 87-106.
Tüxen, R. 1937. Die Pflanzengesellschaften Nordwestdeutschlands. Mitt. Flor.-Soz. Arbgem. Nieder Sachsen 3: 1-170. Reprint 1970. Cramer-Lehre.
Tüxen, R. 1971. Vorläufige Liste von Mitarbeitern am Prodromus der europäischen Pflanzengesellschaften. Vegetatio 22: 265-268.
Westhoff, V. & E. van der Maarel. 1973. The Braun-Blanquet approach. In R. Tüxen (ed.), Handbook of Vegetation Science, Part V, (R.H. Whittaker, ed.) Ordination and Classification of Communities, pp. 617-726. Junk, The Hague.
Whittaker, R.H. 1972. Convergence of ordination and classification. In E. van der Maarel & R. Tüxen (eds.). Grundfragen und Methoden der Pflanzensoziologie, Ber. Int. Symp. Rinteln 1970, pp. 39-57. Junk, Den Haag.
Whittaker, R.H. (ed.). Ordination and classification. Handbook of Vegetation Science (Ed. R. Tüxen). Part V. 737 pp. Junk, The Hague.

TRGRPS - AN INTERACTIVE ALGORITHM FOR GROUP RECOGNITION WITH AN EXAMPLE FROM SPARTINETEA*

László ORLÓCI**

Department of Plant Sciences, University of Western Ontario, London, Canada N6A 5B7

Keywords: Algorithm, Clustering, Groups, Hypotheses, Numerical phytosociology, Testing

Introduction

I know that group recognition is a subject of great interest to this group. I realize, of course, that there are preferences for different methods, according to different objectives.

The algorithms used by phytosociologists appear to be of two generic types:

1. The first focuses on homogeneity within the groups. Groups are sought in which the member objects are as similar as possible in terms of given characters. The clustering algorithm, which I described in an earlier paper (Orlóci 1967) with group homogeneity measured in terms of the sum of squares, represents an example in this connection.
2. The second type shifts focus from homogeneity within the groups to discontinuities in the entire sample. The clustering algorithm TRGRPS is of this type. It is formulated around the definition that one group of relevés is discontinuous with another group if none of its neighbourhoods, with given radius, overlap with any of the neighbourhoods with the same radius in the other. The definition and related concepts are explained in the text.

The algorithm is programed in BASIC under file name TRGRPS.BAS. When it is hypothesized that discontinuities subdivide the sample into a given number of groups and the hypothesis is presented to the program with data, TRGRPS.BAS performs the necessary tests automatically and, if groups are found, it produces group descriptors.

*Contribution from the Working Group for Data Processing in Phytosociology, International Society for Vegetation Science. For nomenclature of species see Lausi, Beeftink & Kortekaas (1975).

**The research project, from which this paper summarizes partial results, is supported by a National Research Council of Canada grant. Computer time was provided by the University of Western Ontario.

Relevant concepts

Sample space

It will be assumed that there are N relevés in the sample labelled by symbols A, B, C, etc. It will be assumed also that in the associated sample space the relevés are represented by points in the line,

The relative placements of the points to one another are important. The assumption of such a simple unidimensional arrangement is not necessary, but convenient, because it can simplify matters when the concepts are explained without detracting from the generality of the conclusions.

Nearest neighbour relevés

The length of the line segment, connecting any two points in sample space, is proportional to differences in the species composition of the relevés. The line segments represent distances in p-dimensions. Each dimension in p-space corresponds to a different species in the sample. It is noted that differences need not be measured in terms of a distance function. Any symmetric measure of dissimilarity is permissible. The point nearest to another point in terms of any such distance or dissimilarity is said to be that point's nearest neighbour. In the example shown, A is nearest neighbour to B, E to F, etc.

The neighbourhood of a relevé

Each relevé may or may not have other relevés in its neighbourhood within a given radius (r) for which the neighbourhood is defined. At $r = d(\mathrm{B, C})$, where $d(\mathrm{B, C})$ is the distance of point B from C, the following neighbourhoods of relevés exist:

{AB}	for relevé A
{ABC}	for relevé B
{BC}	for relevé C
{DE}	for relevé D and E
{F}	for relevé F

It is noted that {AB} and {BC} are overlapping neighbourhoods; *i.e.* relevé B is possessed in common by both. {ABC}, {DE} and {F} are non-overlapping. Neighbourhoods that are non-overlapping are said to be discontinuous.

Clustering by neighbourhoods

Given $r = 0$ as the initial state of the neighbourhood radius in a clustering procedure, as many discontinuous neugbourhoods can be recognized in the sample A, B, C, D, E, F as there are non-identical relevés,

{A}, {B}, {C}, {D}, {E}, {F}

Now, let r be incremented by a constant, small quantity in steps until the condition $d(A, B) \leqslant r < d(B, C)$ is satisfied. Here, $d(A, B)$ represents the smallest distance other than zero in the sample and $d(B, C)$ the second smallest distance. At such a radius,

{AB}, {C}, {DE}, {F}

are recognized as discontinuous neighbourhoods. If r is further incremented, so that it will satisfy the condition $d(B, C) \leqslant r < d(E, F)$, where $d(E, F)$ is the third smallest distance in the sample, the discontinuous neighbourhoods are

{ABC}, {DE}, {F}

Further increments in r, until $d(E, F) \leqslant r < d(C, D)$ is satisfied, yield the discontinuous neighbourhoods,

{ABC}, {DEF}

Finally, when $r \geqslant d(C, D)$, only a single group

{ABCDEF}

is recognized. All neighbourhoods at a given r under the given constraints satisfy the original definition of discontinuous groups stated in the introduction. In other words, none of the neighbourhoods in any one group overlaps with any neighbourhood in the other. It is evident from what has been said that the procedure of clustering by neighbourhoods uses single link fusions similarly to Sneath's (1957) method of single linkage clustering.

The algorithm TRGRPS

TRGRPS incorporates clustering by neighbourhoods and various other devices which enable the user to test hypotheses about the existence of discontinuous groups of revelés. The algorithm regards a group of relevés as discontinuous if none of its neighbourhoods overlap with any neighbourhood in other groups. The flow pattern of operations and decisions is described below:

a. Present data; specify a hypothesis for the number of groups R; compute distances between relevés.
b. Specify initial value of neighbourhood radius and its increments.
c. Cluster relevés and count the number of groups. Designate this number by k.
d. If $k = R$ then accept the hypothesis.
   If $k < R$ then modify the hypothesis to less than R groups and continue with step b.
   If $k > R$ then increment neighbourhood radius and continue at c.

I realize that based on sampling considerations tests of the kind as in this algorithm should preferably be probalistic. However, the algorithm as I present it here does not incorporate probabilistic considerations. I plan to discuss these aspects in a separate eassay presently in the preparatory stage.

A related algorithm

TRGRPS evolved through modifications of the original Jancey (1974) algorithm which I abstracted (Orlóci 1975) under the name TRGRUP. I stressed a nearest neighbour property of the groups in the original descriptions of TRGRUP, although I incorporated the idea of clustering by neighbourhoods in the program TRGRUP.BAS. The present paper corrects this anomaly by recognizing that although the condition that every member in a discontinuous group must have its nearest neighbour within the group is necessary it is by no means sufficient because when this condition is satisfied neighbourhoods can overlap between the groups. The groups with contents

{AB}, {C}, {DEF}

represent an eclatant example in this connection. In these, the nearest neighbour property is satisfied; every member in each group has its nearest neighbour within its own group. Yet two groups, {AB} and {C}, appear continuous because the neighbourhoods of relevés A, B and C overlap at the implied radius $d(E, F)$.

TRGRPS.BAS differs from TRGRUP.BAS by the specific constraint that none of the relevés are allowed to remain unclustered when the hypothesis is true. This means that for the hypothesis of R groups to be accepted in TRGRPS, the sample must divide into exactly R groups, and not just R noda plus a number of points that float unclustered among the groups. Programs TRGRUP.BAS and TRGRPS.BAS yield identical results when the group structure is strong in the sample.

Running the program TRGRPS.BAS

The program uses the BASIC programing language, written in conversational mode for processing from a teletype console. After the RUN command, the program requests values for the following parameters:

Number of species (P)
Number of relevés (N)
Minimum group size (S)
Number of groups (R)
Initial value of neighbourhood radius (C)
Increment in neighbourhood radius (D)

Values for P and N are requested first, and after P and N, values for S and $R.S.$ can be set to one if no hypothesis specifies *a priori* minimum group size. An initial value for R derives from the hypothesis. Alternate values, small or large, may be assigned to R in successive runs until a solution is accepted. It is noted that running time and R are inversely related.

Data input is from DISK file RAWD which contains P sets of N numbers. The data in computer core are stored in array **D** which has P rows and N columns. Data input is followed by computation of relevé distances. The program uses the Euclidean distance. However, if so desired, any symmetric function of a relevé dissimilarity may replace the Euclidean distance. The distances are stored in array **A** which has N^2 elements.

After the elements of **A** are generated, a search begins for the smallest value. The message initial value of linkage parameter is printed followed by display of the smallest distance value. A request to confirm this value is displayed next. The user responds by typing either the same value, or another value, which then serves as the initial value for C. The program determines a maximum of 50 values, each representing a difference between pairs of the smallest distance values in **A**, ordered according to size, to assist in the choice of a value for the increment D of C. Any value can be chosen. However, it must be kept in mind that a large value of D shortens the computation time required, while a small value increases precision. A zero is not permissible for D.

Once the program gets underway, the sample is searched for discontinuous groups. When exactly R groups are found, the group descriptors are printed and the execution of the program stops. If R groups are not found a new search begins for $R-1$ groups.

Printout of results

Examples are given in the next section. When a solution is found under a current hypothesis for R, as modified by parameters S, C, D, the following descriptors are printed:

a. A value of C representing the current size of the neighbourhood radius r at the last fusion. This can be used as a global measure for coherence in the groups.
b. A label for each group.
c. Relevé labels in each group.

Results of repeated runs at different R can be used to trace a hierarchical group structure of relevés in the sample.

Results from Spartinetea

The results presented here derive from the data in Table 1 representing saltmarsh relevé section f (Van der Maarel, Orlóci & Pignatti 1976, Lausi, Beeftink & Kortekaas 1976). Inspection of the raw data reveals that an initial hypothesis of 5 relevé groups may be tenable with S set equal to 5. When TRGRPS is run a solution is indeed found at this value of R (see Table 2).

Table 1. Sample of 50 relevés from Spartinetea. The Braun-Blanquet scale is replaced by the following scores in the table: 1 → r, 2 → +, 3 → 1, 5 → 2, 7 → 3, 8 → 4, 9 → 5. Relevé order follows syntaxonomic affiliation.

Species	Relevé
Triglochin maritima	00000200000000000000000000000000000000000002000000
Agrostis stolonifera	00000000000000200000000000000000000000000000000000
Spartina alterniflora	00000000000899900000000000000000000000000000000000
Spartina maritima	79889533330000000000000000000000000000000000000000
Spartina townsendii	00000000000000000000789985533333333353335575525555
Spartina patens	00000000000000009999000000000000000000000000000000
Phragmites australis	00000000000000002025200000000000000000000000000000
Puccinellia maritima	00000872530002300000000088775003522023030050000000
Scirpus maritimus	00000000000005000000000000000000000000000000500000
Juncus acutus	00000000000000000000200000000000000000000000000000
Juncus maritimus	00000000000000000552000000000000000000000000000000
Atriplex hastata	00000000000000000000000000000000000000008797900000
Halimione portulacoides	00000233220000000000220000220220020022530030035500
Salicornia radicans	00000778890000000000000005000000000000000000000000
Salicornia fruticosa	00000000000000000000200000000000000000000000000000
Arthrocnemum glaucum	00000000000000000200000000000000000000000000000000
Salicornia europaea	00020502200000000000000003555388777025230005033533
Suaeda maritima	000205520000000000000200022232232222887890025055555
Spergularia media	00000000000000000000000003000000000000000000000000
Oenanthe lachenalii	00000000000000000030200000000000000000000000000000
Limonium vulgare	00030000000002003522000000000000000200000000000000
Plantago crassifolia	00000000000000000000200000000000000000000000000000
Plantago maritima	00000000000000000000000000000000000000000002000000
Aster squamatus	00000000000000000002000000000000000000000000000000
Aster tripolium	00302003230022500002002000322332320352303323053353
Inula crithmoides	00000000000000002203000000000000000000000000000000
Sonchur arvensis	00000000000000002200000000000000000000000000000000

Table 2. Results from program TRGRPS.BAS. The number of groups hypothesized is 5. See the main text for description of program.

```
TRGRPS          16:15          22-Nov-75
PROGRAM NAME -- TRGRPS
==================================
NUMBER OF SPECIES ?27
NUMBER OF QUADRATS ?50
MINIMUM GROUP SIZE REQUIRED ?5
NUMBER OF GROUPS REQUIRED ?5
INITIAL VALUE OF NEIGHBOURHOOD RADIUS = 1
CONFIRM BY TYPING  1 ELSE TYPE DESIRED VALUE: ?1
SELECT A VALUE FROM THE LIST BELOW
TO SERVE AS THE INCREMENT IN LINKAGE PARAMETER:
 1  2
TYPE SELECTED VALUE:
 ?2
 5 --GROUPS
MAXIMUM NEIGHBOURHOOD RADIUS 67
GROUP 1
 1  2  3  5  4
GROUP 2
 6  7  8  10  9
GROUP 3
 11  12  14  15  13
GROUP 4
 16  19  20  17  18
GROUP 5
 21  22  25  24  23  27  28  29  30  34  31  32  35  33  36  37  49
 50  46  47  48  38  40  39  26  41  42  43  45  44

TIME:  177.74 SECS.
```

The correspondence of the groups at the 5−group level in TRGRPS and the units in conventional syntaxonomy is most remarkable. The first group, and the second, are subassociations (*typicum* and *salicornietosum*) of *Spartinetum maritimae*. The third group represents *Spartinetum alterniflorae*, the fourth *Spartinetum patensis*, and the fifth *Spartinetum townsendii*.

A hierarchical pattern of fusions, derived by repeated runs of TRGRPS. BAS with different values of R, is traced in Fig. 1. Interestingly, the first fusion does not occur between the subassociations of *Spartinetum maritimae*, notwithstanding expectations, but between *Spartinetum maritimae salicornietosum* and *Spartinetum townsendii*. With these the other subassociation, *typicum*, fuses only in the second step, followed by fusions with *Spartinetum alterniflorae*, and lastly, *Spartinetum patensis*.

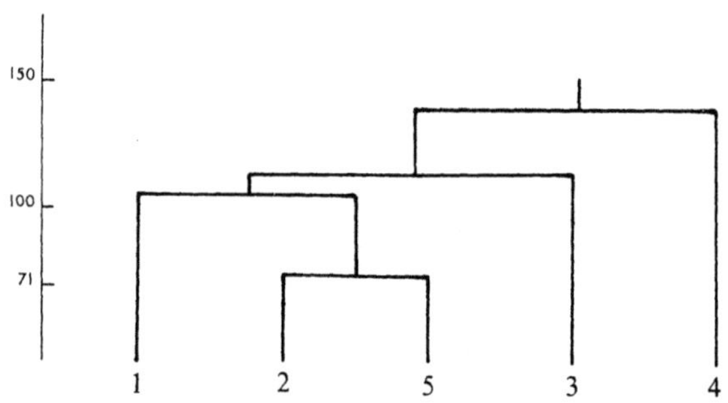

Figure 1. Hierarchical structure in Spartinetea (Table 1) at the level of 5 groups. Vertical scale indicates values of the neighbourhood radius. Symbols at the base designate syntaxonomic units recognized in TRGRPS. Legend to symbols: 1 - *Spartinetum maritimae typicum*, 2 - *Spartinetum maritimae salicornietosum*, 5 - *Spartinetum townsendii*, 3 - *Spartinetum alternifoliae*, 4 - *Spartinetum batensis*. Numbers 1 to 5 identify groups in the computer printouts at the 5 groups level in Table 2.

Discussion

It is evident intuitively that the relevés in a discrete group must exhibit a density pattern that permits their recognition as members in a coherent group. Coherence, the discreteness of groups, is associated with criteria of non-overlapping neighbourhoods in algorithm TRGRPS. When applied to a sample of *Spartinetea*, the results of a comparison of the groups suggest that the method of group recognition in TRGRPS resembles the method of group recognition in traditional syntaxonomy.

The algorithm TRGRPS should be a useful tool in the hands of phytosociologists, interested in the construction of syntaxonomic systems and in the identification of a parsimonious path through neighbouring groups for which hierarchical relationships are to be shown.

Summary

TRGRPS can detect groups or signal if discrete groups cannot be found in a sample. The present paper elaborates on the concepts, describes the algorithm and provides illustrations from syntaxonomy. A computer program (TRGRPS.BAS) is available from the author upon request.

References

Jancey, R.C. 1974. Algorithm for detection of discontinuities in data sets. Vegetatio 29: 131-133.

Lausi, D., W.G. Beeftink & W.M. Kortekaas, 1976. Survey of salt marsh relevés selected for the working group for Data Processing in Phytosociology, to be published in Vegetatio.

Maarel, E. van der, L. Orlóci & S. Pignatti 1976. Data Processing in Phytosociology: retrospect and anticipation. Vegetatio 32: 65-72.

Orlóci, L. 1967. An agglomerative method to classify plant communities. J. Ecol. 55: 193-205.

Orlóci, L. 1975. Multivariate Analysis in Vegetation Research. Dr. W. Junk b.v. Publishers, The Hague, 276 pp.

Sneath, P.H.A. 1957. The application of computers to taxonomy. J. Gen. Microbiol. 17: 201-226.

DEUX APPLICATIONS EN PHYTO-ÉCOLOGIE DE LA NOTION DE LIMITE 'OPTIMALE'**

J.L. GUILLERM[1], B. GAUTHIER[2] & F. ROMANE[1]

1 Centre d'études phytosociologiques et écologiques L. Emberger (C.N.R.S.) B.P. 5051-34033 Montpellier CEDEX (France)
2 Université Laval-Faculté des sciences, de l'agriculture et de l'alimentation, département de phytologie Cité universitaire, Québec, P.Q., GIK7P4, Canada

Mots-clef:
Hétérogénéité (Heterogeneity), Limite optimale (Optimal limit)

Introduction

La recherche de limites dans une suite d'événements ordonnés est un problème fréquent en écologie et c'est dans ce but que Godron (1972) propose une méthode pour chercher la limite 'optimale'. Elle a d'abord été appliquée à l'étude de la structure horizontale de la végétation, en exploitant les observations recueillies le long d'un transect établi dans une formation végétale (Godron 1972, Corre 1975, B. Thiébaut, en préparation). Après avoir rappelé le principe des calculs, deux applications seront présentées.

- l'une au niveau spatial, pour tenter de préciser, à une échelle moyenne (1/250000) les limites des régions écologiques,
- l'autre à partir d'échantillons répartis dans le temps, à savoir des relevés effectués dans les stades post-culturaux récents, pour chercher les ruptures les plus marquées de cette évolution.

Rappel du principe de la méthode

Nous ne raisonnerons ici que sur le cas de N échantillons alignés dans l'espace, en considérant la présence ou l'absence des espèces dans chaque échantillon. Ceux-ci peuvent être, par exemple, des segments réguliers répartis le long d'un transect. Considérons d'abord le cas d'une seule espèce présente dans F échantillons. L'ensemble des positions des F présences et des N-F absences constitue un 'cas de figure', appelé 'complexion' (Godron 1971). Avec ces F présences sur N échantillons, il existe C_N^F 'complexions différentes'*. Si chacun de ces cas est considéré comme ayant la même chance de survenir, la probabilité d'avoir la complexion effectivement réalisée est égale à $1/C_N^F$.

Il est souvent convenu que l'hétérogénéité $H(E)$ de la végétation pour l'espèce observée est égale à (Godron 1971):

$$H(E) = \log_2 C_N^F$$

Si une limite sépare en deux parties la suite d'événements considérés, (ici, des segments le long d'une ligne) il est alors possible d'envisager une hétérogénéité 'à droite', $H(E_d)$, qui est l'hétérogénéité de la suite d'évènements situés à droite de la limite, ainsi qu'une hétérogénéité 'à gauche', $H(E_g)$, hétérogénéité de la suite d'évènements situés à gauche de la limite. La différence entre l'hétérogénéité de l'ensemble et la some des hétérogénéités des deux sous-ensembles peut-être prise comme une mesure de la valeur de cette limite (figure 1):

$$V(L) = H(E) - (H(E_d) + H(E_g))$$

La limite L, de la division fournissant une valeur maximum pour V(L) est appelée 'optimale'.

Nous avons raisonné jusqu'à présent dans le cas d'une complexion donnée avec F présences et N échantillons, or, il existe C_N^F complexions possibles.

** Contribution from the Working Group for Data-Processing in Phytosociology, International Society for Vegetation Science.
Nomenclature d'après P. Fournier (1961); Les quatres flores de France.

* C_N^F = combinaisons de N objects pris F à $F = \frac{N!}{F!\,(N-F)!}$

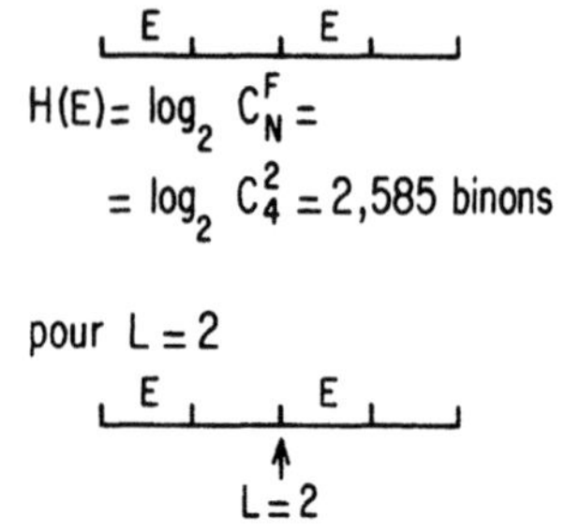

$$V^i(L) = H(E) - [H(E_d) + H(E_g)]$$

$$V^i(2) = \log_2 C_4^2 - (\log_2 C_2^1 + \log_2 C_2^1) = 0{,}585 \text{ binon}$$

Fig. 1. Exemple de calcul de la valeur d'une limite (L = 2) dans le cas d'une complexion i donnée.

Pour une limite donnée L, l'espérance mathématique de cette valeur est égale à (Godron & Bacou 1974):

$$V_e(L) = \frac{\sum_{i=1}^{i=C_N^F} V^i(L)}{C_N^F}$$

$V^i(L)$ étant la valeur pour une césure L de la complexion i; cette valeur de l'espérance mathématique passe par un maximum pour la césure située au milieu de la suite des évènements ordonnés, et s'annule aux deux extrémités.

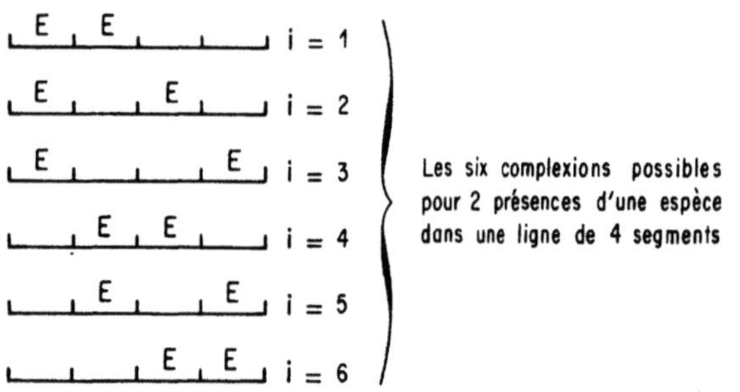

$$V_e(L) = \left[\sum_{i=1}^{i=C_N^F} V^i(L)\right] / C_N^F$$

$$V_r^i(L) = \frac{V^i(L)}{V_e(L)}$$

Fig. 2. Calcul de l'espérance de la valeur d'une limite et de la valeur relative de la limite.

Dans l'étude de la structure de la ligne, ce qui nous intéresse, c'est surtout la limite particulièrement éloignée de la valeur attendue, et il parait donc logique de remplacer $V^i(L)$, pour chaque limite, par le rapport (figures 2 et 3):

$$V_r^i(L) = \frac{V^i(L)}{V_e(L)}$$

Jusqu'ici, nous avons envisagé le cas de la présence ou de l'absence d'une seule espèce, or, dans l'étude de la végétation, ce sont généralement la présence ou l'absence

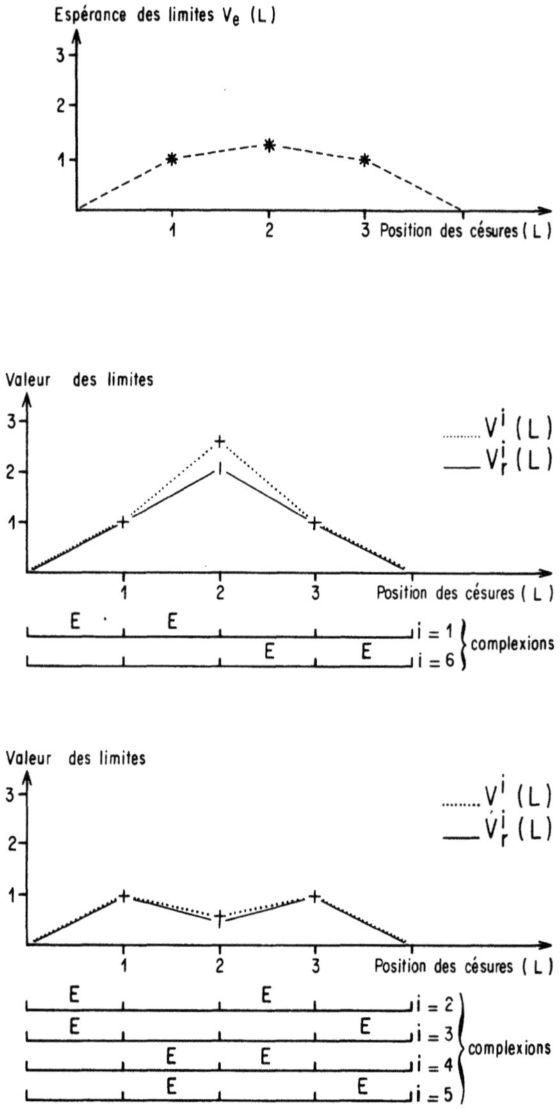

Fig. 3. Représentation de $V^i(L)$, $V_e(L)$ et $V_r^i(L)$ dans un cas simple.

de k espèces qui sont notées. Pour juger de la valeur d'une limite, nous avons finalement retenu comme critère celui de la moyenne des valeurs relatives de cette limite pour chaque espèce*:

$$\bar{V}_r(L) = \frac{\sum_{j=1}^{k} V_r(L)}{k}$$

Précisons bien qu'il s'agit là d'une première approximation. En effet, il n'est pas tenu compte ainsi des relations qui peuvent exister entre les espèces pour lesquelles cette moyenne est calculée (L. Orlóci, comm. verb., 1975).

D'un point de vue pratique, ces calculs ont été programmés au C.E.P.E. Louis Emberger sur un ordinateur CII 10010, en Fortran modifié, par A. M. Bacou ainsi q'en langage *APL* (par M, Godron et F. Romane) sur un ordinateur de la série IBM 360. Le programme en Fortran a été utilisé pour l'étude des régions écologiques à moyenne échelle, celui en APL pour l'étude de l'évolution de la végétation.

Ces deux programmes fournissent, pour chaque césure possible:

- la moyenne calculée sur toutes les classes de variables ou sur toutes les espèces de $V(L)$,
- la moyenne calculée sur toutes les classes de variables ou sur toutes les espèces de $V_e(L)$,
- la moyenne calculée sur outes les classes de variables ou sur toutes les espèces de $V_r(L)$.

En outre, celui en *APL* fournit pour chaque césure, les espèces ayant la plus forte valeur $V_r(L)$, ce qui permet de connaître les espèces jouant un rôle important dans la limite.

Les applications

Limites de régions écologiques

La notion de région écologique a été définie pour la première fois par Emberger (1930) et reprise récemment dans le Code écologique du C.E.P.E. (Godron et al. 1968) et discutée par Long (1974).

Les régions écologiques présentées dans l'Atlas régional du Languedoc-Roussillon (E. Le Floch et al. 1973) ont été déterminées sur la base de la confrontation des données (Joly 1960, Godron 1971) provenant essentiellement de quatre documents cartographiques:

- la carte hypsométrique,
- la carte géologique qui, à l'échelle envisagée, peut aussi apporter des informations sur la lithologie et le bilan hydrique des sols,
- la carte des précipitations,
- la carte des étages et des séries de végétation.

En superposant ces documents, on constate que certaines limites sont communes à plusieurs thèmes.

Les limites communes aux quatre documents constituent des limites écologiques très fortes; elles séparent des régions nettement distinctes ou des portions d'une région écologique donnée, isolées dans une autre région.

Lorsque la superposition des quatre documents de base laisse apparaitre la coïncidence des limites de trois thèmes, il a été estimé que cela était encore suffisant pour discriminer des régions écologiques. Des subdivisions d'une même région écologique sont établies lorsque les limites de deux thèmes seulement sont en coïncidence. Le choix des variables retenues dépend des documents cartographiques disponibles et ils sont généralement les mêmes que ceux utilisés pour bâtir un échantillonnage stratifié en vue d'un inventaire phyto-écologique, à la même échelle. Le choix des variables est relativement arbitraire et discutable.

Pour mieux connaitre la valeur de ces limites, nous avons repris à l'échelle du 1/250000, une partie de la coupe du littoral méditerranéen (Montpellier) à l'Aubrac (Nasbinals) publiée dans l'Atlas régional.

Sur une ligne allant de la région du Pic St-Loup au début du Causse de Sauveterre, nous avons réparti 64 segments de 1,25 km chacun. Pour chaque segment, les présences ou les absences de une ou plusieurs classes des 5 variables suivantes ont été notées:

- carte de la végétation actuelle (in Atlas régional du Languedoc-Roussillon),
- carte des séquences de végétation (in Atlas régional du Languedoc-Roussillon),
- carte topographique au 1/500000 de la France (I.G.N., 1969),
- carte des précipitations, d'après H. Gaussen,
- carte des formations géologiques (in Atlas régional du Languedoc-Roussillon).

Notons que dans le cas présent, les cartes utilisées ont été établies indépendamment, ce qui permet d'éviter en partie, au moins, certaines relations factices (carte des précipitations établie à partir d'une carte de végétation par exemple). Cependant, l'interprétation des résultats devra être faite avec précaution.

Le détail des observations faites sur les cartes est présenté, sous forme de tableau, (tableau 1, légende en annexe) et de coupe (figure 4).

* Dans la suite du texte $\bar{V}_r(L)$ sera noté $V_r(L)$.

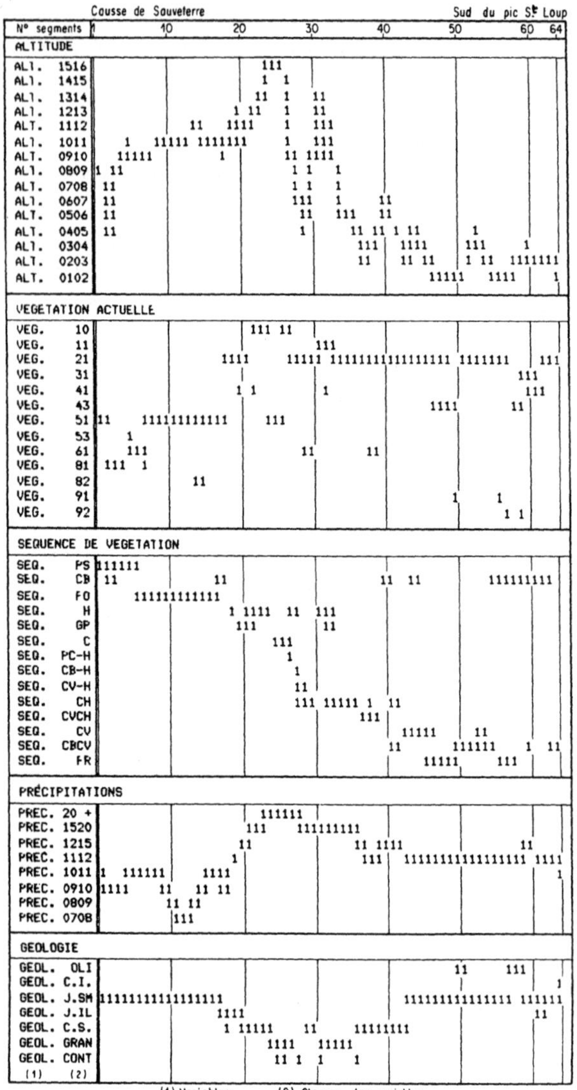

(1). Variables (2). Classes des variables

N.B. Liste des codes des classes des variables en annexe

Tableau 1

Résultats pour une variable: les séquences de végétation

Le diagramme de la figure 5 représente pour l'ensemble des 63 limites possibles, L, les valeurs de $V(L)$, de l'espérance de cette valeur $V_e(L)$ ainsi que celles du rapport $V_r(L)$. La courbe des valeurs de $V_r(L)$ présente 7 pics correspondant à des changements dans le thème étudié. Ces changements ont été notés *C* sur le diagramme et numérotés de *C*1 à *C*7 par ordre de valeur décroissante de la valeur $V_r(L)$ pour chacun d'eux.

Pour essayer de mieux faire saisir la signification de ces changements, les observations relatives aux séquences ont été reprises dans la figure 6 où pour chacun des changements *C*1 à *C*7, il a été fourni une explication par le début ou par la fin d'une ou plusieurs séquences, ou la combinaison de ces deux processus.

Résultats pour les cinq variables retenues

Les valeurs réelles des limites pour chaque variable sont données dans le tableau 2. La moyenne calculée sur toutes les classes des variables montrent 7 changements (figure 7). Nous avons présenté sur la coupe étudiée (figure 4) la correspondance entre les limites obtenues, les régions écologiques et la contribution des différentes variables. D'une manière générale, les césures délimitent des sous-régions écologiques dans les régions données par l'Atlas régional.

Sur les 7 changements mis en évidence par la technique de la limite 'optimale', 3 correspondent à des 'frontières' de régions écologiques ou de subdivisions de ces régions, fournies par l'Atlas régional (*C*1, *C*3, *C*4). Un autre (*C*2) peut être considéré comme correspondant à une des frontières de l'Atlas, si l'on admet que la limite entre deux des régions écologiques mérite d'être déplacée (par suite d'une erreur de tracé peut-être).

Quant aux trois autres (*C*5, *C*6, *C*7) ils ne correspondent pas à des frontières déjà détectées dans la carte des régions

Tableau 2. Valeur $V_r(L)$ pour les variables étudiées et valeur moyenne sur l'ensemble des variables

Variables	Valeur du changement[1]	n° de la coupure[2]	Valeur du changement[1]	n° de la coupure[2]
	(1)	(2)	(1)	(2)
Altitude	C1 : 3,248	34	C6 : 1,617	53
	C2 : 2,784	41	C7 : 1,419	8
	C3 : 2,126	28	C8 : 1,300	13
	C4 : 2,005	3	C9 : 0,757	63
	C5 : 1,760	22		
Végétation	C1 : 2,701	17	C6 : 2,129	39
	C2 : 2,635	27	C7 : 2,121	56
	C3 : 2,373	7	C8 : 2,100	46
	C4 : 2,294	32	C9 : 2,067	58
	C5 : 2,155	5		
Séquence	C1 : 2,641	39	C5 : 2,351	17
	C2 : 2,615	36	C6 : 2,197	24
	C3 : 2,568	33	C7 : 1,952	6
	C4 : 2,367	27		
Précipitation	C1 : 4,584	18	C5 : 1,780	4
	C2 : 3,834	36	C6 : 0,906	60
	C3 : 3,471	28	C7 : 0,867	63
	C4 : 3,190	42		
Géologie	C1 : 7,756	63	C4 : 2,378	56
	C2 : 2,804	43	C5 : 2,013	17
	C3 : 2,634	49	C6 : 1,028	28
Valeur moyenne pour l'ensemble des variables	C1 : 2,734	36	C5 : 1,731	7
	C2 : 2,585	42	C6 : 1,634	53
	C3 : 2,423	18	C7 : 1,490	63
	C4 : 2,323	27		

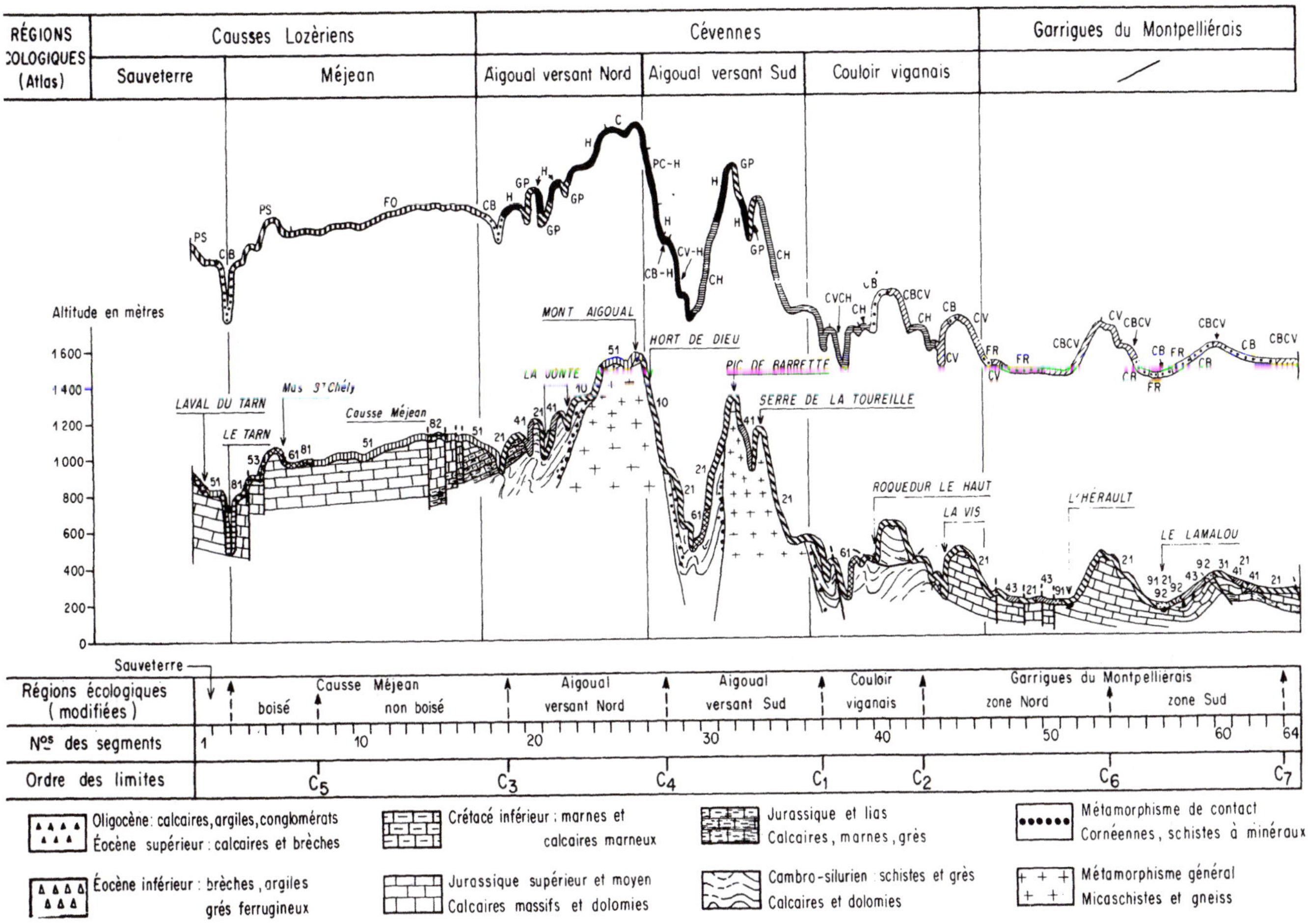

Fig. 4. Coupe suivant le transect retenu pour l'étude des régions écologiques (géologie, végétation actuelle, séquence de végétation, altitude).

écologiques. Nous allons rapidement discuter tous ces changements.

Le premier changement (C1) correspond principalement à une variation de la géologie qui se traduit par le passage du schiste au granite, donc des sols à bilans hydriques très différents. Ce premier changement apparait dans l'Atlas régional comme la frontière entre deux subdivisions de la région écologique des Cévennes (entre couloir du Viganais, au sud, et Massif du Liron, au nord).

Le deuxième changement (*C*2) correspond aussi à deux substrats différents (calcaires et schistes). Cette modification du substrat peut paraitre plus importante qu'un passage du schiste au granite mais il ne faut pas oublier que le Chêne vert peut-être dominant aussi bien sur calcaire que sur schiste dans cette zone (la séquence à Chêne vert se retrouve sur schiste et sur calcaire) ce qui 'tamponne' peut-être la valeur de la limite. Ce changement *C*2 ne correspond à aucune frontière entre les régions écologiques proposées dans l'Atlas. Cependant, après une examen attentif de la carte des régions écologiques, il semble que la frontière entre la région des garrigues du Montpelliérais et celle du couloir du Viganais doive être déplacée un peu plus au nord, ce qui la ferait correspondre avec *C*2. La frontière entre la Massif de l'Aigoual (granite et schiste) et le Causse Méjean (calcaire) correspond à *C*3, et celle entre le Massif de l'Aigoual (sommet et versant nord de ce massif) et Massif du Liron (englobant le versant sud de l'Aigoual) est en concordance avec *C*4.

Les changements *C*5, *C*6 et *C*7 quant à eux ne correspondent à aucune frontière entre les régions écologiques ou leur subdivisions, dans l'Atlas régional mais, dans chacun des cas, ces limites nous ont paru justifiées. Pour

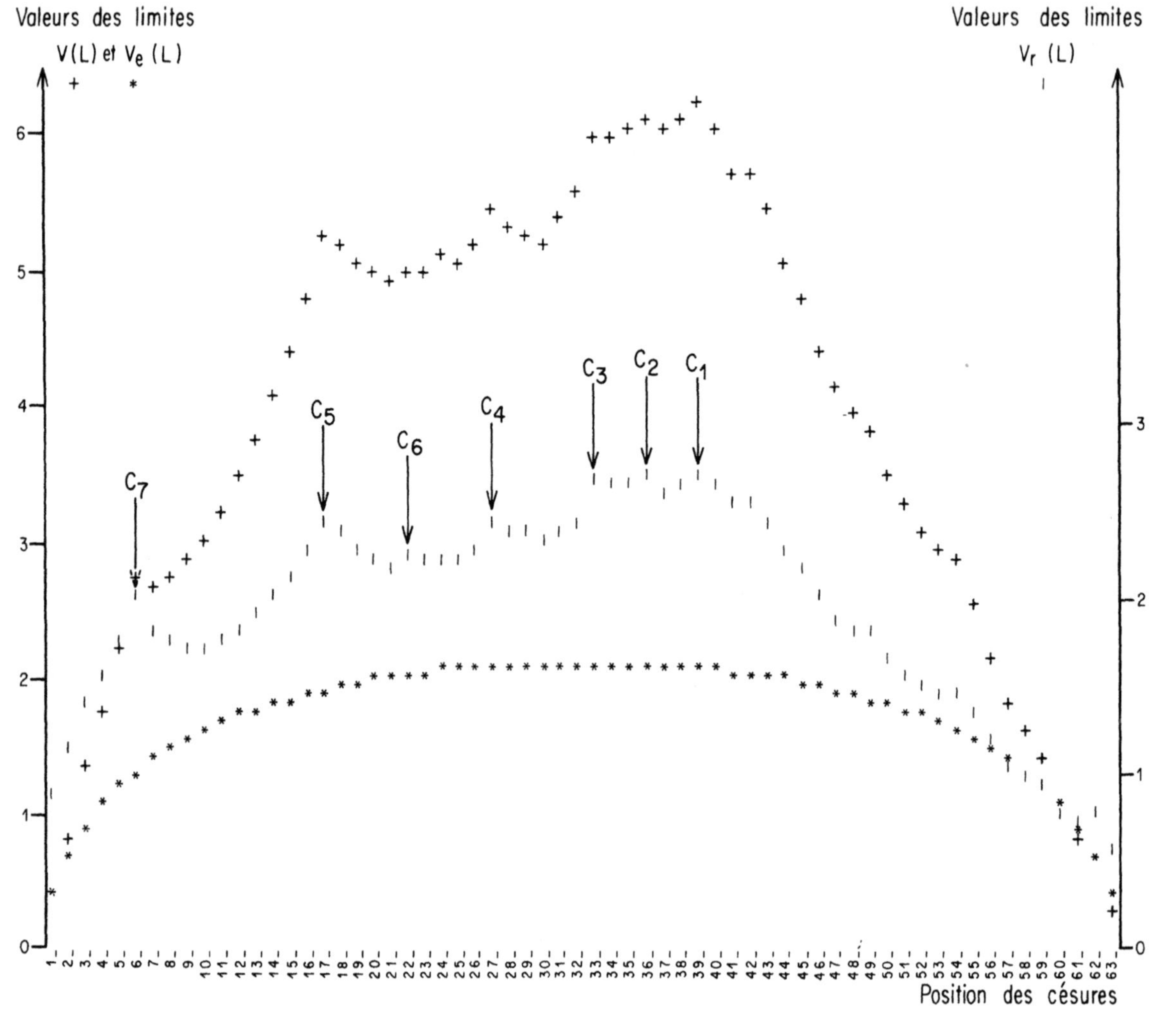

Fig. 5. Etude des régions écologiques. Valeurs des limites pour la variable séquence de végétation.

*C*5, il s'agit effectivement de deux aspects du Causse Méjean, très homogène du point de vue de la géologie, mais non de la végétation. Les auteurs de la carte des régions écologiques n'avaient pas estimé nécessaire de marquer ce changement qui est ici retenu par la technique que nous avons utilisée. Le changement *C*6 ferait une distinction entre les zones au nord du Pic St. Loup et celles au sud. Ceci nous parait justifié dans la mesure où au nord de ce massif, le Chêne blanc semble jouer un rôle plus important qu'au sud. Le changement *C*7 est peut-être dû à l'apparition d'un étage géologique limité au dernier segment sud du transect (sédiments tertiaires dominants), et il annonce la zone du Pin d'Alep correspondant à un climat plus chaud et un peu plus sec.

Pour terminer, il faut noter ici qu'une frontière entre deux régions écologiques ne correspond à aucun changement mis en évidence par la technique de la limite optimale: celle entre Causse de Sauveterre et Causse Méjean. Ceci s'explique très bien car cette limite, entre deux régions très semblables, est la vallée étroite du Tarn qui, avec la longueur retenu (1,25 km) pour chaque segment, apparait très mal.

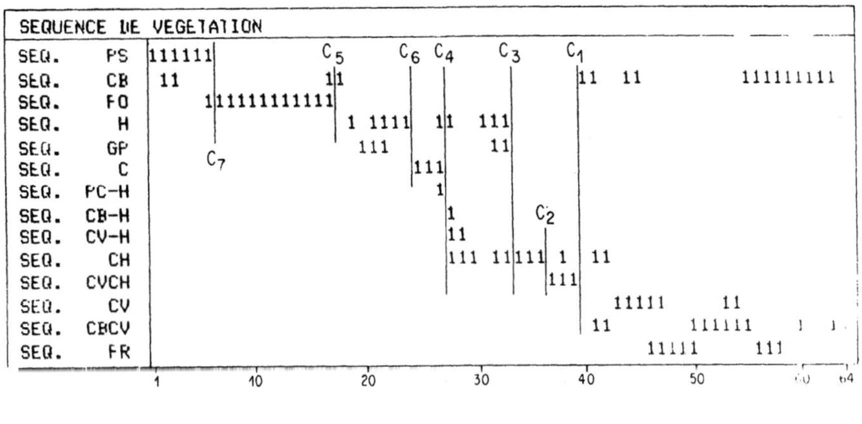

Rappel des observations et position des changements

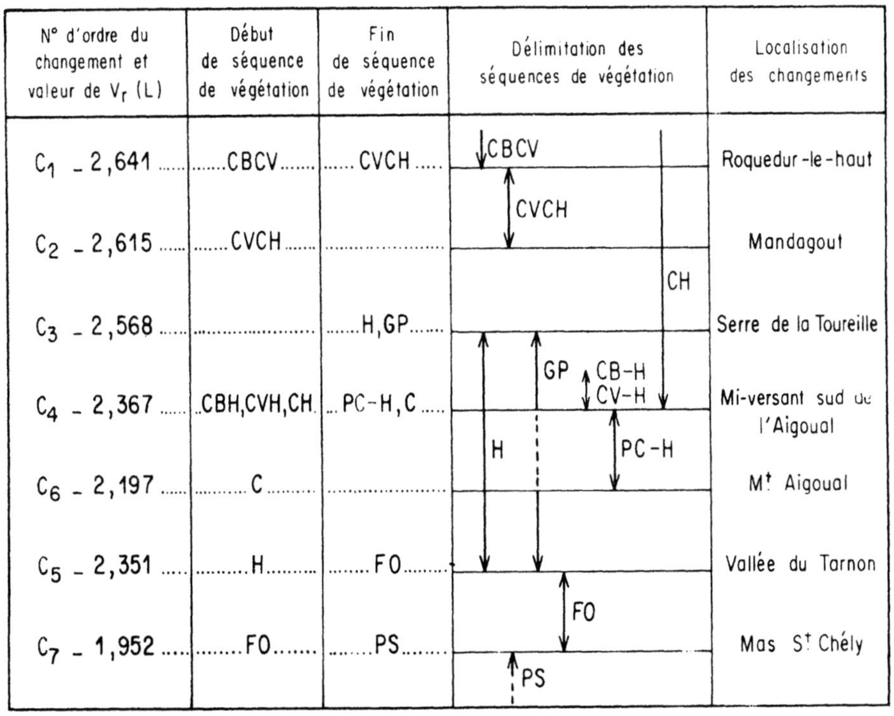

Interprétation des changements

Fig. 6. Etude des régions écologiques. Interprétation des résultats pour la variable séquence de végétation.

Recherche des limites optimales dans une série de stades post-culturaux

A la suite d'un inventaire phyto-écologique réalisé dans les stades post-culturaux du Bas-Languedoc par J. L. Guillerm, une série de relevés appartenant aux mêmes milieux mais comportant différents degrés d'abandon (vigne, vigne abandonnée, friche, lande) ont été analysés pour rechercher les meilleures limites entre stades. L'application de la notion de limite optimale est donc faite dans ce

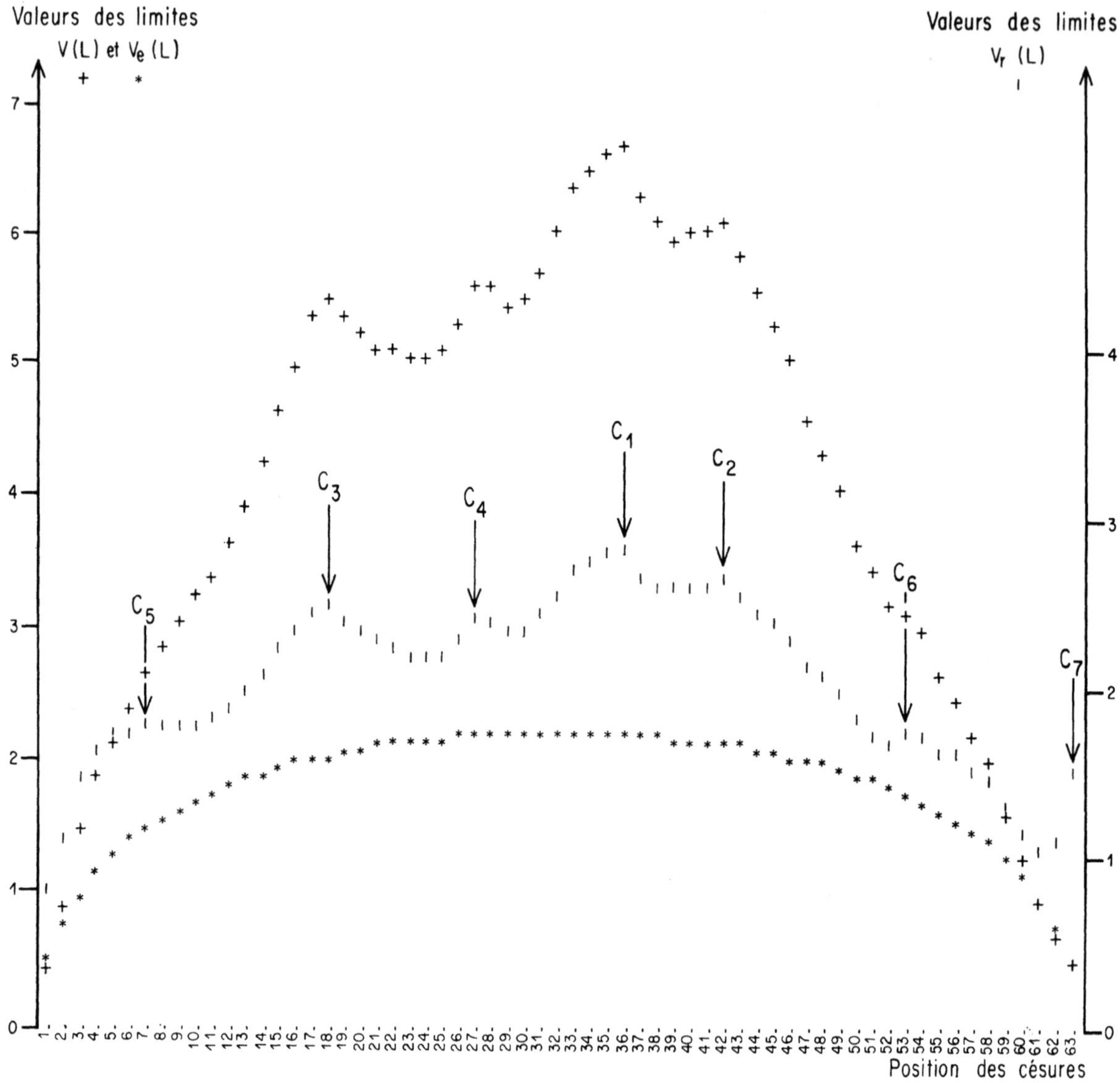

Remarque Pour faciliter la lecture du dessin les échelles pour V(L), $V_e(L)$ et $V_r(L)$ sont différentes.

Fig. 7. Etude des régions écologiques. Valeur des limites pour l'ensemble des cinq variables.

cas sur des échantillons répartis dans le temps et non dans l'espace.

Les données de base sont constituées par l'observation de la présence ou de l'absence d'une espèce dans un état d'évolution. Les principales espèces caractéristiques des stades pour une séquence sur calcaire dolomitique (Roquedur le Haut) sont données dans le tableau 3.

Les différentes valeurs de limites: $V(L)$, $V_e(L)$, $V_r(L)$ sont représentées sur la figure 8 et font apparaitre deux césures, la première entre la vigne et la vigne abandonnée, la deuxième plus faible entre la pelouse 2 et la pelouse 3. La première coupure correspond à l'abandon cultural car ce changement sépare la vigne de tous les autres stades post-culturaux. La deuxième coupure marque une évolution nette entre les pelouses étudiées. Si les pelouses 1 et 2 sont peu différentes, la pelouse 3, par contre, annonce déjà les landes avec une chute forte du pourcentage d'espèces annuelles.

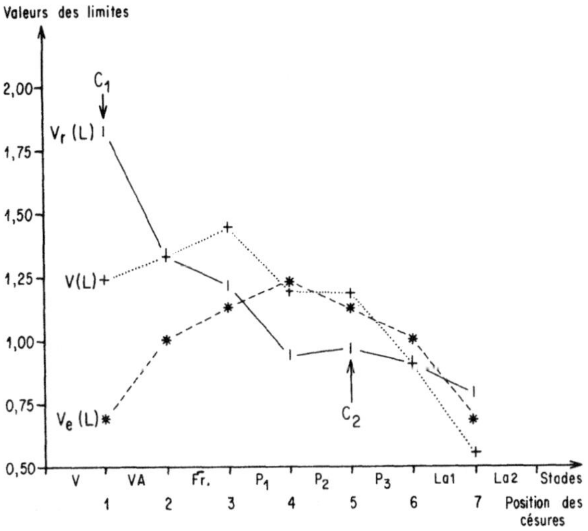

Fig. 8. Etude des tsades postculturaux. Valeur des limites pour l'ensemble des espèces.

Les 35 espèces, soit 24% des espèces recensées, uniquement présentes dans la vigne et contribuant à la première coupure, sont les suivantes:

V	VA	FR	P1	P2	P3	La1	La2
XXX							

Convolvulus arvensis	*Plantago psyllium*
Avena sterilis	*Trifolium stellatum*
Chenopodium album	*Chenopodium botrys*
Trifolium repens	*Mentha rotundifolia*
Bromus mollis	*Fumaria parviflora*
Rumex obtusifolius	*Potentilla reptans*
Crepis foetida	*Silene nocturna*
Chondrilla juncea var. *latifolia*	*Convolvulus sepium*
Bromus macrostachys	*Senecio gallicus*
Vulpia myuros	*Equisetum maximum*
Tragopogon dubium	*Mentha pulegium*
Anagallis phoenicea	*Verbena officinalis*
Reseda phyteuma	*Equisetum campanulatum*
Senecio silvaticus	*Valerianella coronata*
Papaver rhoeas	*Anthoxanthum odoratum*
Chrysanthemum leucanthemum	*Melilotus sulcatus*
Heliotropium europeum	*Avena barbata*
Saxifraga tridactylites	

Quatre espèces, soit 3%, sont absentes de la vigne mais présentes de la vigne abandonnée à la lande La2:

Tableau 3. Caractères principaux de la végétation des stades étudiés à Roquedur le Haut (Cévennes) sur calcaire dolomitique

V	- vigne à *Anagallis arvensis* ssp. *phoenicea*, *Chondrilla juncea*, *Saxifraga tridactylites*, *Plantago psyllium*, *Senecio gallicus*, *Chenopodium botrys*, avec 24% de vivaces, 76% d'annuelles
V.A	- vigne abandonnée à *Euphorbia cyparissias*, *Rubus caesius*, *Koeleria phleoides*, *Chondrilla juncea*, *Erigeron canadense*, *Psoralea bituminosa*, avec 30% de vivaces, 70% d'annuelles
Fr	- friche à *Psoralea bituminosa* (50% de recouvrement), *Holcus lanatus*, *Origanum vulgare*, avec 43% de vivaces, 57% d'annuelles et l'espèce ligneuse: *Thymus vulgaris*
P1	- pelouse à *Brachypodium phoenicoides* (60% de recouvrement), *Ononis repens*, *Psoralea bituminosa*, avec 60% de vivaces, 40% d'annuelles et une espèce ligneuse: *Genista scorpius*
P2	- pelouse (très semblable à P1) à *Brachypodium phoenicoides* (55% du recouvrement), *Psoralea bituminosa*, *Dactylis glomerata*, avec 63% de vivaces, 37% d'annuelles et une espèce ligneuse: *Genista scorpius*
P3	- pelouse à *Brachypodium phoenicoides* (65% du recouvrement), *Bromus erectus*, *Psoralea bituminosa*, *Origanum vulgare*, *Dactylis glomerata*, *Poa pratensis*, avec 88% de vivaces, 12% d'annuelles et deux espèces ligneuses: *Thymus vulgaris*, *Satureia montana*
La1	- pelouse à *Brachypodium phoenicoides* (60% du recouvrement) envahie par *Pteridium aquilinum* (40% du recouvrement) avec 85% de vivaces, 15% d'annuelles et 4 espèces ligneuses: *Genista scorpius*, *Pistacia terebinthus*, *Clematis flammula*, *Quercus pubescens*
La2	- lande de *Pteridium aquilinum* (80% du recouvrement) avec 90% de vivaces, 10% d'annuelles et 4 espèces ligneuses: *Artemisia campestris*, *Thymus vulgaris*, *Fumana ericoides*, *Lavandula latifolia*

V	VA	FR	P1	P2	P3	La1	La2
	XX	XX	XX	XX	XX	XX	XX

Psoralea bituminosa	*Euphorbia nicaeensis*
Artemisia vulgaris	*Ononis natrix*

Pour la deuxième coupure, les 4 espèces présentes de la vigne à la pelouse P2 et absentes ensuite, soit 3%, sont les suivantes:

V	VA	FR	P1	P2	P3	La1	La2
XXX	XXX	XXX	XXX	XXX			

Vulpia ciliata	*Verbascum sinuatum*
Arenaria serpyllifolia	*Holcus lanatus*

Sur les 143 espèces recensées, le 'poids' de la première coupure est de 27%, tandis que nous n'avons plus que 3% pour la seconde. Le changement de la composition floristique est brutal dès l'abandon cultural, puis beacoup plus progressif par la suite. L'implantation des espèces vivaces ligneuses dépend de la présence de semenciers, du stock de graines du sol et de l'environnement.

Conclusion

Il pouvait sembler *a priori* ambitieux d'utiliser la notion de limite 'optimale' pour les deux applications que nous proposons. C'est qu'en effet, cette technique avait été proposée et notée pour exploiter des observations floristiques recueillies le long de transect de quelques mètres ou dizaine de mètres de longueur, dans les communautés végétales. Pour nous, dans un des cas, il s'agit d'un transect de 64 km de long, les observations effectuées n'étant plus floristiques mais relatives au milieu, dans l'autre, d'observations non plus réparties dans l'espace mais dans le temps.

Il semble pourtant que cette technique soit fort fructueuse dans chacun des cas, en n'oubliant pas toutefois que toutefois que certains problèmes restent à résoudre comme celui des corrélations possibles entre les différentes variables prises en considération pour rechercher les changements.

Summary

The optimal limit conceived permits the appraisal of the value of the limits that one has imposed on a series of ordered events, frequently of segments (observing the presence or absence of a species) distributed along a transect in order to study the horizontal structure of the vegetation. The value of a limit is defined by the difference between the heterogeneity of the whole line and the sum of heterogeneities of its parts to the right and to the left of the limit envisaged. The heterogeneities H are themselves defined for F presences of a species E over N segments of a line by:

$$H\,(E) = \log_2 C_N^F$$

This value is elsewhere corrected, taking into account the expectation of the value of a limit.

In a collection of several species, it is possible to take as a limit the mean of the limits obtained for each species. This mean is a first approximation that does not take into account the possible relations between these species.

This concept of a limit has been applied here in two cases:

The first, at a spatial level to determine the limits of ecological regions at a scale 1/250000.

In this application the data consist of observations, made along a transect of 80 km long divided into 64 equal segments, of the classes of five variables (altitude, actual vegetation, successional status of the vegetation, precipitation, geology). The results obtained permit the confirmation of the limits that had been proposed previously, and also, recognition of new ones which had been overlooked.

The second, at an abstract level, to search for clear differences between stages in the evolution of abandoned cultivated land. This is concerned with samples distributed in time, i.e. with relevés that have been recorded in different postcultural stages. Here the segments are post-cultural stages and the stages are ordered according to the period elapsed since abandonment. The example given shows the importance of the method in recognizing stages due to abandonment after cultivation (in this case viticulture).

Résumé

La notion de 'limite optimale' a été proposée pour l'étude de la structure horizontale de la végétation à l'aide de transects divisés en segments.

La valeur d'une limite est estimée à partir de la différence entre l'hétérogénéité de l'ensemble de la ligne et la somme des hétérogénéités de la partie de la ligne à droite et de la partie de ligne à gauche de la limite envisagée. Ces hétérogénéités sont mesurées par $H(E) = \log_2 C_N^F$ (F = nombre de présences de l'espèce E envisagée, N = le nombre de segments du transect).

Cette notion a été appliquée ici à deux problèmes:

- l'un au niveau spatial pour tenter de préciser à l'échelle du 1 250000, les limites de régions écologiques,
- l'autre pour étudier l'évolution dans le temps d'une culture après abandon.

Bibliographie

Atlas régional du Languedoc-Roussillon. 1969-1973. Nombreuses cartes thématiques (direction R. Dugrand), Berger Levrault, Paris.

Corre, J.J. 1975. Etude phyto-écologique des milieux littoraux salés en Languedoc et en Camargue. Thèse, Université des Sciences et techniques du Languedoc, Montpellier, 179 p. + annexes.

Emberger, L. 1930. La végétation de la région méditerranéenne, essai d'une classification des groupements végétaux. Rev. Gén. Bot., p. 648.

Gaussen, H. Carte de la pluviosité annuelle des Alpes, du bassin du Rhône et de la Corse au 1/500000. Ministère des travaux publics. (sans date).

Godron, M. 1971. Essai sur une approche probabiliste de l'écologie des végétaux. Thèse, Université des Sciences et techniques du Languedoc, Montpellier, 247 p.

Godron, M. 1972. Echantillonnage linéaire et cartographie. Investigacion Pesquera, 36(1): 171-174.

Godron, M. et al. 1968. Code pour le relevé méthodique de la végétation et du milieu. Principes et transcription sur cartes perforées. C.N.R.S. Paris, 291 p.

Godron, M. & A.M. Bacou. 1974. Sur les limites 'optimales' séparant deux parties d'une biocénose hétérogène. C.E.P.E. Louis Emberger C.N.R.S. Montpellier, document dactylographié, 8 p.

Godron, M. & J. Poissonet. 1972. Quatre thèmes complémentaires pour la cartographie de la végétation et du milieu. (Séquence de végétation, diversité du paysage, vitesse de cicatrisation de la végétation, sensibilité de la végétation). Bull. Soc. Lang. de géogr., 6(3): 329-356.
Institut géographique national. 1969. Carte de la France au 1/500000. Feuilles de Lyon et Marseille.
Joly, F. 1960. Une carte au 1/1000000 de l'utilisation du sol au Maroc. Com. Nat. géogr. Rabat, 18 cartes, 114 p.
Le Floc'h, E., G. Long, J. Poissonet & M. Godron. 1973. Notice des cartes de la végétation et de son écologie pour la région Languedoc-Roussillon. In Atlas régional du Languedoc-Roussillon, Univ. Paul Valéry, Montpellier, 35 p.
Long, G. 1974. Diagnostic phyto-écologique et aménagement du territoire. Masson Edit. (Paris), t1, 252 p.
Thiebaut, B. 1975. Etudes des hêtraies de l'arc montagneux périméditerranéen de la vallée du Rhône à celle de l'Ebre. Contribution n 2. Structure horizontale, en mosaïque des hêtraies de la Montagne Noire. Institut de Botanique, Montpellier, 11 p.

Annexe

Tableau des observations relatives aux cinq variables retenues pour l'étude régionale et légende. (ci-dessous).

Légende

1) variable *altitude* (ALT). Code:
1516 altitude comprise entre 1500 et 1600 m,
1415 altitude comprise entre 1400 et 1500 m,
0102 altitude comprise entre 100 et 200 m,

x

– le premier chiffre indique le type de formation
1 – ligneux hauts denses
2 – ligneux hauts clairs
3 – ligneux très clairs
4 – ligneux bas
5 – herbacés
6 – ligneux hauts et ligneux bas
8 – ligneux bas et herbacés
9 – ligneux hauts, ligneux bas et herbacés
– le deuxième chiffre le degré d'artificialisation
0 – 'climax'
1 – degré d'artificialisation extensif
2 – degré d'artificialisation demi-extensif
3 – degré d'artificialisation intensif
ex: 41, ligneux bas à degré d'artificialisation extensif (landes, garrigues, maquis)

3) variable *séquences de végétation* (SEQ)

La séquence de végétation est définie par M. Godron et J. Poissonet (1972) comme étant 'les stades d'évolution dont la situation est prévisible, en fonction des principaux critères visibles *in situ*'.

Chaque séquence est nommée d'après le dernier stade prévisible.

Le code adopté est le suivant:
Séquence à:

PS	*Pinus silvestris* L. (Pin sylvestre)
CB	*Quercus pubescens* Willd (Chêne blanc)
FO	*Festuca ovina* L. (Fétuque ovine)
H	*Fagus silvatica* L. (Hêtre)
GP	*Cytisus purgans* (L) Benth. (Genêt purgatif)
C	*Calluna vulgaris* (L) Hull (Callune en altitude)
PC-H	*Pinus montana uncinata* Ram. et *Fagus silvatica* L. (Pin à crochet, Hêtre)
CB-H	*Quercus pubescens* Willd et *Fagus silvatica* L. (Chêne blanc, Hêtre)
CV-H	*Quercus ilex* L. et *Fagus silvatica* L. (Chêne vert, Hêtre)
CH	*Castanea sativa* Miller (Chataignier)
CVCH	*Quercus ilex* L. et *Castanea sativa* Miller (Chêne Chataignier)
CV	*Quercus ilex* L. (Chêne vert)
CBCV	*Quercus pubescens* Willd et *Quercus ilex* L. (Chêne blanc, Chêne vert)
FR	*Fraxinus oxyphylla* M. Bieb, *Ulmus campestris* L. et *Quercus pubescens* Willd (Frêne, Orme champêtre, Chêne blanc)

4) variable *précipitations* (PREC)
20 +: plus de 2000 mm
15 20: de 1500 à 2000 mm
12 15: de 1200 à 1500 mm
11 12: de 1100 à 1200 mm
10 11: de 1000 à 1100 mm
09 10: de 900 à 1000 mm
08 09: de 800 à 900 mm
07 08: de 700 à 800 mm

5) variable *géologie* (GEOL)

OLI	Oligocène (calcaires, argiles, conglomérats) Eocène supérieur (calcaires et brèches)
C.I.	Crétacé inférieur (marnes et calcaires marneux)
J.SM	Jurassique supérieur et moyen (calcaires massifs et dolomies)
J.IL	Jurassique et lias (calcaires, marnes, grés)
C.S.	Cambro-silurien (schistes et grés, calcaires et dolomies)
GRAN	Granites
CONT	Métamorphisme de contact (cornéennes, schistes à minéraux).

A SYSTEM FOR CODING PLANT SPECIES FOR DATA-PROCESSING IN PHYTOSOCIOLOGY*

Sandro PIGNATTI
Istituto ed Orto Botanico, Università, Trieste, Italy I 34100

Keywords:
Coding system, Data-Processing, Phytosociology, Plant species, Nomenclature

Introduction

A first attempt to code plant names for use in phytosociological tables has been made during 1969. A mimeographed paper (4th Circular, Appendix 3 of the Working Group for Data Processing in Phytosociology) gave an account of the method proposed. Since this time, the method has been used subsequently for different tasks (storage of information, ordination, elaboration) without serious incovenience and it has become more and more complete. It is now known to members as the *Trieste Code*.

Although some problems remain unresolved (chiefly in the coding of Cryptogams) it seems now opportune to discuss the criteria on which the code is based and to describe methods used to establish the link between the scientific name of the plant and the code number. The present paper adds further to the proposals in the 4th Circular of 1969, which has had only limited distribution among colleagues in the Group, with addition of results obtained since 1969. A short abstract has been published on the same topic in Italian (Pignatti 1973).

The problem of coding

The floristic information in a phytosociological survey is represented by the names of species which were observed in the relevés. The species following the Linnean nomenclature are identified by three elements: (1) the name of the genus to which the species belongs, (2) the specific attribute and (3) the name of the author(s); e.g. *Convolvulus arvensis* L., *Calystegia sepium* (L.) R. Br., etc. In this way, although the name could be easily stored in a computer file or printed, the manipulation of complete names may yet meet with serious difficulties.

Our first experiment involved an elaboration of data on the phenology of European beech-woods (Lausi & Pignatti 1973) with a complex of nearly 900 specific binomial. An IBM 1620 computer automatically sorted the binomial in alphabetical order and gave to each of them a progressive number. However, at the end, synonyms and typographical errors too, became coded. Even the dominant tree was sorted in two distinct forms – *Fagus sylvatica* l. (modern orthography) and *Fagus silvatica* L. (old usage). This gives an idea of the difficulties occurring by the direct use of the scientific nomenclature when processed by computer. Clearly, it seems desirable to substitute the complete scientific name with a code based on a system different from the Linnean nomenclature.

To code plant names there are three different possibilities:
(1) Coding *ad hoc* without a logical basis. A correspondence is established between plant names and series of numerals or letters, e.g. an alphabetical list corresponding to progressively larger numbers. This method can be used only in a well circumscribed problem without possibility to extend into any other.
(2) Of systems of coding, based on logical principles, it is possible to distinguish two kinds -
(2a) Closed systems in which all elements are conceived a priori, containing a finite number of possibilities, and used only for the flora of a relatively large region.
(2b) Open systems in which only general rules are conceived a priori with feasibility of extension to any number of possibilities and all possible data, even the unexpected.

* Contribution from the Working Group for Data-Processing, International Society for Vegetation Science
** The author thanks G. Cristofolini, E. Feoli and D. Lausi, Trieste, for stimulating discussions, E. van der Maarel, Nijmegen, for critical revision of parts of the manuscript, and Mrs. M. Sauli-Pizzulin, Trieste, for technical assistance.

It seems very difficult to extend the well-established concept of closed and open thermodynamic systems even to informational systems. In the latter case, a closed system corresponds to a compact and homogeneous construct with the pretention of foreseeing all possibilities; it shows a static character (e.g. a flora, a dictionary). An open system is based on a flow of information and shows a dynamic character (e.g. a herbarium). Indeed the difference between closed and open information systems is quantitative rather, than qualitative.

Systems of coding

Several possibilities have been studied:

(a) Ehrendorfer's (1967, 1973) Liste der Gefäßpflanzen Mitteleuropas.

This list gives a complete numerical code of all species occurring in Germany, Austria, Eastern France, Western Czechoslovakia, Switzerland, Northern Italy, and Northern Jugoslavia – all-in-all nearly 960 genera with ca 5000 species. This list is alphabetically ordered; taxonomy and nomenclature are up-to-date and refer to the taxa mostly used in plant sociological tables. This system has been developed for a study of the distribution of plant species belonging to the flora of Central Europe. Our first idea was to investigate if this system could also be used for the coding of plant sociological records from Europe. But then it became clear that this would be impractical for the following reasons:

1) Each genus is labelled with a 3 digit number, the maximum possible entry being 999. However, 960 spaces are already occupied, leaving 38 open for future additions. This number is too small and hinders the extension from the Central European code to a code of broader relevance. The system can be regarded as a closed one.

2) Besides the alphabetical arrangement, which is convenient for finding plant names, the system excludes information on the taxonomy of the genera; direct investigations of higher taxa (family, etc.) are therefore impossible. This is perhaps of secondary importance, but we think that a system must principally store as much information as possible.

3) It must further be remembered that the largest genera (*Carex, Alchemilla* and *Rubus*) possess 2 numbers instead of 1; this can given difficulties.

4) Also the code for the species is unsatisfactory for our work; principally species are coded with 2 digits in sequence to the code number of the genus, e.g. 006 = *Achillea*; 00601 = *Achillea atrata*. Ehrendorfer (1967) suggests a system with three taxa around the specific level, i.e. aggregates such as 00609 = *Achillea millefolium*, normal species such as 00618 = *Achillea nana* and small species such as 00611 = *Achillea collina*. In all three cases, the code has 5 digits; from the coded number it is impossible to know the level of the corresponding taxon. The coded aggregate includes many small species but numbers of aggregates and of small species form a linear succession, so that the information of this hierarchic arrangement is lost. In many cases the specific binomial of the aggregate is equal to the binomial with which one of the included small species is indicated. This is the case of the 'eu-subspecies'. However they have different numbers; e.g. *Achillea millefolium* aggregate 00609 and *Achillea millefolium* small species 00613.

This brings further difficulties into focus because in many cases it is impossible to distinguish in plant sociological tables between the binomial used for the aggregate and for the small species. We can conclude that Ehrendorfer's (1967) code is a coding *ad hoc* to a very large extent, which may well serve its proper task, but which cannot be used for plant sociological records. This is not a criticism of the system itself, but only of our attempt to use it for another task, for which it was not planned. Yet, it can be used as a provisional code, but only in Central Europe and with caution because of the above cited difficulties. The study of this code was particularly important for us in suggesting basic concepts in regard with a more elaborate system.

(b) The Flora Europaea

The materials are taxonomically orded at all levels; nomenclature and the concepts of species and genus are updated and can be easily applied to plant sociological tables. Three taxonomic levels are used around the species, *i.e.* the whole species, the species group (almost corresponding to the 'aggregate' of Ehrendorfer (1967) and the subspecies (in Ehrendorfer (1967) not always recorded).

Up to now only volumes I–III (*Pteridophyta, Gymnospermae, Archichlamydeae* and the first half of *Gamopetalae*) have been published; the following 2 volumes will appear within 2-5 years. Besides, this work contains in a more or less complete description of the species and the subspecies of vascular plants growing in Europe. For this reason, it is a closed system on a global level.

Unfortunately the Flora Europaea lacks continuous numbering of genera and species. A taxon of the lowest order (sub-species) can be identified only by citing the code numbers of all taxa to which it is subordinated (see Table 1).

For example, coding *Silene vulgaris* subsp. (c) *commutata* requires a series of 12 digits, *i.e.* 3-057-028-056-02. The possibility of coding information is enormously great but only a small part is really exploited for the estimated total number of 20.000 European species. Of the 1.10^{12} possible places only 2.10^4 are effectively used; ca 999.999.980.000

Table 1. Estimated number of taxa and number of digits required to code taxa in the system of Flora Europaea

	Estimated maximum number of taxa	Digits required to code taxa
Divisions	3	1
Families in each division	300	3
Genera in each family	200	3
Species in each genus	300	3
Subspecies in each species	25	2*

* In Flora Europaea subspecies are not numbered, but indicated with letters of the alphabet. To eliminate these letters we can think of substituting them by the corresponding numbers from 01 to 25.

places are left unused. In the free places we can introduce new taxa and possibly extend the system to cover the World flora. In this way we can get a partially open system. We think that no division will have more than 999 families, no family more than 999 genera and perhaps no genus more than 999 species. The number of divisions is only 3; thus 7 places remain free for the coding of thallophytes.

However, at this point, a serious difficulty must be recognized. All taxa, as mentioned above, are taxonomically ordered, but only taxa of the European flora are numbered. When a new taxon is added, it obtains the number which follows the last number of the European system; *e.g.* the genus *Nigella* has 12 European species numbered as 001, 002, 003, . . . etc.; if we add *N. oxypetala* and *N. latisecta* (Asia), these species will be coded 013 and 014. The non-European taxa will thus be coded without any order and its numerical sequence will reproduce only the temporal sequence of its appearance in a coded plant sociological relevé. This will be especially disturbing in the more comprehensive units. At the species level, it is rather unimportant whether the species of a genus are partly taxonomically ordered and partly in disorder because one is generally not interested in for instance the number of *Carex*-species with bisexual spikes (sect. *Vignea*) present in a set of relevés. But already with genera, this can produce serious difficulties, at least in larger families.

A case in point is the *Poaceae* including predominantly small herbaceous plants, but its tribe *Bambuseae* comprises plants with a different kind of non- or hardly herbaceous habit. If the genera are ordered taxonomically, we can directly obtain a separation of *Bambuseae* from other grasses. Similarly, we can distinguish spinous *Asteraceae* or *Apiaceae* in a study of pastures. With families we find still more difficulties. Families of the *Angiospermae* will be numbered continuously from the *Salicaceae* to *Orchidaceae* – but each new so far uncoded family must follow the *Orchidaceae*. E.g. *Menispermaceae*, more related to the *Ranunculaceae than to the Orchidaceae*, would still follow the latter. This will make it much more difficult to perform numerical manipulations of the taxonomical composition of different types of vegetation, a field which is just opening exactly by the use of computers. (cf Van der Maarel 1972)

We can conclude, that a code of 12 digits for each singular taxonomical unit is rather extensive but has the advantage that it stores much taxonomical information (and perhaps all information pertaining to the study of the vegetation). The disadvantages are that it is phytogeographically closed, only one part is actually available, and the completion of the lacking parts on *Gamopetalae* and *Monocotyledonae*, which would be necessary for our purpose, may require many years of work.

It is possible to look for other possibilities, such as a direct numbering of genera, combined with the numbering of species in each genus (thus with an estimated number of 3.000 genera, a code of 4 + 3 + 2 digits) or, a numbering of volume, page and species (thus a code of 1 + 3 + 2 + 2 digits), or a direct linear numbering of the species (5 + 2 digits). In these cases we can strongly reduce the number of required digits (see Table 2), but the system becomes too much closed; e.g. a simple comparison with neighbouring plants (for instance between salt-marsh vegetation of Spain and Morocco) is impossible in this way.

The results can be summarized as follows: methods 2, 3 and 4 (Table 2) are closed and cannot be used; method 1 can be possibly used but it must first be completed through very difficult taxonomical work. Method 1 can be regarded as the most open, but the extension to other floras will raise

Table 2. Methods of coding and digits needed to code taxa. Explanation to symbols: 1 - system in Flora Europaea; 2 - numbering genera and species in each genera; 3 - numbering of volume, page and species; 4 - direct linear numbering of species

	Estimated maximum number of taxa	Digits required for coding taxa (1)	(2)	(3)	(4)
Divisions	3	1	-	-	-
Families in each division	300	3	-	-	-
Genera in each family	200	3	4	-	-
Species in each genus	300	3	3	5	2
Subspecies in each species	25	2	2	2	2
Volumes	5	-	-	-	1
Pages	400	-	-	-	3
Total numbers of digits required		12	9	7	8

problems with the taxonomical order in each unit; this can be particularly disturbing in families and genera.

(c) Alphabetic systems

The binomial nomenclature for plant and animal species was perhaps one of the most original ideas of Linnaeus; binomals have a minimum of complexity but a high information content and are easily remembered. Is it possible to use efficiently the classical nomenclature in computer manipulations? This was repeatedly proposed, particularly by Ehrendorfer (1967) and also by Lieth & Moore, (1971) by Holzner & Stockinger (1972) and others. Ehrendorfer (1967) uses 6-7 letters for the genus and 3-6 letters for the species; proper combinations of initial and final letters avoid confusions between similar names (e.g. Centau'a = *Centaurea*; Centau'm = *Centaurium*). In most cases such shortened names permit to recognize immediately the original binomial. Lieth & Moore (1971) print 4 letters for the genus and 4 for the species. These systems are infinitely open, but generic or specific names do not contain information of the taxonomical position of the corresponding taxa. The only possible order is the alphabetical, which is absolutely for scientific purposes irrelevant.

Nevertheless, it must be remembered that an important task of the coding is to avoid computer errors in the presence of synonyms. When we read in a first table *Quercus petraea* and in a second table *Q. sessiliflora* we immediately remember that both binomials designate the same species, but the computer treats them as completely distinct. The situation does not change, if the computer works with shortened synonyms. The alphabetic system does not give a general solution of the coding problem either, but only for handling single tables or local floras, in which only one author is involved and where is no possibility of error because of synonyms.

(d) Gould's International Plant Index.

This work can represent an appropriate system for coding of plant sociological records, but up to the present time only a key for family names and a key for authors of plant genera are available. However, it must be remembered that Gould (1962) suggests a very complicated code:

4 digits for order
1 letter (this can be neglected)
4 digits for the family
5 digits for the genus
3 digits for the species
1 digit for the variety
—
17 digits

It seems that 17 digits for each taxon will make the punching and coding very tedious.

(e) The Trieste Code

Dalla Torre & Harms (1900–1907) published the *Genera Phanerogamarum*, a work in two parts, with a taxonomical and an alphabetical enumeration of the genera of Phanerogams. All genera of the World flora are recorded; a system derived from this list can therefore be regarded, although closed, a global one. The taxonomical enumeration follows the system of Engler (1964) (which has not been considerably modified since its first presentation in 1892) but with the enumeration of Monocotyledons at the beginning of Angiosperms. Recent floras show indeed a tendency to place Monocotyledons after Dicotyledons, but this is not disturbing. A serious disadvantage is that the Pteridophyta are not recorded. Genera are numbered from 1 (*Cycas*) to 9629 (*Thamnoseris*) *i.e.* with 4 digits. Families are numbered 1 to 280 in an independent series and also genera within each family. The maximum is reached by the *Asteraceae* with nearly 900 genera. Synonyms are listed. This work, reprinted in 1963, is found in most botanical libraries. It is used in most herbaria for the general ordering of specimens. And importantly, it is a well-known book to plant taxonomists.

Two possibilities have been examined: (1) adoption of the direct numbering of genera (as in herbaria), each genus coded with 4 digits, or (2) the combined numbering of families (3 digits) and genera (3 digits) using 6 digits for each genus. The first possibility is more efficient and contains full information on the taxonomy of genera. It can also be extended to families but with sub-programs (e.g. genera 101-451 are *Poaceae*, etc.). The second possibility offers lower efficiency (2 digits more for each plant-name), but presents direct storage of taxonomical information on both genera and families. Indeed, it must be remembered, that the family is used only exceptionally in elaborations of plant sociological data. Thus we opt for the direct numbering of genera using four digit numbers from 0001 to 9629.

Species can be added to this system with 3 more digits. Possibly no genera exist with more than 999 species; and also if the largest genera (*Solanum, Carex*) have more than 999 species, it may well be possible to go on for a century before such a number of coded species of a single genus in plant sociological tables will be reached.

For the infraspecific taxa it was originally planned to add 2 digits more after the code of the species, but this idea was subsequently abandoned. This point is discussed further on with reference to the relations between scientific nomenclature and code numbers.

The complete code numbers of species are determined in two ways, depending on whether the species is included or not in the three published volumes of Flora Europaea.

1. For species recorded in Flora Europaea Volumes I-III the code number is determined as follows:

Arabis pauciflora (Grimm) Garcke – code of the genus (Dalla Torre & Harms p. 188): 3001; code of the species (I: 291): 004. With the combination 3001004 it is possible to distinguish *A. pauciflora* from every other species.

Silene cucubalus Wibel – This binomial is a synonym of *Silene vulgaris* (Moench) Garcke (see Flora Europaea I: 168), *i.e.* the 56th species of this genus. The polymorphic species includes 8 subspecies and the coder must judge the subspecies to which the binomial *S. cucubalus* must be referred on the basis of the taxonomical views of the author of the releves. If the author means the (a) subspecies (*S. vulgaris* subsp. *vulgaris*), the definitive coding combination is 2490056, otherwise see the discussion on infraspecific taxa. This adjustment of plant sociological data is very important and in most cases requires the collaboration of an expert taxonomist to avoid wrong interpretations.

Melandryum album (Miller) Garcke is now under *Silene* as the 105th species. The genus is coded as *Silene* (2490) and the species as 105, i.e. 2490105.

Honkenya peploides (L.) Ehrh. is considered in Dalla Torre & Harms as the 13th Section of the genus *Alsine* = *Minuartia*). This genus is coded as *Minuartia* (2441); the species at the end of the European species of *Minuartia, i.e.* as 058. We write for the complete code 2441058.

2. For taxonomical groups not included in Flora Europaea Vol. I-III (*Gamopetalae* p.p. and *Monocotyledones*), the procedure to find the code number of the genus is the same as in the first case (see in Dalla Torre & Harms). The code number of species has been chosen arbitrarily by us mostly on the basis of alphabetical lists of species occurring in Southern and Southwestern Europe; this is incongruous regarding the system of the Flora Europaea, where species are in a taxonomical order and not in an alphabetical order, but this has caused no difficulties up to date. New taxa have been added successively without any order, whenever they occurred in a plant sociological table. Some examples:

Scabiosa argentea L. – 8546001
Chrysanthemum segetum L. – 9341001
Tanacetum alpinum (L.) Sch.-Bip. – 9341810.

Implementation of the coding

Up to 1975, the complete flora of Italy and all names recorded in the list of Ehrendorfer (1967) have been coded, at the Botanical Institute of Trieste, in total ca 9000 plant names. A first card is punched with code number and complete scientific name, a second card with code number and coded biological information such as life-form, chorological groups, polyploidy, etc. The species occurring in phytosociological relevés are indicated only with their code number and, if necessary, the computer, can recall and print scientific name and biological information.

The Flora of Italy, which appear in print in 1976, contains code numbers of ca 5600 species that are known from the country. In this way, we hope that within a short time the data will be available for those who might be interested.

Because of its geographical position, the Italian flora, contains the majority of species of central Europe (ca 95%) and also many species of the Mediterranean flora. Consequently, it will be a relatively easy task to extend the coded list to the flora of the neighbouring countries. If necessary, an alphabetical list of coded names of the species occurring in western Europe, say in the states of the European Economic community, can be obtained and published. Listing can be accomplished automatically on a computer.

Some problems of coding

An open problem: code for the cryptogams

The work of Dalla Torre & Harms lists the phanerogams. Only three groups of cryptogams generally occur in plant sociological tables, i.e. *Lichenes, Bryophyta* and *Pteridophyta*. For these groups are assigned the places 9700, 9800 and 9900.

In *Pteridophyta* it is possible to obtain a complete correspondence with phanerogams, i.e. the 45 genera occurring in Europe were numbered from 9900 to 9944 in the same order in which they are listed in Flora Europaea, vol. I; 55 places are free for successive additions. Species have the same code number of Flora Europaea with 3 digits, e.g. *Huperzia selago* (L.)Bernh. subsp. (a) *selago* is coded as 9900001. The system is open, but the places are not sufficient to allow an extension to the World flora. Even the massive inclusion of data from a fern-rich flora, such as the Japanese, can rapidly exhaust the 55 free places. In this case it will be necessary to study more complex systems such as the system in Christensen's (1906) Index Filicum and in other modern work. Between 9900001 and 9999999 about 10^5 places are available, which are more than sufficient for the complete World flora, estimated to include up to 10,000 species of Pteridophyta.

For *Bryophyta* and *Lichenes* only an *ad hoc* coding has been considered: species are identified with 7 digits – 3 for the general group and 4 for the species, following the enumerations in standard floras.

Table 3. Code and source for enumeration of Cryptogams

Code of the group		Source for enumeration of species
Lichenes	970	Poelt, J. - Schlüssel europäischer Flechten. Lehre 1969
Hepaticae	981	Muller, K. - Die Lebermoose in Rabenhorsts Kryptogamenfl. 3. Aufl. 1951-57
Musci - *Sphagnales*	982	Van der Wijk, R. et al. - Index Muscorum. Vol. I-V. 1959-69
Musci - (other groups)	983	

Examples for coding: *Cetraria islandica* (L.) Ach. 9700120, *Plagiochila asplenioides* Dum. 9810214, *Sphagnum acutifolium* Ehrh. 9820026, *Mnium undulatum* (L.) Hedw. 9830594.

This coding is not homologous with the systems used for vascular plants, but this seems to involve no important difficulties. In vascular plants the necessity to distinguish genera and species is a practical one and a scientific problem such as generic diversity. Such problems are meaningless in *Bryophyta* and *Lichenes,* whose genera are built on other basic concepts and can be hardly compared to genera of vascular plants. Moreover, in most of the plant sociological relevés *Lichenes* and *Bryophyta* are incompletely investigated so that computations would bring uncertain results.

Up to now only vegetation types poor in mosses and lichens have been coded, so that this cryptogamic component is of little consequence. We can imagine that in other vegetation types such as *Oxycocco-Sphagnetea* the problem of coding may become much more complicated.

Nomenclatural changes and code-numbers

Taxonomy, as a science, is continuously changing, whereas Code Trieste as any coding system is necessarily static. This contradiction is a source of few but subtle problems.

The views on the status and the limitation of genera have not remained unchanged after Dalla Torre & Harms, and will change as well in the future. Broad genera are divided into smaller, or conversely, small genera are lumped into broader ones; new genera are described, old genera are recognized as wrong and discarded. The same can happen with species. The static system can develop into dynamic one when it is continuously revised and adapted to the most advanced taxonomical views, but this is possible only with a great deal of difficulty. Are such revisions really necessary? It must be remembered that here, a system for coding information in plant sociological records is studied, not an all-inclusive coding system for plant science. Major points can be separately proposed:

1. *Changes in nomenclature* are without any effect. The numerical code is a stabilizing factor for the taxonomical nomenclature!
2. *Changes in the delimitation of genera* are important only in research on the generic diversity of communities. This topic is very interesting but only occasionally investigated.
3. *The 'genus' concept* in Dalla Torre & Harms is more or less as broad as in modern work and can give a satisfactory basis for geobotanical investigations; with small adaptions it is possible to uphold the system without loosing the information derived from modern taxonomical work.

This last point can be discussed in more detail. The flora of Europe and neighbouring areas is in a quasi climax status as to the number of genera. E.g. *Caryophyllaceae* from the flora of Europe comprise 36 genera according to Dalla Torre & Harms and 37 genera in Flora Europaea. 33 are cited in both works, 3 only in Dalla Torre & Harms, and 4 genera, recorded as such in Flora Europaea, are cited in Dalla Torre & Harms as subgenus or section. Major differences can possibly be expected in tropical floras with a tendency towards an increasing number of genera, but these floras are not yet important for the elaboration of relevés.

In a more practical sense, we can consider fusion and division of genera as a more or less personal matter or mood of the time; in this respect, the opinion of Dalla Torre & Harms is a compromise that can be accepted with just as much justification as some of the prevailing opinions of taxonomists in present time. Moreover, because the opinion of Dalla Torre & Harms is founded on the last general survey of the complete World flora, the genera can be judged in a homogeneous way.

However, in the practical work done coding the whole flora of Italy, it was attempted to harmonize the opinion of Dalla Torre & Harms with modern taxonomical views using the following methods: If several related genera of Dalla Torre & Harms are unified in a single genus, only the code number of the last one is used and the other code numbers for genus are neglected. If a single genus of Dalla Torre & Harms is split in several genera the species of each of them are listed under inverse order of the 5th digit, e.g. *Satureja* (7305) was divided in *Satureja* a.s. (7305001 . . . etc.), *Calamintha* (7305901 . . . etc.), *Micromeria* (7305801 . . . etc.), *Clinopodium* (7305701 . . . etc.) and *Acinos* (7305601 . . . etc.). Major cases of split genera include *Cytisus, Genista, Satureja, Scirpus*.

Minor problems

Scientific plant names are punched completely and without necessity to apply any variation. All letters of the Latin alphabet are available on the computer and other symbols, such as hyphen in *Legousia speculum-veneris* also. In the computer printout plant names are usually typed in capitals. In the taxonomical use, however, only the first letter of the genus-name is capitalized (in old publications also the first letter of some specific epithet). The standardization of letters is not source of trouble, on the contrary, it avoids errors caused by the distinction between normal and capital types. The cards with code number and scientific name are labelled with an F and are regarded as the basic cards of the system. Synonyms have been partly punched on cards indicated with the letter S, but at this moment it is not clear if the punching of synonyms has sufficient utility. In the case of nomenclaturalchanges it is sufficient to change the letter that characterizes the card, the code number itself remains unchanged.

The citation of Author names is well standardized in the classical taxonomical literature, such as L. for Linnaeus, Ten. for Tenore, or Lam. for Lamarck. In flora Europaea some citations have been enlarged (*e.g.* Reichenb. or Dumort. instead of Rchb. and Dum.). This was introduced to avoid mistakes, but can be neglected in the computerized nomenclature. If the complete scientific denomination is firstly punched without error, it is impossible that any error can successively appear.

On the contrary, a very important problem is to reduce the length of the complete scientific denomination. On the card 80 places are available of which the first 7 are used for the code number (two more places are free for possible additions of informations on infraspecific taxa) and the last for the code F of basic cards. These leave 70 places free for the complete scientific name, including Author name. This number is generally sufficient, but it is desirable that the standarized names be as short and compact as possible for using them in the tables. As it is not possible to shorten the latin names, the only way is to write author names in a short form. Therefore, we write in general author names as short as possible. Only surnames are indicated, given names and other indications are neglected, *e.g.* for both G. Don. and G. Don. fil. we write Don., for the geobotanist it is irrelevant if the species has been described by the father or the son. More troublesome are double words which have been shortened to the extreme, *e.g.*

Sibthorp et Smith	= S. et S.
Waldstein et Kitaibel	= W. et K.
Mertens et Koch	= M. et K.
Grenier et Godron	= G. et G.
Humboldt, Bonpland et Kunth	= H., B., K.
Hunter, Porta et Rigo	= Huter, P, et R.

In these extremes the identity of the author is lost as K. can represent in one case Kitaibel and in the other Koch. The same Author must be cited in two different ways, Godron, if alone as Godr., but in conjunction with Grenier as G. This procedure is somehow unlogical but can be explained admitting that in the first case (one author), a definite person is cited, in the second case (two or three authors) a published botanical work. In any way author names are only an element to distinguish the scientific binomial from similar binomials or synonyms and are not involved in geobotanical elaborations. The definitive form in which they are written cannot present any further difficulty.

The names of some French authors are written with accents and apostrophe which can be printed by the computer only with difficulty. These symbols are neglected, e.g.: L'Her., D'Urv. are printed LHER. and DURV.

The complete set of coded names is recorded on magnetic tape and conserved in the Centro di Calcolo dell'Università di Trieste (Italy). It is available for research use. A great deal of effort was spent to obtain a list free from errors. Any addition to this list can bring new errors in the sense that a single species can be assigned two distinct numbers, or two species can be recorded under the same number. This is a good example of informational entropy. Therefore, it is necessary to consider this list as a central source of information. It must be conserved with caution and it can be modified or enriched only following strictly defined rules.

Scientific nomenclature and code numbers

The scientific nomenclature is a way to express a biological reality in symbolic language. The symbolism accords with the rules of nomenclature.

Code numbers represent a further degree of symbolization characterized by higher efficiency in return of the loss of any semantic value. They require only very few general rules (as exposed above), and become explicit by the internal logic of the computer itself.

The relationships between biological reality and symbolic language can be shortly discussed. The basic operational reality is the species. It is probably impossible to give a general definition of this unit, and correspondingly, it is hardly possible to find two species which are completely equivalent* in a taxonomical sense. Ecological and geographical amplitudes are different in each species, and also the inner variability is more or less unique, differring from every other species. We can speak of narrow species as *Primula palinuri* or broad species as *Primula officinalis*; still broader are groups as *Stellaria media*, whose elementary taxa (*Stellaria neglecta* or *Stellaria pallida*) can be regarded also as narrow species, and the cycle is complete. The term 'species' is evidently a generalization. Some authors (cf. Fiori 1923–1929, or Rouy 1893–1913) have a broad concept of species while our concept is more narrow. In the Flora Europaea species are often somewhat broader than in the Flora of Italy but smaller than in Fiori (1923–1929). The first 2600 species of our Flora correspond to about 2550 species in Flora Europaea and only 1950 species in Fiori (1923–1929) Clearly, modern work is more or less standardized around a narrow concept of the species and the differences, as between Flora Europaea and Flora of Italy, are in the range of 1-2% of cases which are regarded as good species by some authors and as infraspecific taxa (mostly subspecies) by others.

Infraspecific taxa are generally listed as subspecies, varieties, forms and so on. In some taxonomical work (e.g. Ascherson & Graebner 1896–1938) a dozen different levels are distinguished. But this hierarchy is mostly avoid of scientific value. Varieties and forms are generally without genetical basis and it does not seem opportune to consider them in plant sociological relevés to avoid subjectivity in the taxonomical information. Subspecies are in many cases better defined, but also in this case a complete coding would seem perfectionism without real advantages. If we accept a current opinion, the one in Flora Europaea, subspecies, in some cases more than 9 to a single species, must also become coded with 2 digits. This means more work, a larger source of error and a loss of efficiency in any elaborations involving code numbers. It seems better to choose a narrower taxonomical concept and regard as species (coded with 7 digits) also the subspecies, when indicated by geographical distrubution, ecological characteristics, or life-form. (In effect most subspecies may be 'good' species but incompletely known and their status is only a provisory one. Otherwise, subspecies can be lacking genetical barriers which characterize species and have only geographical barriers.)

Contrary to these, in special problems it is possible to make use of a broader species concept. In the study of salt marsh vegetation we find the following groups of sociological parallels and taxonomical closely related species:

Atlantic	*Mediterranean*	*Pannonian*
Puccinella maritima	*P. palustris*	*P. limosa*
Limonium vulgare	*L. serotinum*	*L. gmelini*
Aster tripolium	*A. tripolium*	*A. pannonicus*

The corresponding code numbers are:

0384008	0384009	0384010
6351017	6351018	6351012
8900025	8900025	8900026

If we want to differentiate vegetation types in salt marshes on a geographical basis, we must work with the taxa listed above and possibly see if for instance *Aster tripolium* is the same ecotype in Atlantic and in the Mediterranean regions. But, when the investigation is devoted to vicariant associations, with completely distinct species, any comparison becomes impossible: in such cases the syntaxonomy must delineate arbitrary groups of connected species. The connection can exist as geographical vicariance, ecological vicariance, life form, and so on. For these cases subprograms would have to be written.

Conclusions

The combined use of genus numbers (derived from Dalla Torre & Harms (1900–1907) and arbitrary species numbers (partly derived from Flora Europaea) make it possible to identify any European species with a numerical code of 7 digits, which, with some adjustments, can be extended to the World flora. This code, known as the Trieste Code, is adequate for the use in plant sociological work.

The Trieste Code is a provisory one, but it is sufficiently flexible to permit future changes as discussed above. The need of a definitive code is undisputable, but probably it is not a pressing problem in a time of vigorous progresses in plant taxonomy. However, the numerical code is an alternative to the scientific nomenclature which it stabilizes and simplifies. An important point is that classical taxonomy is only the source for the Trieste Code *(Index Siphonogamarum*, species concept, nomenclature) but the Code evolves independently with its own logic and it represents a form of symbolic language of operational value.

Summary

A system for coding plant species is described. The system is in use in the project Data-Processing in phytosociology. The code consists of 7 digits, 4 for the genus and 3 for the species. The genus code is derived from Dalla Torre & Harms, with own ad-hoc numbers for Pteridophyta, Bryophyta and Lichenes. Species codes are established according to Flora Europaea, Vol. I-III, and for species not treated there according to an alphabetical list of species based on S.W. European floras, completed with ad hoc data for other parts of Europe. Ca 9000 species names have been coded at Trieste. The coding system is compared with the systems of Ehrendorfer, Flora Europaea, alphabetic systems and Gould's international plant index. Problems of nomenclatural changes and species concepts and their bearing on the code are discussed.

Zusammenfassung

Ein System für die Codierung von Pflanzenarten wird beschrieben. Das System wird im Projekt der Datenverarbeitung in der Pflanzensoziologie verwendet. Der Code besteht aus 7 Ziffern, 4 für die Gattung und 3 für die Art. Der Gattungscode ist dem System von Dalla Torre & Harms entnommen, mit eigener Numerierung für Pteridophyten, Bryophyten und Flechten. Die Artcodes werden der Flora Europaea Vol. I-III gemäss bestimmt und stützen sich auf eine eigene alphabetische Liste für S- und SW-Europa, mit ad hoc Ergänzungen für das übrige Europa, basiert. Etwa 9000 Arten sind bisher in Trieste codiert worden. Das Code-System wird mit den Systemen von Ehrendorfer, Flora Europaea, alphabetischen Systemen und Gould's international plant index verglichen. Probleme der Änderungen in der Nomenklatur und der Artbegriffe und ihre Bedeutung für den Code werden besprochen.

References

Ascherson, P. & P. Graebner. 1896–1938. Synopsis der mitteleuropäischen Flora. voll. 1-7 and 12, Leipzig.

Christensen, C.F.A. 1906. Index filicum. Hafniae. Suppl. 1-3 (1913–1934).

Crisofolini, G., D. Lausi & S. Pignatti. 1969. Survey of the system for coding of plant sociological records used by the Trieste Group. Working Group for data-processing in phytosociology, Circular 4, Appendix 3. (Mimeographed).

Dalla Torre, C.G. & H. Harms. 1900–1907. Genera Siphonogamarum. Lipsiae, 2 vol. Reprint Wiesbaden (1963).

Engler, A. 1964. Syllabus der Pflanzenfamilien 2. Bd., 12. Aufl. by H. Melchior (ed.). Berlin.

Ehrendorfer, F. 1967. Liste der Gefässpflanzen Mitteleuropas. Graz. 2. Aufl. Stuttgart (1973).

Fiori, A. 1923–1929. Nuova flora analitica d'Italia. voll. 1-2. Firenze.

Gould, S.W. 1962. Family names of the plant kingdom. New Haven and New York.

Greene, D.M. 1972. A taxonomical data bank and retrieval system for a small herbarium. Taxon 21: 621-630.

Hall, A.V. 1972. Computer-based data banking for taxonomic collections. Taxon 21: 13-25.

Heywood, V.H. 1975. Discussion of 'Löve & Löve: Cytotaxonomical Atlas of the Slovenian Flora'. OPTIMA Newsletter 2: 31-35.

Lausi, D. & S. Pignatti. 1973. Die Phänologie der europäischen Buchenwälder auf pflanzensoziologischer Grundlage. Phytocoenologia 1: 1-63.

Lieth, H. & G.W. Moore, 1971. Computerized clustering of species in phytosociological tables and its utilization for field work. In: G.P. Patil, E.C. Pielou & W.E. Waters (ed): Spatial patterns and statistical distributions. Statistical Ecology vol. 1 Penn. State Univ. Press p. 403-422.

Löve, A. & D. Löve. 1974. Cytotaxonomical Atlas of the Slovenian Flora. Lehre.

Maarel, E. van der. 1972. Ordination of plant communities on the basis of their plant genus, family and order relationships. In: E. van der Maarel & R. Tüxen (eds.): Basic problems and methods in phytosociology, p. 183-190. Junk, Den Haag.

Morris, J.W. & O.A. Leistner. 1975. Computerization of the National Herbarium, Pretoria. Taxon 24: 261-270.

Pignatti, S. 1973. Problemi di codifica dei dati floristici in fitosociologia. Not. Fitosoc. 7: 17-20.

Rouy, G.C.C. 1893–1913. Flore de France. voll. 1-14. Asnières, Paris et Rochefort.

Shetler, S.G., J.H. Beaman, M.E. Hale, L.E. Morse, J.J. Crockett & R.A. Creighton. 1971. Pilot data processing systems for floristic information. In: Cutbill J.L. (ed), 'Data processing in Biology and Geology'. pp. 275-310. London and New York.

Stockinger, J.J. & W.F. Holzner. 1972. Rationelle Methode zur Auswertung pflanzensoziologischer Aufnahmen mittels Elektronenrechner. In: E. van der Maarel & R. Tüxen (ed): Grundfragen und Methoden in der Pflanzensoziologie p. 239-261. Junk, Den Haag.

Tutin, T.G. & al. 1964–1972. Flora Europaea. vol. 1-3. Cambridge.
Watson, L. 1971. Basic taxonomical data: the need for organization over presentation and accumulation. Taxon 20: 131-136.

A NUMERICAL CLASSIFICATION OF EUROPEAN SPARTINA COMMUNITIES*

Wil M. KORTEKAAS[1], Eddy VAN DER MAAREL[1] & Wim G. BEEFTINK[2]**

1. Division of Geobotany, University of Nijmegen, Toernooiveld, Nijmegen, The Netherlands
2. Delta Institute for Hydrobiological Research, Yerseke, The Netherlands***

Keywords:
Classification, Numerical methods, Numerical syntaxonomy, *Spartina* communities, Syntaxonomy

Introduction

This contribution is based on two earlier internal reports (Kortekaas & Van der Maarel 1972, 1973): the first one presented at the Colloquium during the Symposium of the International Society for Vegetation Science, at Todenmann near Rinteln, March, 1972; the second at the conference of the Working Group for Data-Processing, Prague, September 1973.

The project described here is aimed at comparing a classical syntaxonomy of *Spartina* communities with the results of numerical treatments. Concurrently with the numerical analyses by W. M. Kortekaas and E. van der Maarel at Nijmegen, W. G. Beeftink at Yerseke, demonstrated and discussed the outcome of the classical approach which he published (Beeftink & Géhu 1973) as the first volume of the new Prodrome series for the European plant communities 'Prodromus der europäischen Pflanzengesellschaften' (Prodrome des Groupements végétaux d'Europe). About 500 relevés were analysed and four associations with sixteen subassociations in one alliance, one order and one class, *Spartinetea maritimae*, were distinguished.

The idea of our approach was to produce a numerical classification on the basis of a set of relevés overlapping as much as possible with the set used for the Prodrome study.

* Contribution from the Working Group for Data-Processing in Phytosociology, International Society for Vegetation Science. Nomenclature follows the Trieste coding system (Pignatti, this issue, and Lausi, Kortekaas & Beeftink, next issue).

** Part of this work was supported by the Netherlands Organization for the Advancement of Puve Research.

*** Communication number 4[illegible].

Material and methods

The Nijmegen material used in the analysis consists of 576 relevés with at least one *Spartina* species with a Braun-Blanquet combined estimation value of at least 1 in each. This material has been selected from about 2200 relevés, stored on punched cards and computer tapes; 1300 forming selection a (see Van der Maarel, Orlóci & Pignatti 1976) and 900 stored at Nijmegen from additional, mostly unpublished *Spartina* relevé tables. It should be noted that it was obvious from the beginning that many relevés thus chosen may not normally be considered as a '*Spartinetum*'. Nevertheless, the sample facilitates establishment of boundaries between the *Spartina* communities towards other communities. The lower Braun-Blanquet value 1 was effective to reduce the total size of the selection.

The relevés were treated with various methods of the CLUSTAN program developed and described by Wishart (1969). This program is accessible on disc via the IBM 370-158 computer of the University of Nijmegen. The method RELOC appeared most promising and will be briefly described here.

The objective of RELOC is a relocation of misclassified relevés within the initial clusters, which have to be specified by the user before the relocation process. These clusters may be the result of a random grouping of the original relevés or a preliminary classification. Each relevé is considered in turn and its similarities with all clusters are computed. A relevé removed from the initial cluster is attached to the one with which it has the highest similarity. A threshold minimum similarity value is chosen at which a relevé is fused with a cluster. The program achieves optimum clustering depending on the homogeneity of the basic material, the number of clusters demanded, and the threshold value. In combination with the relocation process, hierarchical fusions may be performed.

The similarity measure used by us is the similarity ratio (Wishart 1969),

$$S = \frac{\sum x_i y_i}{\sum x_i^2 + \sum y_i^2 - \sum x_i y_i}$$

where x_i is a score of species i in relevé x and y_i a score of species i in relevé y.

This measure is rather similar to the Jaccard index and could be considered as a quantitative analogue of the qualitative Jaccard formula,

$$S_J = \frac{c}{a+b+c}$$

In the Jaccard formula, c is the sum of the lesser values for each species. According to our experience values of S are about 20% higher than the quantitative values of S_J, and about 10% higher than the equally similar Sørensen values.

$$S_s = \frac{2c}{a+b}$$

We started RELOC with 34 clusters, obtained from preliminary CLUSTAN analyses, viz. HIERAR and Group analysis. A threshold value of 0.70 seemed reasonable considering that average similarity of not more than 0.80 may be found between samples of one and the same stand (cf. Curtis 1959). After the first relocation cycle 62 relevés could not be assigned to any cluster because of the threshold value adopted. They were placed in a remainder group. These remainder relevés were then considered as potential noda for additional clusters. From this set 49 clusters were obtained after fusion on the threshold level, i.e. 0.70. Together with the 34 initial clusters they formed a set of 83 clusters for the next relocation cycle. During this cycle some relocation took place, whilst no relevés were placed in a remainder group.

Then a series of fusions and subsequent relocations were performed. At the 81 cluster phase one relevé was placed in the remainder group. Since we wished each relevé to be classified in a cluster that is homogeneous at the chosen level, i.e. 0.70, the procedure was stopped at that phase and the result was thus a system of 82 clusters.

To obtain an idea of the hierarchical structure of the clusters, an agglomerative clustering was performed. In the CLUSTAN set, program HIERAR is available for this purpose. Since this program would take too much computer time, a complete linkage analysis (cf. Sokal & Sneath 1963) was performed long-hand. This clustering was terminated at the 0.30 similarity level where only 12 clusters were left.

Interpretation of the clusters

Selection of Spartineta clusters

From an inspection of the clusters it appeared that nearly all clusters showed a constant occurrence of one or more species and were thus easily to characterise. Then some provisional rules were set up in order to select the *Spartineta* clusters. (These rules are entirely arbitrary and they have no relation to the numerical procedure. They do relate though to the problem of delimitating species-poor plant communities with dominance or sometimes codominance of species).

1. At least one *Spartina* species should constantly occur in the relevés with a minimum Braun-Blanquet value of 2.
2. If one constant companion species occurs, the Braun-Blanquet value of *Spartina* should exceed the value of this species with at least 2 scale values.
3. If two or more constant companions occur, the Braun-Blanquet value of *Spartina* should exceed the value of this species with at least 2 scale values.

Based on the use of these rules 20 clusters were recognised. A new dendrogram of these 20 clusters has been constructed (Fig. 1). The other clusters were removed from consideration. Within the selected groups two categories of clusters may be distinguished:

1. Clusters in which all relevés conform to one of the above mentioned rules.
2. Clusters in which one of the rules can be applied to a portion of the relevés.

In the synoptic Tables as presented in Tables 1–3, the relevés which do not satisfy the rules were omitted. The three *Spartina patens* clusters have also been left out of the general dendogram and the synoptic tables, because two of them represent transitional types and the remaining typical cluster would contain only three relevés.

Syntaxonomical ranking

The 20 clusters were syntaxonomically ranked according to the following considerations: It is reasonable to expect a certain relation between the homogeneity level within a cluster and the syntaxonomical rank that cluster should obtain. Ellenberg (1956) stated that the average similarity of relevés within an association is between 0.25 and 0.50 and that subunits should be distinguished on levels over 0.50. These are Jaccard values. According to our experi-

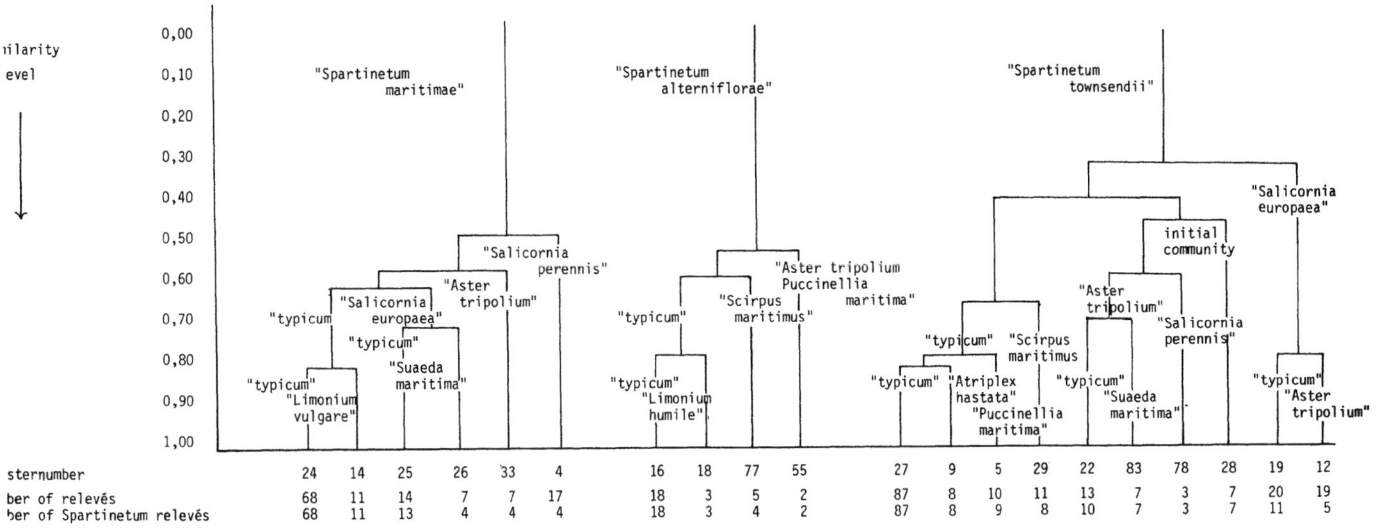

Fig. 1. Agglomerative clustering of *Spartina* communities. Resulting clusters are interpreted in terms of the existing syntaxonomical hierarchy. See text.

ence the Sørensen values are 10% higher, in which case the interval would be 0.30–060. Sørensen (1948) found that his grassland groups on a similarity level between 40 and 50 (Sørensen values) corresponded with the alliance level. Looman & Campbell (1960) calculated Sørensen values of > 0.70 within subunits belonging to one grassland association, whereas values between subunits were all <0.50. Hofmann & Passarge (1964) presented group affinity values between various woodland associations and subassociations. Within associations values were mostly > 0.60, between typical subassociations of related associations values tended to be between 0.30 and 0.50. Raabe (1952) obtained affinity values (Kulczińiski coefficient) between associations and alliances of weed, salt marsh and alpine communities. Within-alliance values were between 0.40 and 0.50, between-alliance values were 0.20 to 0.40. Within salt marsh associations values varied more widely, from 0.30 to 0.80, and here the influence of geographical distance between local representatives of associations was evident. These and other similarity level indications have been discussed by Westhoff & Van der Maarel (1973).

For our purpose we take values between 0.40 and 0.60 as association level, values between 0.61 and 0.70 as subassociation level and values from 0.71–0.80 as variant level. Of course, these values are again arbitrary; the main purpose of the ranking is to facilitate a direct comparison with the units of the Prodrome. We shall therefore speak of associations, subassociations and variants, but use quotation-marks to indicate that they were derived by our scheme based on the given similarity levels.

Fig. 1 shows three groups of clusters, characterised by respectively *Spartina maritima*, *Spartina alterniflora* and *Spartina townsendii*. They have been considered as representatives of three associations, '*Spartinetum maritimae*', '*Spartinetum alterniflorae*' and *Spartinetum townsendii*'. The three groups will be discussed. One should realise that:

1. At present no algae have been recorded on punched cards. Thus in our study algae could not be used as differential taxa.
2. At present no subspecies have been recorded on punched cards and thus used in characterizing the units.

'*Spartinetum maritimae*'

General structure of the dendogram:

Cluster 24 represents the central cluster here; clusters 24 and 14 are fused at a similarity level of 0.81, 25 and 26 at 0.69; clusters 24, 14, 25 and 26 are fused at 0.61 while cluster 33 and 4 remain separated down to below the 0.61 level.

Syntaxonomical interpretation:

The following interpretations are based upon the rules presented above for the clusters identified by numbers:

24 'subassociation *typicum*', 'variant *typicum*'*

14 'subassociation *typicum*', 'variant with *Limonium vulgare*'

* We use the indication typicum here, although we realise that we usually deal with 'a-typical' species-poor forms, which should rather we called 'inops' (cf. Westhoff & Van der Maarel 1973).

Table 1. Synoptic table of *Spartina maritima* communities on 'variant' and 'subassociation' level. See Fig. 1 for interpretation of clusters. For each species % presence and range of Braun-Blanquet values in each cluster is indicated

	"variant" level						"subassociation" level					
Cluster	24	14	25	26	33	4	24-14	25	26	33	4	Total
Spartina maritima	100^{2-5}	100^{2-5}	100^{3-5}	100^{3-4}	100^{4-5}	100^{4}	100^{2-5}	100^{3-5}	100^{3-4}	100^{4-5}	100^{4}	100^{2-5}
Limonium vulgare	24^{+}	100^{1-2}	15^{+}				34^{+-2}	15^{+}				28^{+-2}
Salicornia fruticosa	7^{+-1}	56^{r-2}					14^{r-2}					11^{r-2}
Salicornia europaea	19^{+}	9^{+}	100^{1-2}	50^{+-1}	50^{+-1}	50^{+}	18^{+}	100^{1-2}	50^{+-1}	50^{+-1}	50^{+}	32^{+-2}
Spartina townsendii	15^{r-+}		8^{+}	100^{1-2}			13^{r-+}	8^{+}	100^{1-2}			14^{r-2}
Aster tripolium	9^{+}	18^{+}	8^{1}		100^{1-2}	25^{+}	10^{+}	8^{1}		100^{1-2}	25^{+}	13^{+-2}
Salicornia perennis						100^{3}					100^{3}	4^{3}
Puccinellia maritima	4^{+}		38^{+}		25^{+}		4^{+}	38^{+}		25^{+}		9^{+}
Halimione portulacoides	1^{+}		8^{+}			50^{+}	1^{+}	8^{+}			50^{+}	5^{+}
Suaeda maritima	6^{+-2}		15^{+}				5^{+-2}	15^{+}				7^{+-2}
Puccinellia palustris	7^{+-1}	27^{+-1}					16^{+-1}					8^{+-1}
Triglochin maritima			8^{+}					8^{+}				1^{+}
Juncus maritimus		9^{1}					9^{1}					1^{1}
Number of relevés	68	11	13	4	4	4	79	13	4	4	4	104

25 'subassociation with *Salicornia europaea*'
26 'subassociation with *Spartina townsendii*'
33 'subassociation with *Aster tripolium*'
4 'subassociation with *Salicornia perennis*'

Table 1 presents the synthetic data on the various subassociations and variants. It follows from this table that each of the syntaxonomic subunits is characterised by exactly one differential species. The subassociation with *Spartina townsendii* almost reaches the variant level. Should a lower similarity level such as 0.50 be used for the distinction sub-associations, two subunits would remain, the first including clusters 24, 14, 25, 26, 33 and the second including cluster 4. The first of these two units cannot be characterised, however, by a constant differential species and would then be called '*typicum*'. When a still lower level would be accepted, the whole dendrogram would represent one single subassociation with five variants.

'*Spartinetum alterniflorae*'
General structure of the dendrogram:
Cluster 16 is the central cluster; clusters 16 and 18 are fused at a level of 0.75; both cluster 77 and 55 remain separated below 0.61.
Syntaxonomical interpretation:
Based on the rules presented above, the following interpretations can be made for the clusters:
16 'subassociation *typicum*', 'variant *typicum*'
18 'subassociation *typicum*', 'variant with *Limonium humile*'
77 'subassociation with *Scirpus maritimus*'
55 'subassociation with *Aster tripolium* and *Puccinellia maritima*'

Table 2 presents the synthetic data for the four subunits; the subassociations are clearly characterised, the third one even by two constant differential species: *Aster tripolium* and *Puccinellia maritima*. Should the 0.50 similarity level be accepted for the subassociation level only one subassociation '*typicum*' would remain.

'*Spartinetum townsendii*'
General structure of the dendrogram:
Cluster 27 is the general cluster; clusters 27, 9 and 5 are fused at a similarity level of 0.82; clusters 27, 9, 5 and 29 form one cluster at 0.69; clusters 22 and 83 are fused at 0.70 and fused with 78 at 0.61; 28 is fused with clusters 22, 83, 78 at 0.48; clusters 27, 9, 5, 29 form one group with 22, 83, 78, 28 at 0.42; 12, 19 form an own group at 0.78 (cluster 19 is close to 27, 9, 5); 12, 19 are fused with the large cluster of 27, 9, 5, 29, 22, 83, 78, 28 at 0.34.
Syntaxonomical interpretation:
According to the established rules, we can make the following interpretations:
27 'subassociation *typicum*', 'variant *typicum*'
9 'subassociation *typicum*', 'variant with *Atriplex hastata*'
5 'subassociation *typicum*', 'variant with *Puccinellia maritima*'

Table 2. Synoptic table of *Spartina alterniflora* communities. See Fig. 1 and Table 1 for further explanation.

	"variant" level				"subassociation" level			
Cluster	16	18	77	55	16-18	77	55	Total
Spartina alterniflora	100^{4-5}	100^{3-4}	100^{4-5}	100^{4-5}	100^{3-5}	100^{4-5}	100^{4-5}	100^{3-5}
Limonium humile	6^{+}	100^{1-2}			19^{+-2}			15^{+-2}
Scirpus maritimus			100^{2}			100^{2}		15^{2}
Aster tripolium	28^{+-1}	33^{+}		100^{1-2}	29^{+-1}		100^{1-2}	30^{+-2}
Puccinellia maritima	6^{+}			100^{1-2}	5^{+}		100^{1-2}	11^{+-2}
Agrostis stolonifera				50^{+}			50^{+}	4^{+}
Spergularia media				50^{1}			50^{1}	4^{1}
Juncus maritimus		33^{+}			5^{+}			4^{+}
Atriplex hastata				50^{+}			50^{+}	4^{+}
Suaeda maritima				50^{+}			50^{+}	4^{+}
Salicornia europaea		33^{2}			5^{2}			4^{2}
Number of relevés	18	3	4	2	21	4	2	27

Table 3. Synoptic table of *Spartina townsendii* communities. See Fig. 1 and Table 1 for further explanation

	"variant" level									"subassociation" level						
Cluster	27	9	5	19	12	22	83	29	78	27-9-5	12-19	22-83	29	78	Total	(28)
Spartina townsendii	100^{3-5}	100^{5}	100^{4-5}	100^{3-5}	100^{4-5}	100^{3-5}	100^{3-5}	100^{4-5}	100^{4-5}	100^{3-5}	100^{3-5}	100^{3-5}	100^{4-5}	100^{4-5}	100^{3-5}	100^{1-2}
Atriplex hastata	14^{r-+}	100^{1-3}	56^{+-1}			50^{+}	14^{+}	75^{+-2}		23^{r-3}		35^{+}	75^{+-2}		24^{r-3}	
Puccinellia maritima	9^{r-+}		100^{1-2}		20^{+}	20^{r-+}	43^{+}	25^{+-1}		17^{r-2}	6^{+}	29^{r-+}	25^{+-1}		18^{r-2}	
Salicornia europaea	16^{+}		11^{+}	100^{1-2}	100^{1-2}	30^{r-+}	86^{+-1}			15^{+}	100^{1-2}	53^{r-+}			27^{r-2}	29^{+}
Aster tripolium	13^{r-+}	43^{+}	33^{+-1}	56^{r-+}	100^{1-2}	100^{1-2}	76^{+-1}	37^{r-+}		17^{+-1}	69^{r-2}	76^{+-2}	37^{r-+}		32^{r-2}	29^{+}
Suaeda maritima	7^{+}	14^{+}	22^{+}	56^{+}	60^{+}	10^{+}	100^{1-2}			9^{+}	56^{+}	47^{+-2}			18^{+-2}	14^{+}
Scirpus maritimus	2^{r-+}	14^{+}	11^{1}			20^{+}		100^{1-3}		4^{r-1}		12^{+}	100^{1-3}		9^{r-+}	
Salicornia perennis	1^{+}								100^{2}	1^{+}				100^{2}	3^{+-2}	
Halimione portulacoides	7^{+-1}		11^{+}	9^{1}		10^{1}	71^{+-1}		33^{2}	7^{+-1}	6^{1}	35^{+-1}		33^{2}	10^{+-2}	
Limonium vulgare	2^{+}	29^{+}		9^{+}		10^{+}	29^{+}			4^{+}	6^{+}	18^{+}			5^{+}	
Triglochin maritima	2^{r-+}	14^{+}				20^{r-+}	14^{1}			3^{r-+}		18^{r-+}			4^{r-+}	
Plantago maritima	1^{+}		11^{+}			20^{+}				1^{+}		12^{+}			2^{+}	
Spergularia media						10^{+}				1^{+}		6^{+}			1^{+}	
Glaux maritima	1^{+}									1^{+}					1^{+}	
Spartina maritima	1^{r}									1^{r}					1^{r}	
Agrostis stolonifera						20^{+-1}						12^{+-1}			2^{+-1}	
Phragmites communis								12^{r}					12^{r}		1^{r}	
Number of relevés	87	7	9	11	5	10	7	8	3	103	16	17	8	3	147	7

22 'subassociation with *Aster tripolium*', 'variant *typicum*'
83 'subassociation with *Aster tripolium*', 'variant with *Suaeda maritima*'
19 'subassociation with *Salicornia europaea*', 'variant *typicum*'
12 'subassociation with *Salicornia europaea*', 'variant with *Aster tripolium*'
29 'subassociation with *Scirpus maritimus*'
78 'subassociation with *Salicornia perennis*'
28 initial community of *Spartina townsendii*

The last cluster does not satisfy item 1 of the rules presented and cannot therefore be referred to as a '*Spartinetum*'. It nevertheless is still retained for comparison as initial cluster.

Table 3 presents the synthetic data for the ten subunits. Three out of four subassociations are well characterized, but the subassociation *asteretosum tripolii* is not entirely satisfying. Now the two clusters forming this subunit, 22 and 83, are fused at 0.61, almost the lower limit of subassociations. One should therefore consider these two clusters rather as subassociations, characterised by *Aster tripolium* and *Suaeda maritima* respectively.

At the subassociation level of 0.50 three subassociations would remain, but again the characterisation would then be less satisfactory: 27, 9, 5, 29 would form '*typicum*', and 22, 83, 78 would form a floristically loose grouping with *Aster tripolium* as its (weak) differential species. The group 12, 19 only fuses below 0.40 and would properly fall outside the association category. However, a large number of transitional relevés occur in this cluster, which caused the low fusion level. In its purified førm the cluster would certainly qualify as an association.

Comparison of clusters with syntaxonomical units of Spartinetea maritimae in the Prodrome

The Prodrome's system

The Prodrome treats four associations in one alliance *Spartinion maritimae* (order *Spartinetalia maritimae*, class *Spartinetea maritimae*). The associations are: *Spartinetum maritimae*, *Limonio-Spartinetum maritimae*, *Spartinetum alterniflorae* and *Spartinetum townsendii*. The *Limonio-Spartinetum* is not found in the Nijmegen system. This N. Adriatic association is characterised only by *Limonium vulgare* ssp. *serotinum* as differential taxon. These subspecies occurs in mediterranean *Spartinetum maritimae*, while in the atlantic *Spartinetum maritimae* the ssp. *pseudo-limonium* can be found. Up till now no taxa below the rank of species have been distinguished in the storage files of the Working Group, and for that reason no clusters with the subspecies serotinum could be expected. It should however be doubted, that a cluster would separate on the association level. According to Pignatti (personal communication) the occurrence of this *Limonium* taxon in *Spartina* communities is only marginal. Thus both on numerical and local phytosociological grounds the existense of the association *Limonio-Spartinetum maritimae* could be questioned.

Spartinetum maritimae (Table 4)

In both systems five subassociations are found of which four are similar. The subassociation *salicornietosum fruticosae* is not represented in the Nijmegen system while the '*spartinetosum townsendii*' is not distinguished by the Prodrome. The '*spartinetosum townsendii*' cluster fuses at

Table 4. Comparison of numerical system of *Spartina maritima* communities with syntaxonomical system of *Spartinetum maritimae*

Prodrome \ Nijmegen: "Syntaxa" / "subassociations" / "variant"	Mean species number	Total species number	"typical": "typical"	"typical": "*Limonium vulgare*"	"*salicornietosum tripolii*"	"*asteretosum tripolii*"	"*salicornietosum perennis*"	"*spartinetosum townsendii*"	Total	
Syntaxa			1,9	2,1	2,7	2,8	3,3	2,5	2,3	Mean species number
			10	11	9	4	5	3	13	Total species number
typical*	1,3	6	68 / 66	11						
*salicornietosum strictae**	2,7	6			13 / 18					
asteretosum tripolii	3,0	6				4 / 16				
salicornietosum perennis	3,1	9					4 / 16			Number of relevés
salicornietosum fruticosae	2,6	3					5	4		
Total	2,5	9							104 / 121	

* Division of these subassociations into variants are based on algae species. At present no algae have been recorded on punched cards. Thus in our study algae could not be used as differential taxa.

the 0.61 level with clusters 24, 14, 25 and is almost a variant. The 'variant with *Limonium vulgare*' within the 'subassociation *typicum*' is only present in the Nijmegen system. The relevés of this unit originate from the N. Adriatic and have been assigned to the *Limonio-Spartinetum maritimae maritimae* in the Prodrome.

Comparison of the synthetic table on page 8 of the Prodrome with our Table 1 indicates (see our Table 4) that the mean species numbers in both systems are the same, both in the total association and in the subassociations, except in the subassociation *typicum*, where the Nijmegen value is higher; the total species numbers in the tables are 6 (Prodrome) and 10 (Nijmegen): the total relevé numbers are about equal when we compare our material with the table in the Prodrome manuscript. In the final version of the Prodrome nearly 30 new relevés were taken into account (most of them from the 1972 excursion to Portugal of the International Society for Vegetation Science and most of them assigned to subassociation *typicum*). This makes a direct comparison more difficult. Still one conclusion is permitted: when we compare our cluster 24 with the *typicum* from the Prodrome (our cluster 4 contains most of the relevés with *Limonium vulgare* ssp. *serotinum* and is not considered) we find about equal relevé numbers, 68 and 66. However, 28 Prodrome relevés were new, i.e. they were added to the Prodrome set later on, whilst the number of relevés assigned to this syntaxon in the Prodromus manuscript table was only 49. This may be explained by the deletion of relevés during the completion of the Prodrome in order to purify the synthetic table. Interestingly enough the results of both attempts are still very similar.

Spartinetum alterniflorae (Table 5)

There are three corresponding subassociations, *typicum asteretosum tripolii* and *scirpetosum maritimae*, of which the subassociation typicum has similar subdivisions in both systems. The *scirpetosum maritimae* is only subdivided by the Prodrome into the variants *typicum* and *Atriplex hastata*. The third subassociation is characterised by one differential species in the Prodrome and by two in the Nijmegen system – *Aster tripolium* (Prodrome) and *Aster tripolium* plus *Puccinellia maritima* (Nijmegen) respectively. In the Prodrome *Puccinellia maritima* is used to characterise a variant within the *asteretosum tripolii*.

Table 5. Comparison of numerical system of *Spartina alterniflora* communities with syntaxonomical system of *Spartinetum alterniflorae*

Prodrome \ Nijmegen "Syntaxa"				"subassociation": "typical", "variant": "typical"	"subassociation": "typical", "variant": "*Limonium humile*"	"*Aster tripolium* and *Puccinellia maritima*"	"*scirpetosum maritime*"	Total	
				1,4		5,0	2,8	1,9	Mean species number
				6		7	2	11	Total species number
subassociations	**variants**	**Mean species number**	**Total species Number**						
typical	(- typical (- with Limonium (humile	1,1	3	22 / 21					Number of relevés
asteretosum tripolii	(- typical (- with *Puccinellia* (*maritima*	3,0	7			8 / 2			
scirpetosum maritimae	(- typical (- with *Atriplex* (*hastata*	3,0	5				5 / 4		
Total		2,3	10					35 / 27	

The Nijmegen unit has only two relevés. Comparison of the origin of the relevés shows that a number of relevés in our cluster 16 ('typical subassociation', 'typical variant') must have been classified as subassociation *asteretosum* in the Prodrome. The numerical clusters 16 and 55 seem to be not well-distinguished indeed.

The explanation of discrepancies is easy, but requires some additional information on the method described in part 1 of this contribution. The test set of 576 relevés has been subjected to various clustering techniques. (The results of all the treatments are not relevant for the present study and will not be discussed here). In order to obtain a more realistic comparison between techniques bound to presence-absence data (e.g. association analysis according to Williams & Lambert 1959) and quantitative approaches the combined estimation values *r* and + of all species have been omitted. On the whole this seems reasonable when dominants like the *Spartina* species characterise a plant community. In the case of the *Spartinetum alterniflorae asteretosum* most of the relevés which were differently classified contained *Aster tripolium* with Braun-Blanquet value + only.

The mean species number in the total association is somewhat higher in the Prodrome system (see Table 5). The subassociations have similar species numbers except the second one (Prodrome 3.0, Nijmegen 5.0). Total species numbers are for the association 10 (Prodrome) and 11 (Nijmegen); for the subassociations *typical* 3 (Prodrome) and 6 (Nijmegen), *asteretosum tripolii* 7 (both systems), *scripetosum maritimae* 5 (Prodrome) and 2 (Nijmegen). Again the differences are not too great.

Spartinetum townsendii (Table 6)

Both systems distinguish similar subassociations. In the subassociation *typicum* corresponding variants are found. However, the division of the *salicornietosum strictae* and the *asteretosum tripolii* is different in the two systems. The *salicornietosum strictae* from the Prodrome does not have variants, the corresponding cluster of the Nijmegen system has a variant with *Aster tripolium*, while the *asteretosum tripolii* of the Prodrome has a variant with *Salicornia europaea*. In the Nijmegen subassociation *asteretosum Suaeda maritima* is used to characterise a variant.

It is interesting to remark that the subassociation *asteretosum* did not appear in an earlier draft of the Prodrome and was apparently distinguished on the suggestion of the results of the numerical approach. This is thus a first example how a numerical classification can

Table 6. Comparison of numerical system of *Spartina townsendii* communities with syntaxonomical system of *Spartinetum townsendii*

Prodrome \ Nijmegen		"Syntaxa"	"subassociations"	"typical"				*"salicornietosum strictae"*		*"asteretosum tripolii"*		*"Salicornietosum perennis"*	*"scirpetosum maritimae"*	Total	
			"variant"	"typical"	*"Puccinellia maritima"*	*"Atriplex hastata"*		"typical"	*"Aster tripolium"*	"typical"	*"Suaeda maritima"*				
subassociations	variants	Mean species number	Total species number		2,0				3,4		4,3	2,7	3,5	2,6	Mean species number
					15				7		13	3	6	17	Total species number
typical	(- typical (- *Puccinellia maritima* (- *Atriplex hastata*	1,3	5		89	9 100	7								
salicornietosum strictae		2,2	5					11 47	5						
asteretosum tripolii	(- typical (- *Salicornia europaea*	3,4	7							10 74	7				Number of relevés
salicornietosum perennis		2,8	6									3 8			
scirpetosum maritimae		3,4	5										8 17		
Total		2,6	9											147 246	

indicate syntaxa hitherto not recognised in classical treatments.

The mean species numbers at the association level are equal (see Table 6). At the subassociation level only in the two last subassociations are these values similar; while the first three have higher values in the Nijmegen system, especially the *salicornietosum strictae* and the *asteretosum tripolii*.

The total species number of the typical subassociation is 5 for the Prodrome and 15 for the Nijmegen system. This set contains 51 relevés with 1 species, 25 relevés with 2 species, 15 relevés with 3 species, so that there are 12 relevés with more than three species, leading to the difference in total species number. The relevé number is about equal for both systems. Only few relevés have been added to the original Prodrome manuscript table.

For the *salicornietosum strictae* the total species numbers are 5 (Prodrome) and 7 (Nijmegen). The Prodrome has 31 relevés more than Nijmegen, most of which were already present in the original manuscript table.

In the *asteretosum tripolii* the total species number is 7 (Prodrome) and 13 (Nijmegen), while the Prodrome has 57 relevés more than Nijmegen. Comparison of the two systems is rather difficult. In the original version of the Prodrome table this syntaxon was not distinguished (see remark above).

The total species number for the *salicornietosum perennis* is 6 (Prodrome) and 3 (Nijmegen), for the *scirpetosum maritimae* 5 (Prodrome) and 6 (Nijmegen). For both syntaxa the Prodrome used more relevés than Nijmegen. The total relevé number at the association level is about 100 relevés higher in the Prodrome.

Remarks on the Spartina patens clusters

The Prodrome has not taken up *Spartina patens* communities in the *Spartinetea maritimae*. The reason for that is as follows. The *Spartinetea maritimae* were delimited by the Prodrome authors so as to have only those *Spartina* taxa as character-taxa for the lower syntaxa which belong to Mobberley's (1956) taxonomic group II. *Spartina patens* is not assigned to that group. In our study we only had seven *Spartina patens* relevés, spread over three small clusters:

Cluster 17: '*typicum*', consisting of three relevés. Mean species number is 5.3 and total species number 11. Here *Spartina patens* occur with Braun-Blanquet value 5. *Phragmites communis* is present in all relevés (+ to 2).

Cluster 36 and 35: both have two relevés with mean

species number 6.5 and 12 respectively. Total species number is 9 and 14. Cluster 36 and 35 fuse at the 0.68 level and they fuse with '*typicum*' at 0.52. In these two groups *Phragmites australis* (+), *Juncus maritimus* (2 to 3) and *lachenalii* (+ to 1) occur in all relevés, whereas the last cluster also contains species of the *Plantaginion crassifolii* as *Linum maritimum*, *Centaurium tenuiflorum* and *Plantago crassifolia*. Note that the Prodrome mentions the penetration of *Spartina patens* in some indigenous communities of the West Mediterranean.

Discussion

The procedure described in this paper leads to a hierarchical system of vegetation units which can be easily described by constant differential species and compared with each other as to their homogeneity level. The procedure for the recognition of syntaxonomical units, i.e. 'associations', 'subassociations' and 'variants', is admittedly an arbitrary one and it demonstrates the advantages of continuous involvement of the investigator in the classification and it shows a specific and effective use of the computer in automating parts of the job.

The present syntaxonomical system does not include a general rule for the distinction of subassociations and variants. The system has however the advantage that the lower rank units are established in a general and reproducible procedure. The reason for ranking in a syntaxonomical system could have been a variation in the weighting of differential species. This weighting could have synsystematical or synecological reasons, e.g. *Salicornia europaea* is an association character species, therefore a subassociation *salicornietosum* might be considered of more validity than one with *Aster tripolium*. However, we do not recognise any justification for such a different weighting, nor did we find any general motivation for it in textbooks. In fact we think the distinction of lower units proceeds intuitively. At least for the *Spartinetea* the levels chosen here for the subassociation and variant categories seem to be realistic.

We did not consider the difficult problem of higher synsystematical units in *Spartineta*. According to our findings the three associations are floristically discontinuous and should properly be termed classes of their own. However, problems such as the introduction of *Spartina alterniflora* and *townsendii*, which cannot be discussed in the present contribution, should be considered with due weight in a general approach to the syntaxonomy of the *Spartineta*.

The overall similarity between the system of European *Spartina* communities as developed in the Prodrome and the numerical classification system of Nijmegen is obvious. This may be considered encouraging for both approaches. The advantage of the numerical approach is no doubt its easier comprehensibility. Species numbers in the numerical units are slightly higher on the average. This could be understood in such a way that the Prodrome units may have been purified by deleting relevés with additional species. The numerical units are less 'pure', but they reflect the natural variation in the syntaxa in a better way. A disadvantage of the deletion of atypical relevés is that one cannot determine what is considered impure, if the deleted relevés are not presented.

With the help of the numerical approach it is worth trying to fix levels of similarity for the various hierarchically coherent lower units such as the variants and subassociations. It remains to be seen whether such levels should vary in different kinds of plant communities. In conclusion we may state that this study has shown perspectives for a more general use of numerical methods in syntaxonomy, which may be indicated as numerical syntaxonomy (Westhoff & Van der Maarel 1973).

Summary

The classical syntaxonomical treatment of the European *Spartina* communities as published in the series Prodrome of the the European plant communities, is compared with the results of a numerical treatment, based on largely the same set of relevés. 576 relevés, selected from the total salt marsh data set were subjected to agglomerative clustering with relocation with the similarity ratio as similarity measure. The resulting numerical system was compared with the syntaxonomical hierarchy. The correlation between both systems is close. The numerical units are slightly more heterogeneous because no purification occurred, which implies relevés to be left out of consideration. One new syntaxon, *Spartinetum townsendii asteretosum tripolii*, could be suggested from the results of the numerical treatment. Perspectives for the development of a numerical syntaxonomy are stressed.

Zusammenfassung

Die klassische syntaxonomische Bearbeitung der europäischen *Spartina* Gesellschaften, wie im Prodromus der europäischen Pflanzengesellschaften veröffentlicht, wird mit den Ergebnissen einer numerischen Bearbeitung verglichen, und zwar auf Basis eines etwa gleichen Aufnahmematerials. 576 Aufnahmen, aus dem gesammten Salzwiesenmaterial entnommen, wurden mit einem agglomerativen Schwarm-Verfahren auf Basis der similarity ratio bearbeitet. Das resultierende numerische System wurde mit der syntaxonomischen Hierarchie verglichen. Der Zusammenhang zwischen beiden Systemen ist gross. Die numerischen Einheiten sind etwas heterogener, weil keine Tabelle-Bereinigung stattfand und daher Aufnahmen ausser Betracht gelassen wurden. Ein neues Syntaxon, *Spartinetum townsendii asteretosum tripolii* konnte aus den Ergebnissen der numerischen Bearbeitung abgeleitet worden. Die Perspektive für die Entwicklung einer numerischen Syntaxonomie werden betont.

References

Beeftink, W. G. & J.-M. Géhu. 1973. Spartinetea maritimae. In: R. Tüxen (ed.). Prodrome des Groupements végétaux d'Europe. Lieferung 1. J. Cramer Verlag, Lehre. 43 pp.

Curtis, J. T. 1959. The vegetation of Wisconsin. An ordination of plant communities. Univ. of Wisconsin Press, Madison, Wisconsin. XI+657 pp.

Ellenberg, H. 1956. Grundlagen der Vegetationsgliederung. I. Teil: Aufgaben und Methoden der Vegetationskunde. In: H. Walter (ed.), Einführung in die Phytologie IV-1. Stuttgart, 136 pp.

Hofmann, G. & H. Passarge. 1964. Über Homogenität und Affinität in der Vegetationskunde. Arch. Forstwes. 13: 1119–1138.

Kortekaas, W. M. & E. van der Maarel. 1972. A numerical classification of Spartinetum vegetations. Preliminary paper Int. colloquium Rinteln 1972. 9 pp.+tables.

Kortekaas, W. M. & E. van der Maarel, 1973. A numerical classification of Spartinetum vegetations II. Comparison of the computer-based numerical system with the system published in the 'Prodrome des Groupements végétaux d'Europe'. Paper Symp. Working Group for Data-Processing in Phytosociology, Prague september 1973. 5 pp.+tables.

Looman, J. & J. B. Campbell. 1960. Adaptation of Sørensen's K (1948) for estimating unit affinities in prairie vegetation. Ecology 41: 409–416.

Maarel, E. van der, L. Orlóci & S. Pignatti. 1976. Data-Processing in phytosociology, retrospect and anticipation. Vegetatio 32: 65–72.

Mobberley, D. G. 1956. Taxonomy and distribution of the genus Spartina. Iowa State College J. Sci. 30: 471–574.

Raabe, E. W. 1952. Über den 'Affinitätswert' in der Pflanzensoziologie. Vegetatio 4: 53–68.

Sokal, R. R. & P. H. A. Sneath. 1963. Principles of numerical taxonomy. San Francisco XVI+359 pp.

Sørensen, Th. A. 1948. A method of establishing groups of equal amplitude in plant sociology based on similarity of species content. Biol. Skr. K. danske Vidensk. Selsk. (4): 1–34.

Westhoff, V. & E. van der Maarel. 1973. The Braun-Blanquet approach. In R. H. Whittaker (ed.). Handbook of vegetation science. Part 5. Classification and ordination of communities, p. 617–726, Junk, Den Haag.

Williams, W. T. & J. M. Lambert. 1959. Multivariate methods in plant ecology. I. Association analysis in plant communities. J. Ecol. 47: 83–101.

Wishart, D. 1969. Clustan Ia. User manual. St. Andrews Computing Centre.

UNE UTILISATION POSSIBLE DE L'ARBRE DE PORTÉE MINIMALE EN PHYTO-ÉCOLOGIE*

F. ROMANE[1], J. L. GUILLERM[1] & G. WAKSMAN[2]

1. Centre d'études phytosociologiques et écologiques Louis Emberger (C.N.R.S.), B.P. 5051, 34033 Montpellier CEDEX, France
2. actuellement: Compagnie générale de géophysique, 6 rue Galvani, 91301 Massy, France

Keywords:
Factor analysis, Minimum spanning tree, Ordination

Introduction

La recherche de groupes d'espèces indicatrices des différents milieux d'une région naturelle a été abordée par de nombreuses méthodes et à différents niveaux de synthèse. C'est ainsi que dans l'étude des relations entre la végétation spontanée du vignoble du Minervois et son milieu (Guillerm et al. 1973a) nous avons utilisé d'abord la méthode des profils écologiques (Gounot 1958, Godron 1968 et Guillerm 1969) puis celle de l'analyse factorielle. La confrontation des résultats obtenus par différentes techniques se révélant souvent fructueuse, il ne nous a pas paru inutile d'en appliquer une nouvelle aux observations que nous avons recueillies. Il s'agit de celle de l'arbre de portée minimale, exposée par Gower & Ross (1969). (Reprenant la traduction de 'Minimum Spanning Tree' nous avons préféré utiliser 'arbre de portée minimale' à 'arbre de longueur minimale'.) D'ailleurs, selon ces deux derniers auteurs. la comparaison des résultats qu'elle fournit avec ceux obtenus par l'analyse factorielle est pleine d'intérêt.

Dans l'essai présenté ici, notre souci était d'obtenir une représentation synthétique des espèces basée sur leur écologie. Pour cela, nous n'avons pas retenu l'une des distances proposées par Gower & Ross (1969), mais celle utilisée dans l'analyse factorielle des correspondances, ce qui a permis de quantifier la dissemblance entre les espèces du point de vue des conditions écologiques qui régissent leur apparition et leur développement.

* Contribution of the Working Group for Data-Processing, International Society for Vegetation Science.
Nomenclature d'après 'les quatre flores de France', de P. Fournier (éditions P. Lechevalier, 1946, 1091 p.).

Rappel sur la méthode retenue

L'arbre de portée minimale

Il existe plusieurs possibilités pour représenter schématiquement une matrice de dissimilarités; une de ces possibilités est celle dite des 'arbres' (Escoufier 1972) qui consiste à représenter l'ensemble des éléments pour lesquelles la matrice a été établie, par un ensemble de segments de droites (arêtes) joignant deux éléments tels que (Gower & Ross 1969, Briane et al. 1974):

- un chemin existe entre deux quelconques de ces éléments (graphe connexe),
- ces segments ne forment pas de boucles (circuit).

Pour un tel arbre, la portée (ou longueur) est égale à la somme des longueurs de tous les segments formant l'arbre. A une matrice donnée de dissimilarité, correspondent plusieurs arbres possibles dont un a une portée minimale, c'est celui qui est appelé 'arbre de portée minimale'. Si dans certains problèmes, comme celui consistant à relier un ensemble de villes par des lignes téléphoniques (Prim 1957) ou des terminaux d'ordinateurs (Loberman & Weinberger 1957) l'intérêt de ce critère de longueur minimale apparait clairement, il n'en est pas de même dans notre cas. Notons cependant que cette méthode avait déjà été utilisée dans des problèmes de taxonomie (Czekanowski 1909, Florek et al. 1951, cf. Gower et al. 1969, Briane et al. 1974 et Roux 1971).

L'arbre de portée minimale peut être considéré comme une méthode de classification au sens de Dagnélie (1968) à partir du moment où l'on décide de couper tout segment supérieur à une longueur donnée, ce qui fournit des groupes d'éléments. Dans les dendrogrammes habituellement construits ceci revient à retenir tous les groupes formés au dessous d'un niveau donné.

La représentation des résultats* sous la forme de l'arbre de portée minimale semble être plus intéressante que celle souvent retenue pour réaliser un dendrogramme. Généralement, dans un dendrogramme, à un niveau donné, 'l'intermédiaire' par lequel un nouvel individu (ou groupe) se raccroche à un autre groupe n'est pas connu, alors que cela apparait ici. Ainsi, au bas de la figure 1 représentant l'arbre de portée minimale pour les observations retenues, il est possible de constater que le groupe *Aira caryophyllea, Silene gallica, Rumex acetosella, Holcus lanatus, Erica arborea, Phleum pratense, Rumex obtusifolius, Trifolium glomeratum* est relié au reste de l'arbre par l'intermédiaire de *Anthemis arvensis*.

La distance

Dans le cas où les observations relatives à la végétation et à son milieu sont disponibles en un même point (relevé) il est alors possible de présenter les données sous forme d'un tableau où les colonnes sont les espèces (ensemble J) et les lignes (ensemble I) les états d'une variable, un élément du tableau étant égal au nombre de fois où une espèce a été rencontrée dans un état d'une variable du milieu. Un tel tableau de données est analysable par une analyse factorielle dont la variante dite 'des correspondances' nous a paru être fructueuse dans ce cas (Romane 1972a). Dans cette analyse factorielle, la distance entre deux espèces peut-être considérée, dans l'espace des espèces, comme une mesure, basée sur les variables du milieu, de la dissimilarité entre ces deux espèces. Cette distance peut s'exprimer ainsi:

$$d^2_{J(I)}(j,j') = \sum_{i=1}^{n} p_{i.} \left[\frac{p_{j/i}}{p_{.j}} - \frac{p_{j'/i}}{p_{.j'}} \right]^2$$

où $d_{J(I)}(j,j')$ = distance entre deux éléments j et j' de l'ensemble J en correspondance avec l'ensemble I;

p_{ij} = probabilité d'apparition de l'élément (i,j);

$p_{i.} = \sum_{j\in J} p_{ij}$ = probabilité d'avoir l'élément i quel que soit j;

$p_{.j} = \sum_{i\in I} p_{ij}$ = probabilité d'avoir l'élément j quel que soit i;

$p_{i/j} = \frac{p_{ij}}{p_{.j}}$ = probabilité d'avoir l'élément i, j étant fixé;

$p_{j/i} = \frac{p_{ij}}{p_{i.}}$ = probabilité d'avoir l'élément j, i étant fixé.

Les probabilités sont estimées à partir des fréquences:

$$p_{ij} = \frac{n_{ij}}{N} \text{ (avec } N = \sum_{i,j} n_{ij})$$

Cette distance est une expression de la différence entre les 'profils écologiques' des deux espèces (Romane 1972b) et répond donc à notre souci de juger la différence entre les espèces à partir de leurs caractères écologiques. Par ailleurs le fait d'utiliser la même distance dans l'analyse factorielle et dans l'arbre de portée minimale nous a paru favorable car cela permet ainsi une meilleure comparaison des deux techniques, et c'est donc cette distance que nous avons retenue.

Les observations phyto-écologiques

Il s'agit d'observations recueillies dans le Minervois de l'Aude (Sud de la France). Dans une première phase, il s'agissait de définir, par une approche phyto-écologique, les différents milieux naturels de cette région. Dans toute la région viticole du Midi de la France, la rareté d'une végétation 'naturelle', ou tout au moins peu influencée par l'homme, avait suggéré à Guillerm (1969), dans une étude similaire, de baser l'approche phyto-écologique sur l'observation de la flore adventice des vignobles en place. Cette même approche a été choisie pour l'étude du Minervois, aboutissant à une soixantaine de relevés phyto-écologiques décrits d'après la codification proposée par Godron et al. (1968), c'est à dire avec une description de la flore et du milieu. Ces relevés ont été déjà analysés par la méthode des profils écologiques et l'analyse factorielle (Guillerm et al. 1973a, b, Waksman 1973); nous représenterons ici les résultats obtenus avec la méthode de l'arbre de portée minimale en nous référant parfois à ces précédents résultats.

* Le programme de calcul a été fourni par M. M. Roux (Institut de Statistique de l'Université de Paris) et R. Tomassone (Centre National de la Recherche Zootechnique, Jouy-en-Josas).

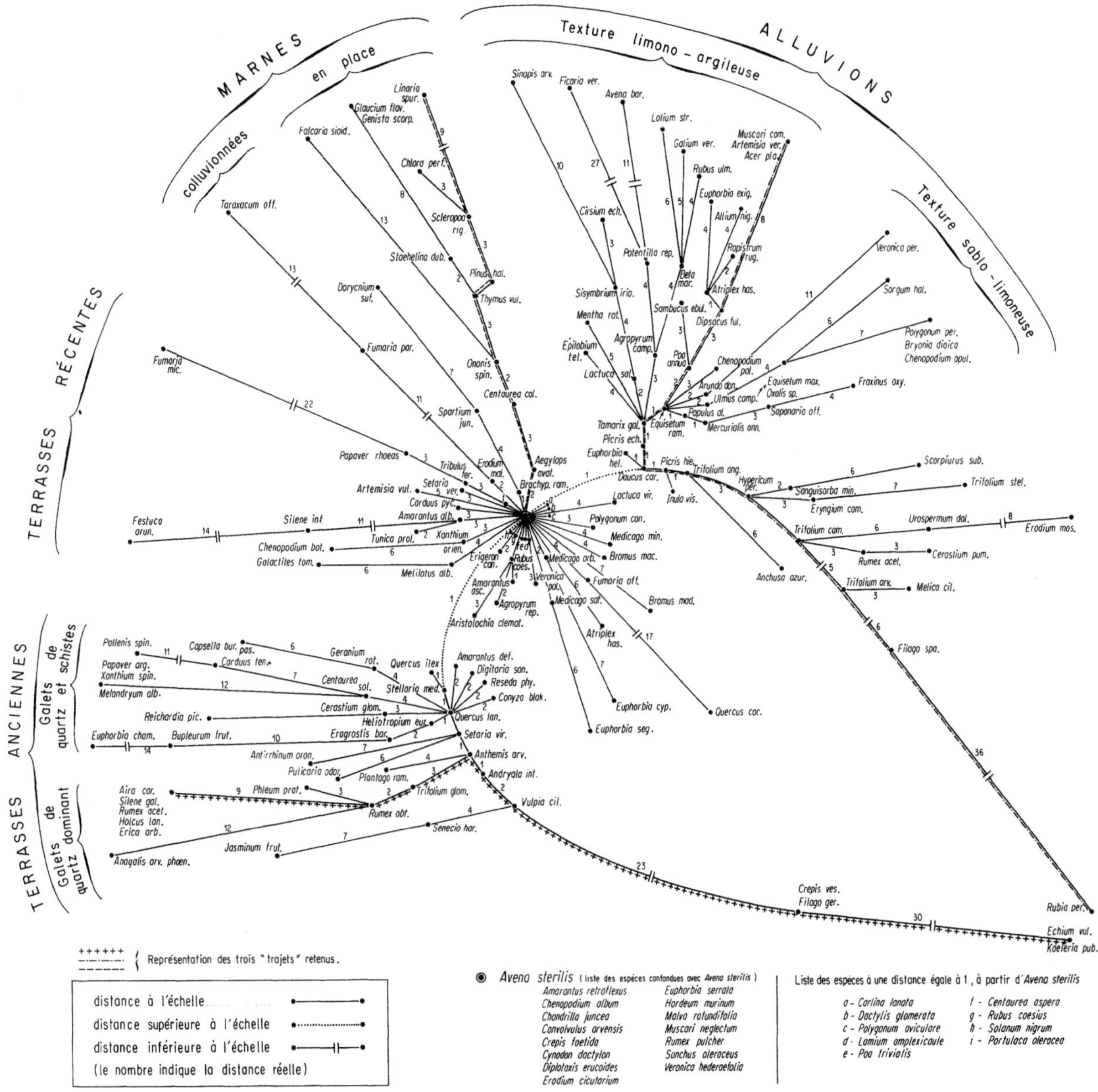

Figure 1 – Représentation des espèces avec la technique de "l'arbre de portée minimale"
(Données du Minervois de l'Aude)

Résultats

Apport de l'arbre de portée minimale

La figure 1 représente l'arbre obtenu pour les espèces des relevés du Minervois. La forme en arc adoptée pour cette représentation est arbitraire et si nous avons rapproché *Rubia peregrina* de *Koeleria pubescens* c'est à partir d'autres considérations que celles fournies par les résultats de l'arbre. Il semble que pour ces données, la 'forme' générale de l'arbre soit un tore et que, par définition, la méthode de l'arbre ne pouvant accepter cette forme, ce tore soit coupé en un point plus 'faible' (M. Roux, comm. verb. 1973).

L'interprétation des 'branches' de l'arbre a pu se faire

aisément grâce aux informations qui avaient été recueillies pour décrire le milieu. Ainsi, à chaque branche ou groupe de branches, nous avons fait correspondre sur la figure 1 quelques caractères dominants du milieu relatif à ce groupe d'espèces. Ce sont des variables édaphiques et géomorphologiques qui ont pu être le plus facilement utilisées pour 'expliquer' ces branches, variables effectivement prédominantes dans le Minervois. Mais pour pouvoir faire une interprétation plus objective de ces branches, il faudrait connaître la participation de chacune des variables aux distances retenues pour la construction de l'arbre, ce qui n'est pas le cas.

Il est intéressant de noter la présence dans la partie centrale de l'arbre, (c'est à dire autour de l'espèce *Avena sterilis*), d'un noyau d'espèces très proches les unes des autres. Ce sont souvent des espèces d'amplitude écologique large qui se retrouvent dand différents milieux cultivés en vignes (*Diplotaxis erucoides, Daucus carota, Chenopodium album, Erodium cicutarium*, etc.).

D'autre part, les observations relatives aux espèces arbustives notées au bord de la vigne où était effectué le relevé (Guillerm et al. 1973a) avaient permis de situer chacun des relevés dans une séquence de végétation (Godron et al. 1972). La distribution de ces espèces arbustives le long des différentes branches de l'arbre de portée minimale fait apparaitre les principales séquences de végétation (Le Floc'h et al. 1973). Les espèces liées à la séquence à *Populus alba* se retrouvent bien sur l'ensemble des alluvions, celle à *Pinus halepensis* sur les marnes et celle à *Quercus lanuginosa* principalement sur les terrasses du quaternaire ancien. Dans l'interprétation des résultats, le fait de pouvoir rattacher une branche à l'ensemble de l'arbre par l'intermédiaire d'une espèce, nous a semblé fort fructueux. Ceci permet de voir un enchainement dans l'organisation de l'arbre, ce qui n'est pas toujours aisé, dans le cas de l'analyse factorielle.

Comparaison avec les résultats d'une analyse factorielle

Pour effectuer cette comparaison, nous avons retenu les résultats d'une analyse factorielle des correspondances entre les espèces et les états des variables (Romane 1972b). Une comparaison objective et complète des résultats obtenus avec chacune des deux techniques nous a paru difficile. Cette comparaison sera donc effectuée à partir de trois exemples, à savoir trois parties de l'arbre dont nous avons reporté le trajet dans le plan des axes 1 et 2 puis celui des axes 1 et 3 de l'analyse factorielle (figure 2 et 3). Le premier trajet retenu, de *Koeleria pubescens* à *Erica arborea*, reporté dans le plan des axes 1 et 2, occupe une zone marginale du nuage de points. Une première constatation peut être faite, c'est que pour les espèces relativement loin d'une extrémité de l'arbre (*Anthemis arvensis*, par exemple) le tracé dans le plan des axes 1 et 2 tend à se rapprocher du milieu du nuage.

Considérons maintenant un deuxième trajet, de *Rubia peregrina* à *Muscari comosum* par l'intermédiaire de *Daucus carota*. Ce trajet reporté dans le plan des axes 1 et 2 traverse tout le nuage de points, mais là aussi, les espèces situées à des extrémités de branche de l'arbre sont sur les bords 'extérieurs' du nuage qui à une forme générale en U; quant aux espèces (*Daucus carota*, par exemple) du 'milieu' de l'arbre, elles se trouvent au centre du nuage de points de l'analyse factorielle. Les espèces situées à des extrémités de l'arbre, c'est à dire au bord extérieur du nuage de points de l'analyse factorielle (axes 1 et 2) sont généralement des espèces d'amplitude écologique faible (dans le cadre de l'échantillon considéré). Ceci est, par exemple, le cas pour *Erica arborea* trouvée sur des terrasses très acides ou encore *Acer platanoides*, dans un milieu très humide au bord de l'Aude (fleuve côtier). Pour d'autres espèces comme *Muscari comosum*, il s'agirait plutôt 'd'imperfections' de l'échantillonnage, car si cette espèce ne se trouve qu'une fois dans l'ensemble des relevés étudiés, cela semble devoir être attribué à la date à laquelle ont été fait les relevés, date peu propice à cette espèce. Elle donne donc l'impression d'être une espèce sténoïque, ce qui n'est pas le cas dans le Minervois.

Le troisième trajet retenu: de *Linaria spuria* à *Avena sterillis* correspond à un trajet avec plusieurs recoupements dans le plan des axes 1 et 2 de l'analyse factorielle. Par contre, si l'on reporte ce même trajet dans le plan des axes 1 et 3, les espèces apparaissent en s'organisant d'une manière très proche de celle obtenue avec l'arbre. Ces trois exemples tendent donc à montrer une bonne concordance des résultats, du moins si l'on considère pour l'analyse factorielle non seulement le plan des axes 1 et 2 mais des axes d'ordre supérieur à 2.

Conclusions

Analyse factorielle et arbre de portée minimum peuvent être considérés comme des techniques de description d'un ensemble d'observations de façon à faire apparaître certains caractères généraux peu apparents sur l'ensemble des données de base. Il semble que pour cet objectif, la technique de l'arbre de portée minimum soit une bonne

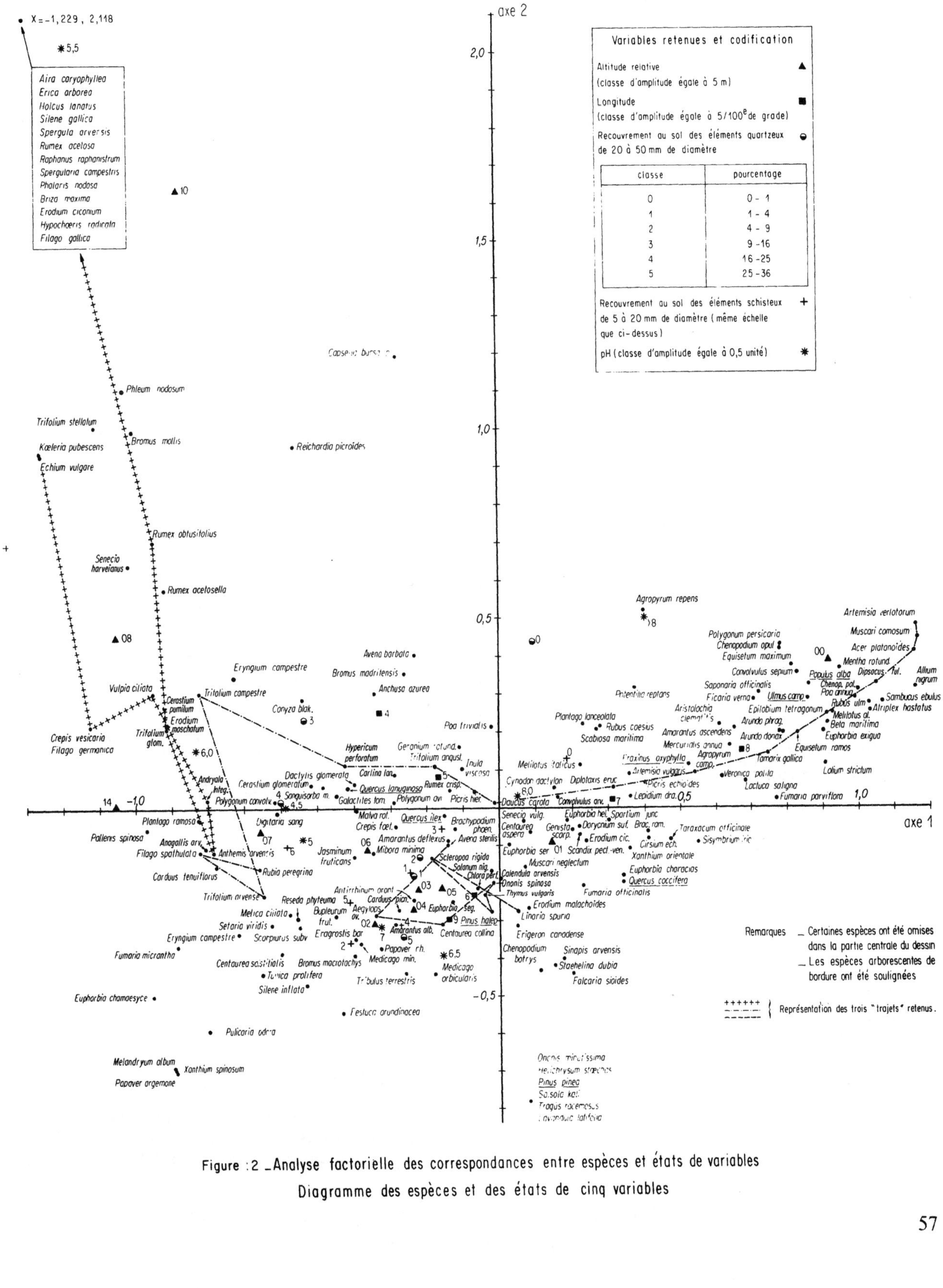

Figure 2 _ Analyse factorielle des correspondances entre espèces et états de variables
Diagramme des espèces et des états de cinq variables

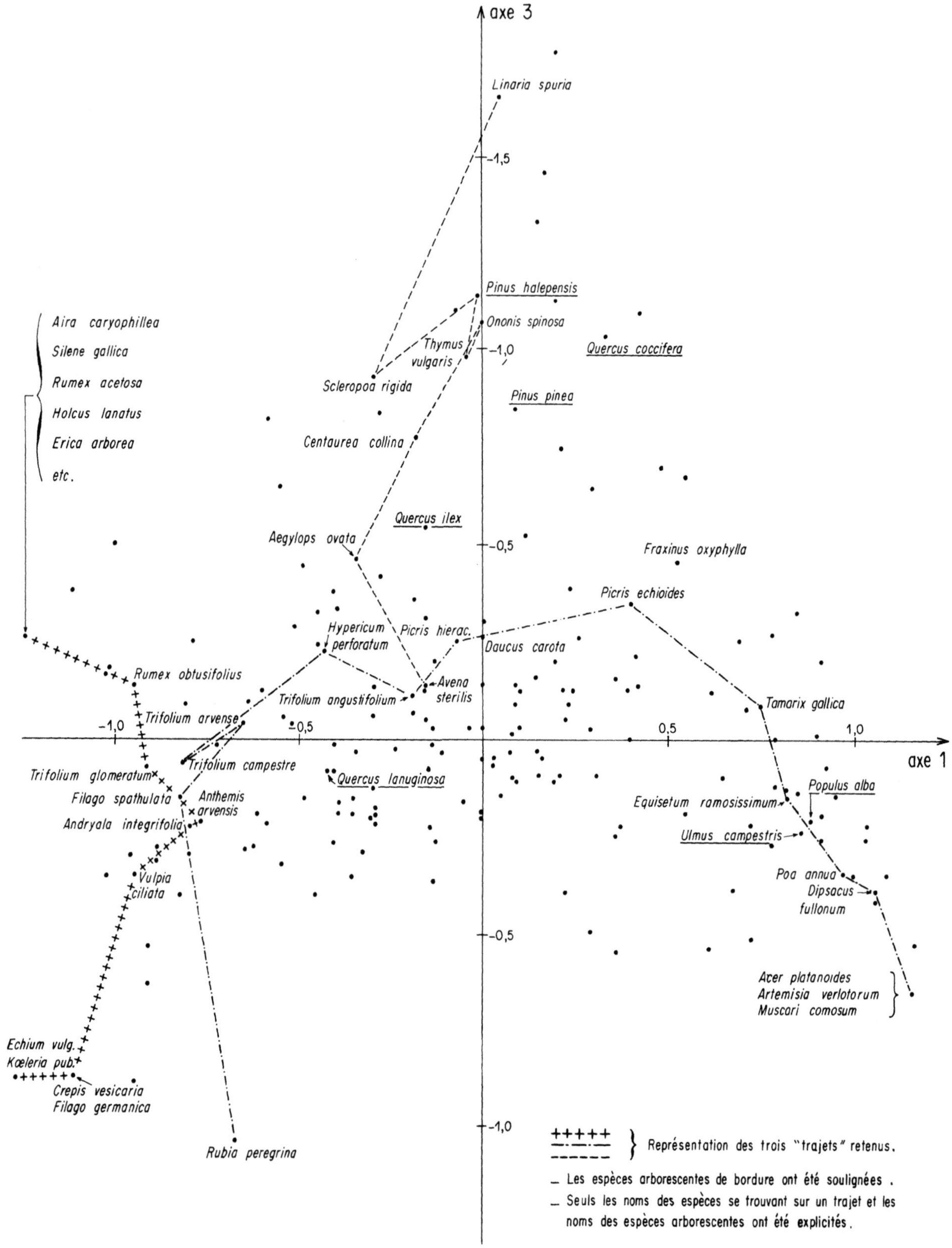

Figure 3 – Analyse factorielle des correspondances entre espèces et état des variables. Diagramme simplifié des espèces

approche avec une interprétation des résultats relativement plus aisée que dans le cas de l'analyse factorielle. Ceci grâce, d'une part à 'l'enchainement' entre les espèces, visible sur l'arbre et d'autre part, à la représentation dans un seul plan des résultats. Dans l'analyse factorielle, il est parfois nécessaire de consulter simultanément plusieurs plans pour avoir une 'image' des observations sans trop de distorsion. L'arbre de portée minimale semble aussi, comme l'analyse factorielle, permettre d'ordonner les espèces entre celles ayant une amplitude écologique faible (généralement aux extrémités des branches, correspondant à des espèces situées sur les bords extrèmes du nuage de points dans l'analyse factorielle) et des espèces d'amplitude écologique plus large (vers le centre de l'arbre ou du nuage de points en analyse factorielle).

Il semble donc qu'en analyse des données pour la phyto-écologie, cette technique de l'arbre de portée minimale puisse être ajoutée à l'ensemble des techniques déjà employées.

Summary

The minimum spanning tree has been envisaged here as a synthetic ordination method of a set of species of phyto-ecological relevés, that is to say samples with descriptions of both the vegetation and its environnent. It was intended to perform this ordination on ecological criteria in order to be able to draw a comparison between the results obtained and those previously acquired by means of a factor analysis of correspondences between the set of species and the set of the states of the variables describing the environnent. The distance for constructing the minimum spanning tree has, then, been implcitly utilized in the factor analysis.

The observations utilised are derived from a phyto-ecological study conducted on the adventive flora of the vineyards of Minervois in southern France. The results furnished by the minimum spanning tree were easily interpretable, emphasizing in particular the importance of geomorphological and edaphic variables for the vegetation of the study zone. The results compare favourably with those acquired by factor analysis; it was formely necessary to note on each occasion that one had to have recourse in certain cases to several plans in order to have available the same information that this method contains in the disposition of the minimum spanning tree.

The minimum spanning tree technique seems to be able to furnish a good image of the information in a collection of observations while its interpretation is easier than that of factor analysis.

Résumé

L'arbre de portée minimale a été envisagée ici comme une méthode d'ordination synthétique d'un ensemble d'espèces provenant de relevés phyto-écologiques, c'est à dire avec description de la flore et de son milieu. Il était souhaitable que cette ordination se fasse sur des critères écologiques, d'une part et que d'autre part on puisse faire une comparaison entre les résultats obtenus et ceux déjà acquis à l'aide d'une analyse factorielle des correspondances entre l'ensemble des espèces et celui des états des variables décrivant le milieu. La distance retenue pour construire l'arbre de portée minimale à donc été celle utilisée implicitement dans l'analyse factorielle.

Les observations utilisées proviennent d'une étude écologique effectuée avec la flore adventice des vignobles du Minervois (sud de la France). Les résultats fournis par l'arbre ont été facilement interprétables faisant ressortir en particulier le rôle important des variables géomorphologiques et édaphiques pour la flore de la zone d'étude.

Ces résultats sont très comparables à ceux acquis par l'analyse factorielle; avant, il faut noter toutefois qu'il faut recourir dans certains cas à plusieurs plans de l'analyse factorielle pour avoir la même 'information' que celle contenue dans la disposition de l'arbre. La technique de l'arbre de portée minimale semblerait donc pouvoir fournir une bonne image de synthèse pour un ensemble d'observations, sa lecture étant d'ailleurs plus aisée que celle de l'analyse factorielle.

Bibliographie

Briane, J. P., J. J. Lazare, G. Roux & C. Sastre. 1974. L'analyse factorielle des correspondances et l'arbre de longueur minimum; exemples d'application. Adansonia 2, 14: 111–137.

Czekanowski, J. 1909. Zur Differentialdiagnose der Neandertalgruppe. Korrespondenzblatt Dt. Ges. Anthropologie 40: 44–47.

Dagnelie, P. 1968. Introduction aux problèmes et aux méthodes de classification numérique. Biométrie-Praximétrie 9: 87–111.

Escoufier, Y. 1972. Analyse de données. Département d'Informatique, Université de Montréal, document INFO/660 L, 101 pp. + annexes.

Florek, K., J. Lukaszewicz, H. Perkal, H. Steinhaus &

S. Zubrzyski. 1951. Sur la liaison et la division des points d'un ensemble fini. Coll. Math. 2: 282–285.
Godron, M. 1968. Quelques applications de la notion de fréquence en écologie végétale (recouvrement, information mutuelle entre espèces et facteurs écologiques, échantillonnage). Oecol. Plant. 2: 185–212.
Godron, M. et al. 1968. Code pour le relevé méthodique de la végétation et du milieu. Principes et transcription sur cartes perforées. Editions CNRS Paris. 292 pp.
Godron, M., J. Lepart & F. Romane. 1972. Etude pour la recherche d'un language informatique commun en phyto-écologie en vue de la création d'une écothèque. Compte rendu CEPE Louis Emberger. 17 pp.
Gounot, M. 1958. Contribution à l'étude des groupements messicoles et rudéraux de la Tunisie. Annales du Service botanique agronomique de Tunisie 31: 1–282.
Gower, J. C. & G. J. S. Ross. 1969. Minimum spanning trees and single linkage cluster analysis. Appl. Statis. 18: 54–64.
Guillerm, J. L. 1969. Relations entre la végétation spontanée et le milieu dans les terres cultivées du Bas-Languedoc. Thèse de spécialité, Faculté des Sciences de Montpellier. 165 pp.
Guillerm, J. L., J. C. Jacquinet & F. Romane. 1973a. Tendances évolutives de la végétation spontanée du vignoble languedocien. Symposium 1973 de l'Association internationale de phytosociologie (comm.), Rinteln/Weser, Allemagne. 12 pp.
Guillerm, J. L., J. C. Jacquinet, M. Marlange & F. Romane. 1973b. Recherches écologiques dans le Minervois de l'Aude. Premiers résultats. Rapport à la Chambre d'agriculture de l'Aude, CNRS/CEPE Louis Emberger, Montpellier, document no. 60. 124 pp.
Le Floc'h, E., G. Long, J. Poissonet, M. Godron et coll. 1972. Notice des cartes de la végétation et de son écologie de la région Languedoc-Rousillon. In: Atlas régional du Languedoc (direction R. Dugrand, Montpellier) Berger-Levrault, Paris, 35 pp.
Loberman, H. & A. Weinberger. 1957. Formal procedures for connecting terminals with a minimum total wire length. J. Assoc. Comp. Mach. 4: 428–237.
Prim, R. C. 1957. Shortest connection matrix network and some generalizations. Bell System Tech. J. 36: 1389–1401.
Romane, F. 1972a. Un exemple d'utilisation de l'analyse factorielle des correspondances en écologie végétale. In: E. van der Maarel & R. Tüxen (eds.), Grundfragen und Methoden in der Pflanzensoziologie. Ber. Int. Symp. Int. Ver. Vegetationskunde. Junk, The Hague, pp. 151–167.
Romane, F. 1972b. Quelques aspects de l'analyse multivariable des données phytoécologiques. Investigación Pesquera 36: 131–139.
Roux, G. 1971. Les problèmes des pelouses à Nardus stricta. Une approche par les méthodes d'analyse numérique. Compte rendu du colloque interdisciplinaire sur les milieux naturels supra-forestiers des montagnes du bassin occidental de la Méditerranée. (Société Botanique de France), Perpignan, 5 et 6 février 1971: 299–314.
Waksman, G. 1973. Contribution à l'étude phyto-écologique du Minervois de l'Aude. Analyse factorielle des correspondances et arbre de portée minimum. Diplôme d'études approfondies, Université des Sciences et techniques du Languedoc, Montpellier. 40 pp.

ON THE RESOLVING POWER OF PRINCIPAL COMPONENT ANALYSIS IN PLANT COMMUNITY ORDINATION*

Enrico FEOLI**

Istituto ed Orto Botanico, Università degli Studi di Trieste, Italy I 34100

Keywords: Centering, Non-centering, Principal component analysis, Resolving power

Introduction

The method of principal component analysis (PCA) is commonly used for the ordination of plant communities (through relevés or synoptic tables). Its diffusion in ecology is due to the work of Goodall (1954), Dagnelie (1960), van Groenewoud (1965), Orlóci (1966, 1973), Austin & Orlóci (1966), Ivimey-Cook & Proctor (1967), van der Maarel (1969), and others. Although PCA is a statistical method, in ecology it serves as an ordination technique in indirect gradient analysis (Whittaker 1967) without much statistical content.

In a PCA, plant communities are ordered basing on co-ordinates obtained by the eigenvalue and eigenvector solutions of a real symmetric matrix; generally a resemblance matrix (Orlóci 1973) of the cross product type. Such co-ordinates may be computed in two different ways (Gower 1966, Orlóci 1966, 1967, 1973, 1975): 1) On the basis of the resemblance of relevés (Q-PCA). 2) On the basis of species correlation, covariance or similarity (R-PCA). The former is generally used when there are more species than relevés, the latter when there are more elevés than species. The results, as component scores of relevés, are mathematically identical in both cases. Therefore, this paper deals only with the former, 1).

A resemblance matrix may be generated by using different coefficients, which may be centered or not. Although, strictly speaking, the term *principal component analysis* applies only to procedures in which the coefficient is centered, being of the covariance type, in the present paper its use will be less restricted. The coefficients have been treated by Orlóci (1972, 1975) and Goodall (1973). The problem of centering and non-centering has been dealt with by Orlóci (1967, 1975) and Noy-Meir (1973). Two types of centering are distinguished: 1) centering by species (variable) and 2) centering by site (relevé or synoptic table).

When the ecologist applies PCA to a data matrix, he should realize that the information content in the data undergoes transformations (Orlóci 1974) through which it assumes a determinate structure in the ordination models. The structure strictly depends on the type of coefficient used. The non-centering coefficients, for instance the coefficients of Jaccard (1901) and Sørensen (1948), lead to models where at least one component is unipolar. This is believed to be a disadvantage in ordinations by Orlóci (1966) and Gower (1966), but an advantage by Noy-Meir (1973). Both these points of view may be accepted depending on what the ecologist would like to accomplish by PCA.

The present work describes the way in which centered and non-centered PCA achieve the discrimination of different objects as distinct points in a scatter diagram.

The data

Two sets of data have been used in the illustrations: 1) 12

* Contribution from the Working Group for Data-Processing in Phytosociology, International Society for Vegetation Science.

** This work was supported by the Italian C.N.R. and the Centro di Calcolo dell'Università di Trieste within the framework of a broader project entitled 'I metodi statistici e le loro applicazioni mediante l'uso dell'elaboratore'.

I am grateful to Dr. E. van der Maarel for his stimulating interest, to Prof. L. Orlóci for reading and correcting the manuscript, to Prof. G. Ghirardi and Prof. G. Alberi for discussions about matrix theory, and to Mr. A. Zampar and Mr. T. Ubaldini for technical assistance.

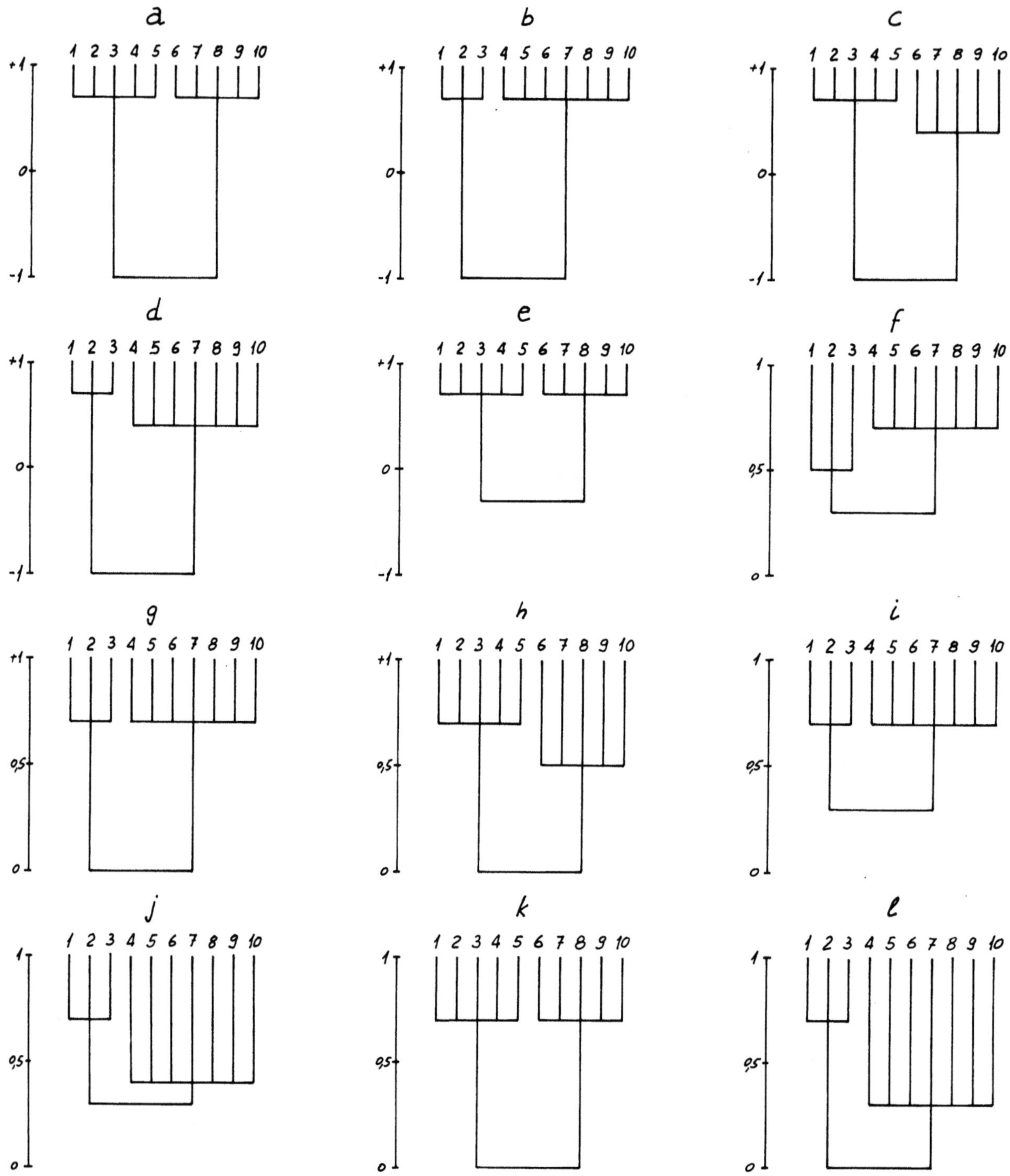

Fig. 1. Dendrograms showing the structure of the artificial matrices described in the text. Numbers across identify relevés or synoptic tables. Vertical scales represent the similarity levels within and between the clusters.

artificial symmetric matrices with 1 as diagonal element. The structures of the matrices may be represented by dendrograms as shown in Fig. 1. Dendrograms a, b, c, d, and e represent centered models, the others non-centered models. 2) 58 relevés of Glauco-Puccinellietalia (Beeftink 1966). This set of data (selection e of the Working Group for Data-Processing, cf. van der Maarel, Orlóci & Pignatti 1976) is also described in Feoli (1977).

Methods

Principal component analysis

Let $\mathbf{X}$ be a data matrix with m species and n relevés. In this work the term PCA is used for all the ordination methods involving the following steps:

a) Generation of a resemblance matrix $n \times n$ of the cross product type. The matrix may be centered or not,

$$\mathbf{X}^T\mathbf{X} = \{s_{jk}\} = \mathbf{S}, \quad j, k = 1, \ldots, n$$

b) Extraction of eigenvalues and eigenvectors of $\mathbf{S}$,

$$(\mathbf{S}' - \lambda_i \mathbf{I})\boldsymbol{\alpha}_i = \mathbf{O},$$

where λ_i is a non-trivial eigenvalue of $\mathbf{S}$ and $\boldsymbol{\alpha}_i$ the corresponding eigenvector, $\mathbf{I}$ is the identity $n \times n$ matrix and $\mathbf{O}$ the vector of n zero components.

c) After standardization of the eigenvectors so that

$$\boldsymbol{\alpha}_j^T \boldsymbol{\alpha}_j = \lambda_j$$

an element of $\boldsymbol{\alpha}_j$, a_{ij} gives the co-ordinate of the ith relevé on the jth principal axis of PCA.

In the present study two coefficients are considered to derive the $\mathbf{S}$ matrix: 1) the similarity ratio coefficient (cf. Westhoff & van der Maarel 1973, Janssen 1975), which is a non-centered one and 2) the weighted similarity coefficient of Orlóci (1966), which centers the data by species. The non-centered PCA was performed by the FACTO programme of IBM (1968) modified by substituting the subroutine CORRE with subroutine SIMILAR which computes the similarity ratio coefficient. The centered PCA was performed by the ORDINA programme (Roskam 1971). PCA of all the artificial matrices was carried out by subroutines EIGEN, TRACE and LOAD of IBM (1968).

Resolving power

The resolving power of any ordination could be estimated by the sum of all the distances between the points of the scatter diagrams:

$$R = \sum_{i,k=1}^{n(n-1)/2} d_{jk}, \quad j, k = 1, \ldots, n \tag{1}$$

where

$$d_{jk} = \sqrt{(x_{ij} - x_{ik})^2 + (x_{hj} - x_{hk})^2}, \; h, i = 1, \ldots, m, \; m < n \tag{2}$$

where j and k are two relevés and x_h and x_i their scores on the hth and ith components. R could be expressed in relative terms. The greater R is, the greater the resolving power of the ordination.

A more rapid estimation of the resolving power may be based on a minimum spanning tree (MST) clustering (Gower & Ross 1969) applied to the scatter diagrams of ordinations where the points are the noda of the graph. The resolving power in this case is expressed by the following formula:

$$R = \sum_{j,k=1}^{c} d_{jk}, \quad j, k = 1, \ldots, n \tag{3}$$

where d_{jk} is defined by expression (2) and c is the number of the edges of MST, $c < n(n-1)/2$. The present work uses expression (3). The relation between the two measures of resolving power are under study.

Mathematical considerations

Transformations, e.g. standardization, normalization, and so on (cf. Noy-Meir et al. 1975), influence the results but without altering the underlying geometrical models. Noy-Meir (1973) states that non-centered PCA puts in evidence discontinuities in a sample better than centered PCA. This fact renders crucial the choice of the ordination method, especially if there are classificatory objectives in hands. All the components of centered PCA are axes passing through the centroid of the hyperspace of relevés (Goodall 1963). The components, all bipolar, fit in order the cloud of relevés along the directions of the maximal dispersion. Since the eigenvectors ($\boldsymbol{\alpha}_j$) are so standardized that $\boldsymbol{\alpha}_j \boldsymbol{\alpha}_j^T = \gamma_j$, each eigenvalue ($\gamma_j$) represents the variance in the frequency distribution of the relevé-scores along the corresponding component.

The non-centered components do not have such geometrical and statistical meaning. In non-centered PCA the components do not pass through the centroid of the cloud of points, but each cluster tends to have its independent set of components. Three theorems are relevant in non-centered PCA: 1) If a data matrix is disjoint, each submatrix has its independent set of eigenvalues and eigenvectors. The proof is given by Noy-Meir (1973). 2) Among similarity matrices of different order with 1 in the diagonal positions and with all the other elements of equal value, the matrix of highest order has the highest eigenvalue. 3) Among similarity matrices of the same order, with 1 in the diagonal positions and with the other elements all equal in each matrix, but different between the matrices, the matrix with elements of greatest value has the greatest eigenvalue. Proofs are given in the Appendix.

Since PCA takes into consideration the eigenvalues in decreasing order of magnitude, it is clear that, if a data matrix is disjoint, the first component is that of the submatrix of highest order or that of the most homogeneous submatrix. Thanks to theorems 2 and 3 and considering perturbation theory, we can state that eigenvalues are functions of order of similarity matrices and of mean homogeneity of the relevés. We can consider the number a in the Appendix as if it were the mean similarity of any non-disjoint symmetric matrix. The deviations from a may be due to perturbation (Wilkinson 1965). Fixing the order of the matrices, the first eigenvalues depend linearly on the mean homogeneity, fixing the mean homogeneity, the first eigenvalues depend linearly on the order of the matrices. For any non-disjoint matrix, the first eigenvalue calculated according to

$$\lambda_1 = 1-a+Na, \qquad (4)$$

where a is the mean similarity of the matrix and N its order, approximates very well the first eigenvalue calculated by the Jacobi method; the greater the standard deviation of the elements of a similarity matrix, the greater the difference between its first eigenvalue and the eigenvalue derived by formula (4). The eigenvalue calculated by formula (4) may be considered as the expected first eigenvalue for a similarity matrix. The difference between this eigenvalue and the real one may be considered as a measure of the variability of the elements of a similarity matrix. The resolving power of centered and non-centered PCA depends on the variability of the similarity coefficients of the resemblance matrices; the higher the variance of the similarities, the higher the resolving power of the ordination. This statement is proved by the application of PCA to artificial matrices. In the scatter diagrams of non-centered PCA, relevés (as vectors) of independent series of vegetation are orthogonal, while in centered PCA they are opposite. This is principally due to the first theorem mentioned and it is the reason why the non-centered PCA always loses about 40–50 % of the resolving power compared to centered PCA.

Illustrations

Artificial matrices

PCA of the artificial similarity matrices produced the esults presented in Table 1. Some important properties may be observed: 1) Elements of clusters, linked at the same level of similarity and having equal similarity with the elements of other clusters, have equal component

Table 1. Results of PCA applied to artificial matrices. Explanations are given in the text

	a	b	c	d	e		f		g		h		i		j		k		l	
λ_i	8.80	8.59	8.23	7.50	5.30	2.30	5.40	2.25	5.2.	2.40	3.80	2.60	5.76	1.90	4.36	1.44	3.80	3.80	3.40	2.40
n_i	1	1	1	1	1	2	1	2	1	2	1	2	1	2	1	2	1	2	1	2
1	0.94	1.00	0.96	1.00	0.73	0.48	0.35	0.53	0.00	0.89	0.87	0.00	0.52	0.72	0.69	−0.57	0.87	0.00	0.00	0.89
2	0.94	1.00	0.96	1.00	0.73	0.48	0.35	0.53	0.00	0.89	0.87	0.00	0.52	0.72	0.69	−0.57	0.87	0.00	0.00	0.89
3	0.94	1.00	0.96	1.00	0.73	0.48	0.35	0.53	0.00	0.89	0.87	0.00	0.52	0.72	0.69	−0.57	0.87	0.00	0.00	0.89
4	0.94	−0.89	0.96	−0.77	0.73	0.48	0.85	−0.16	0.86	0.00	0.87	0.00	0.84	−0.19	0.65	0.26	0.87	0.00	0.70	0.00
5	0.94	−0.89	0.96	−0.77	0.73	0.48	0.85	−0.16	0.86	0.00	0.87	0.00	0.84	−0.19	0.65	0.26	0.87	0.00	0.70	0.00
6	−0.94	−0.89	−0.85	−0.77	−0.73	0.48	0.85	−0.16	0.86	0.00	0.00	0.72	0.84	−0.19	0.65	0.26	0.00	0.87	0.70	0.00
7	−0.94	−0.89	−0.85	−0.77	−0.73	0.48	0.85	−0.16	0.86	0.00	0.00	0.72	0.84	−0.19	0.65	0.26	0.00	0.87	0.70	0.00
8	−0.94	−0.89	−0.85	−0.77	−0.73	0.48	0.85	−0.16	0.86	0.00	0.00	0.72	0.84	−0.19	0.65	0.26	0.00	0.87	0.70	0.00
9	−0.94	−0.89	−0.85	−0.77	−0.73	0.48	0.85	−0.16	0.86	0.00	0.00	0.72	0.84	−0.19	0.65	0.26	0.00	0.87	0.70	0.00
10	−0.94	−0.89	−0.85	−0.77	−0.73	0.48	0.85	−0.16	0.86	0.00	0.00	0.72	0.84	−0.19	0.65	0.26	0.00	0.87	0.70	0.00

Legend to symbols: λ_i eigenvalues, n_i number of components for each of the 12 artificial matrices (a, b, c, d, e, f, g, h, i, j, k, l.), 1, ... 10 component scores for the 10 elements of the 12 artificial matrices (Fig. 1).

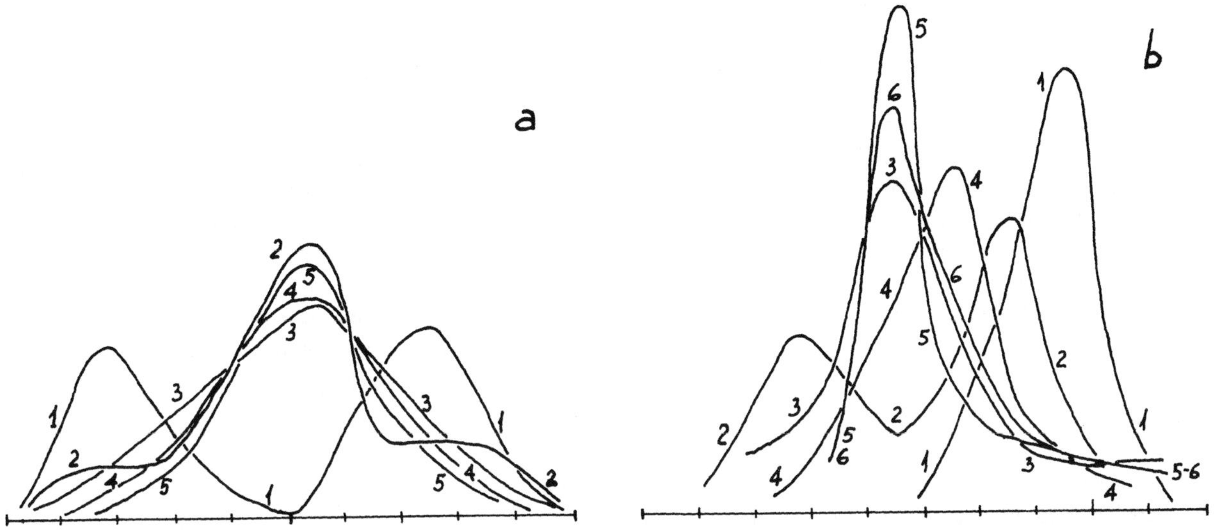

Fig. 2. Frequency distribution of the scores of 58 relevés of *Glauco-Puccinellietalia* along the first five components of centered PCA (a) and along the first six components of a non-centered PCA (b). Explanations are given in the text.

scores. This means that the resolving power of PCA depends on the variability of the similarity measures in the resemblance matrix. 2) Elements of a cluster, linked at a lower level of similarity, have lower component scores than elements linked at a higher level of similarity (Fig. 1 and Table 1 c, d, f, h, j, l). This property explains why it may happen that relevés, although very dissimilar, appear in each other's vicinity as points in an ordination. 3) Both in centered and non-centered PCA the elements of disjoint matrices have higher absolute component scores than the elements of non-disjoint matrices (Fig. 1 and Table 1 a, e and i, g).

Other properties that may be put in evidence on analyzing Fig. 1 and Table 1, agree with the three theorems mentioned above.

Glauco-Puccinellietalia selection

Considering some subsets of relevés based on the presence of species which, according to Feoli (1973) have high diagnostic value, it is shown that non-centered PCA reaches maximally only 65 % of the resolving power of the centered PCA. The loss of resolving power of non-centered PCA is evident in Fig. 2 when the relative frequency distributions of the scores of the relevés along the centered and non-centered components are considered. As regards the interpretations, the scatter diagrams obtained by the components are interpretable in the same way in both cases. The position of the relevés along the first two components of the centered PCA shows a clear distinction between the alliances of the considered order, viz. *Armerion* and *Puccinellion*. The duality of the vegetation explains the bimodality of the frequency distribution of relevé-scores along the first centered component and along the second non-centered component.

The mean similarities of the relevés of *Armerion* and of those of *Puccinellion* are approximately equal, therefore the order of the two submatrices relating to the two alliances plays a fundamental role with regard to the discrimination within the two groups of relevés. It happens that both in centered and non-centered PCA the components, from the first to the last, discriminate within the syntaxonomic groups in decreasing order of their size. In centered PCA the second component discriminates better within the *Armerion* that has the larger group of relevés, while it is the third component that discriminates better within the *Puccinellion*. In non-centered PCA the third component discriminates within the *Armerion*, while the fourth within the *Puccinellion*. The following components discriminate within associations. In this process the ordinations undergo losses of resolving power. PCA may thus be compared to a microscope with components as eyepieces with different resolving power.

Conclusions

In terms of ordination efficiency (Orlóci 1966), non-centered PCA is always less efficient than centered PCA but it is doubtful whether the two methods may be compared in terms of such a measure. Although the mathematics applied to the resemblance matrix is the same in both cases, the underlying geometrical model is completely different. The centered PCA intercepts the axes of maximal variation in the sample of the relevés in the vegetational space, the non-centered PCA intercepts clusters in decreasing order of importance due to size and homogeneity. With regard to the question of whether centering by species is necessary and useful, the present work leads to the conclusion that centered PCA is more appropriate in ordinations while non-centered PCA is more appropriate in cluster seeking.

Summary

A comparison between centered and non-centered principal component analysis is made on the basis of the resolving power of the methods. The results indicate the appropriateness of using a centered PCA when the aim is an ordination of plant communities, and the non-centered PCA when the aim is to elicit the taxonomic structure of a collection.

Appendix

If $\mathbf{A}$ is the resemblance matrix of order N with 1 as diagonal elements and with all other elements equal to a, then the following sequence of steps holds true:

$$\boldsymbol{\alpha} = \begin{bmatrix} 0111\ldots1 \\ 1011\ldots1 \\ 1101\ldots1 \\ \cdots\cdots\cdots \\ \cdots\cdots\cdots \\ 111\ldots.0 \end{bmatrix}$$

$$\begin{aligned} \mathbf{A} &= 1+a\boldsymbol{\alpha}, \\ \boldsymbol{\alpha}^2 &= (N-1)+(N-2)\boldsymbol{\alpha} \\ \mathbf{A}^2 &= (1+a\boldsymbol{\alpha})(1+a\boldsymbol{\alpha}) = 1+2a\boldsymbol{\alpha}+a^2\boldsymbol{\alpha}^2 = \\ &= 1+2a\boldsymbol{\alpha}+a^2(N-1)+a^2(N-2)\boldsymbol{\alpha} = \\ &= 1+a^2(N-1)+(2-2a+Na)\boldsymbol{\alpha}a = \\ &= (1+Na^2-a^2-2+2a-Na)+(2-2a+Na)\mathbf{A} = \\ &= [(N-1)a^2+2a-\mathrm{N}a-1]+(2-2a+Na)\mathbf{A}. \end{aligned}$$

A relevant algebraic theorem states that if a matrix can be expressed in terms of a linear equation, then the roots of the equation are the eigenvalues of the matrix. Therefore,

$$z^2-(2-2a+Na)z-(N-1)a^2-2a+Na+1 = 0,$$

$$z_{1,2} = \tfrac{1}{2}[(2-2a+Na)\pm(4+4a^2+N^2a^2-8a+4Na- -4Na^2+4(N-1)a^2+8a-4Na-4)^{\frac{1}{2}}] =$$

$$= \tfrac{1}{2}(2-2a+Na\pm Na) = \begin{cases} z_1 = 1-a+Na & (z_1 = \lambda_1) \\ z_2 = 1-a & (z_2 = \lambda_2) \end{cases}$$

It is clear that the higher N the greater the first eigenvalue. This demonstrates that if a resemblance matrix is disjoint for submatrices of equal elements but of different order, the highest eigenvalue corresponds to the submatrix of the highest order. If we consider coefficient $b > a$, it is obvious that the eigenvalues will be greater. It is thus also demonstrated that more homogeneous collections have greater first eigenvalues.

References

Austin, M. P. & L. Orlóci. 1966. Geometric models in ecology. II. An evaluation of some ordination techniques. J. Ecol. 54: 217–227.

Beeftink, W. G. 1966. Conspectus of the phanerogamic salt plant communities in the Netherlands. Biol. Jaarb. Dodenaea 30: 325–362.

Dagnelie, P. 1960. Contribution à l'étude des communautés végétales par l'analyse factorielle. Bull. Serv. Carte Phytogéogr. B 5: 7–71, 93–195.

Feoli, E. 1973. Un indice che stima il peso dei caratteri per classificazioni monotetiche. Giorn. Bot. Ital. 107 (6): 263–268.

Feoli, E. 1977. A criterion for monothetic classification of phytosociological entities on the basis of species ordination. Vegetatio 33: 147–152.

Goodall, D. W. 1954. Objective methods for the classification of vegetation. III. An essay on the use of factor analysis. Austr. J. Bot. 2: 304–324.

Goodall, D. W. 1963. Continuum and individualistic association. Vegetatio 11: 297–316.

Goodall, D. W. 1973. Sample similarity and species correlation. In: R. H. Whittaker (ed.), Handbook of vegetation science, Part V: Ordination and classification of communities. Junk, The Hague, pp. 105–156.

Gower, J. C. 1966. Some distance properties of latent root and vector methods used in multivariate analysis. Biometrika 53: 325–338.

Gower, J. C. & G. J. S. Ross. 1969. Minimum spanning tree and single linkage cluster analysis. Appl. Stat. 18: 54–64.

Groenewoud, H. van. 1965. Ordination and classification of some Swiss and Canadian forests by various biometric and other methods. Ber. Geobot. Inst. Rübel Zürich 36: 28–102.

IBM. 1968. System/360 Scientific Subroutine Package (360A-CN-03X). Version III. Programmer's Manual Document H20-0205-3, IBM, White Plains, N.Y. p. 454.

Ivimey-Cook, R. B. & M. C. F. Proctor. 1967. Factor analysis of data from an East Devon heath: a com-

parison of principal component and rotated solutions. J. Ecol. 55: 405–419.
Jaccard, P. 1901. Distribution de la flore alpine dans le Bassin des Dranses et dans quelques régions voisines. Bull. Soc. Vaud. Sci. Nat. 37: 241–272.
Janssen, J. G. M. 1975. A simple clustering procedure for preliminary classification of very large sets of phytosociological relevés. Vegetatio 30: 67–71.
Maarel, E. van der. 1969. On the use of ordination models in phytosociology. Vegetatio 19: 21–46.
Maarel, E. van der, L. Orlóci & S. Pignatti. 1976. Data-processing in phytosociology, retrospect and anticipation. Vegetatio 32: 65–72.
Noy-Meir, I. 1973. Data transformation in ecological ordination. I. Some advantages of non-centering. J. Ecol. 61: 329–341.
Noy-Meir, I., D. Walker & W. T. Williams. 1975. Data transformation in ecological ordination. II. On the meaning of data standardization. J. Ecol. 65: 773–800.
Orlóci, L. 1966. Geometric models in ecology. I. The theory and application of some ordination methods. J. Ecol. 54: 193–215.
Orlóci, L. 1967. Data centering: A review and evaluation with reference to component analysis. Syst. Zool. 16: 208–212.
Orlóci, L. 1972. On objective functions of phytosociological resemblance. Amer. Midl. Natur. 88: 28–55.
Orlóci, L. 1973. Ordination by resemblance matrices. In: R. H. Whittaker (ed.), Handbook of vegetation science, Part V: Ordination and classification of vegetation. Junk, The Hague, pp. 249–286.
Orlóci, L. 1975. On information flow in ordination. Vegetatio 29: 11–16.
Orlóci, L. 1975. Multivariate analysis in vegetation research. Junk, The Hague. 276 pp.
Roskam, E. 1971. Program Bulletin n. 16. Psychologisch Laboratorium, Nijmegen.
Sørensen, T. 1948. A method of establishing groups of equal amplitude in plant sociology based on similarity of species content, and its application to analysis of the vegetation on Danish Commons. Biol. Skr. 5: 1–34.
Westhoff, V. & E. van der Maarel. 1973. The Braun-Blanquet approach. In: R. H. Whittaker (ed.), Handbook of vegetation science, Part V: Ordination and classification of communities. Junk, The Hague, pp. 617–726.
Whittaker, R. H. 1967. Gradient analysis of vegetation. Biol. Rev. 49: 201–264.
Wilkinson, J. H. 1965. The algebraic eigenvalue problem. Oxford University Press, London. 662 pp.

A CRITERION FOR MONOTHETIC CLASSIFICATION OF PHYTOSOCIOLOGICAL ENTITIES ON THE BASIS OF SPECIES ORDINATION*

Enrico FEOLI**

Istituto ed Orto Botànico, Università degli Studi di Trieste, Trieste, Italy I 34100

Keywords: Classification, Monothetic, Ordination, Similarity, Synsystematics

Introduction

Phytosociological systematics according to the Braun-Blanquet approach produces hierarchic classification schemes that are not always justifiable from a formal, logical point of view. The most frequent incongruities are due to the fact that there are species simultaneously characteristic of different syntaxa; and moreover, species which are considered as character taxa of higher syntaxa do not necessarily appear in the phytosociological tables attributed to them. Although intentionally synsystematics uses a procedure of polythetic classification (cf. Moore et al. 1970, Westhoff & van der Maarel 1973), it actually operates rather through a monothetic classificatory process (Williams & Dale 1965). The higher taxa are not defined according to similarity levels between clusters of relevés, but rather, on the basis of the presence of some species which satisfy certain fidelity criteria. Without attempting to evaluate the fidelity concept, which could involve us in a circular argument for the definition of syntaxa, a monothetic, hierarchic systematics, based on the presence of the species, could be considered appropriate depending on personal preference.

If we accept that such a hierarchic systematics of vegetation is appropriate, the classification criterion to be followed is to be formulated in logical terms.

* Contribution from the Working Group for Data-Processing in Phytosociology, International Society for Vegetation Science. Nomenclature follows the Trieste coding system.
** The author is very grateful to Dr. E. van der Maarel for his stimulating interest and to Prof. L. Orlóci for reading and correcting the manuscript.

The aim of the present paper is to formulate a general criterion for this type of systematics. The techniques employed, to derive such a criterion in a most profitable way, include principal component analysis (PCA) and the method of minimum spanning tree (MST) (Gower & Ross 1968, Basagni et al. 1970).

PCA provides a suitable basis for the application of MST. The joint application of the two techniques, PCA for ordination (cf. Goodall 1954, van der Maarel 1969, Orlóci 1973) and MST for classification has been both used to cancel out the weakness that the two techniques may present if separately applied: in PCA, the potential distortions (Orlóci 1974, cf. also Whittaker & Gauch 1973, Kessel & Whittaker 1976) and in MST, the difficulty in finding a suitable disposition for the elements to be classified. PCA may be used for preparation of the data for MST whose execution is then greatly simplified.

The classification criterion

The criterion, here proposed, starts from the fact that every species, in algebraic terms, gives origin to two equivalence classes of phytosociological entities, which may be either relevés or synthetic tables: one class containing the entities possessing that species and the other class lacking it. An optimal situation for a hierarchical classification may be presented as in Fig. 1. If A, B, C, D, E are species, the six relevés in table *a* are easily classifiable as in scheme *c*, after rearrangement as in table *b*. In this case the classification criterion follows the inclusion rule described by:

$$x_i \in \{B\} \wedge \{B\} \subset \{A\} \Rightarrow x_i \in \{A\}$$

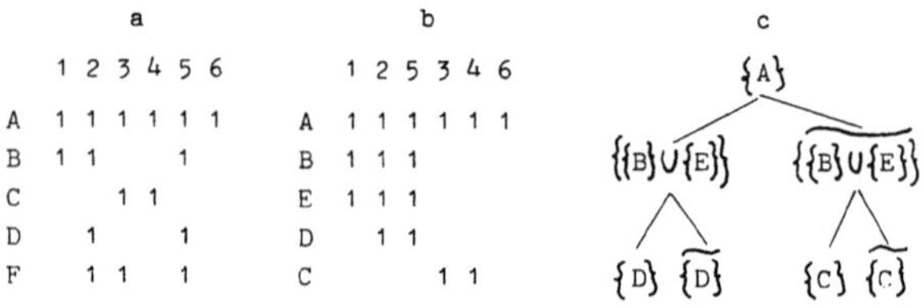

Fig. 1. Illustration of a simple situation for hierarchical classification on the basis of the inclusion criterion.

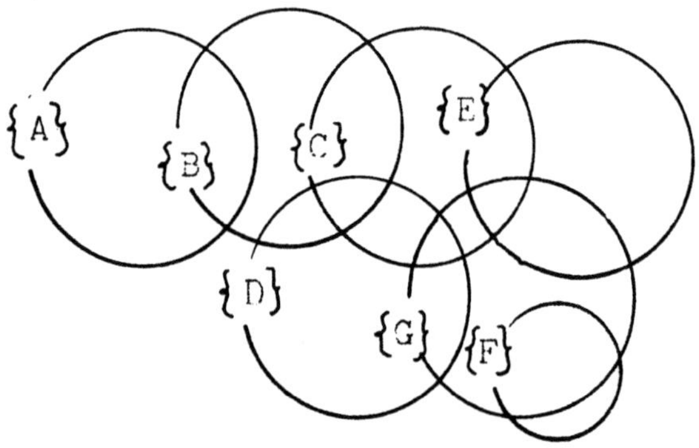

Fig. 2. An example of overlapping classes defined on the basis of species presences. {A}, {B}, {C}, . . . are the classes defined respectively by the presence of species A, B, C, ...

In other words, if x_i is part of {B} and {B} is part of {A} then x_i is part of {A}. The situation as it appears in Fig. 1 may however not often occur in phytosociological collections. A more frequently met situation is shown in Fig. 2.

Based on Fig. 2, the classification would inevitably result in overlapping classes (Williams 1971). It appears that in such a situation a complete hierarchisation of the classes, based only on the inclusion criterion, would be impossible. A more general criterion must be introduced. It may be the following:

$$\text{if } x_i \in \{A\} \text{ then } x_i \in \{B\} \Rightarrow \{A\} \cap \{B\} \text{ (maximal intersection for } \{A\})$$

In other words, if x_i is part of {A}, then x_i is part of {B} if and only if the intersection between {A} and {B} is maximal for {A}. Furthermore if $\{A\} \cap \{B\}$ is the maximal intersection for {A} then {A} coincides with {B} ($\{A\} \equiv \{B\}$).

This criterion introduces in the set of the classes, defined by the species, a relation of equivalence: $\{A\} \equiv \{A\}$; $\{A\} \equiv \{B\} \Rightarrow \{B\} \equiv \{A\}$; $\{A\} \equiv \{B\} \wedge \{B\} \equiv \{C\} \Rightarrow \{A\} \equiv \{C\}$. The criterion thus presented may be defined as a criterion of coincidence according to maximal intersection.

In order to construct a hierarchical system according to this criterion, it is necessary to measure the class intersections. This may be based on the use of, for instance, the Jaccard formula. This formula, in terms of set theory, gives the ratio between the intersection of two sets and their symmetric difference.

Then by calculating the similarity between two species in the phytosociological entities to be classified, we can determine the relative intersection of the classes defined by the presence of the species. Species ordination by PCA may be interpreted as an ordination of the classes based on the species. When MST is applied to a similarity matrix on the species, and carried out on the ordination results of PCA, it will produce a graph of maximal intersection between the classes. The application of the criterion leads to creation of species groups of different hierarchical status. The hierarchy becomes evident on observation of the table of phytosociological entities ordered according to the species groups. It is obvious that the set defined by a group of species with higher presence in the table inevitably represents a high syntaxonomic level because it may include sets defined by groups of species with lower presence.

Example

The material to illustrate the procedure consists of 58 relevés already assigned by the authors to different phytosociological types (Table 1) in the *Glauco-Puccinellietalia*

Table 1. Phytosociological types, considered for the example

Association	Author		Relevé n.
Puccinellietum maritimae	Adriani	(1945)	1,2,3
Artemisietum martimae	,,	,,	4,5,6
Artemisietum martimae facies with *Statice limonium*	,,	,,	7,8,9
Juncetum gerardii Obionetum	Pignatti	(1966)	10,11,12
Juncetum gerardii subass. with *Leontodon autumnalis* and *Odontites litoralis*	Gillner	(1960)	13,14,15
Blysmetum rufi	,,	,,	16,17,18
Puccinellietum maritimae	,,	,,	19,20,21
Puccinellietum distantis	,,	,,	22,23,24
Juncetum gerardii	,,	,,	25,26,27,28
Artemisietum maritimae	,,	,,	29,30,31
Puccinellietum maritimae subass. with *Salicornia europaea*	,,	,,	32,33,34,35
Juncetum gerardii	,,	,,	36,37,38,39
Juncetum maritimi	Corillon	(1953)	40,41,42
Puccinellietum maritimae	,,	,,	43,44,45,46
Junco Caricetum extensae	,,	,,	47,48,49,50
Juncetum gerardii	,,	,,	51,52,53,54
Festucetum arenariae	,,	,,	55,56,57,58

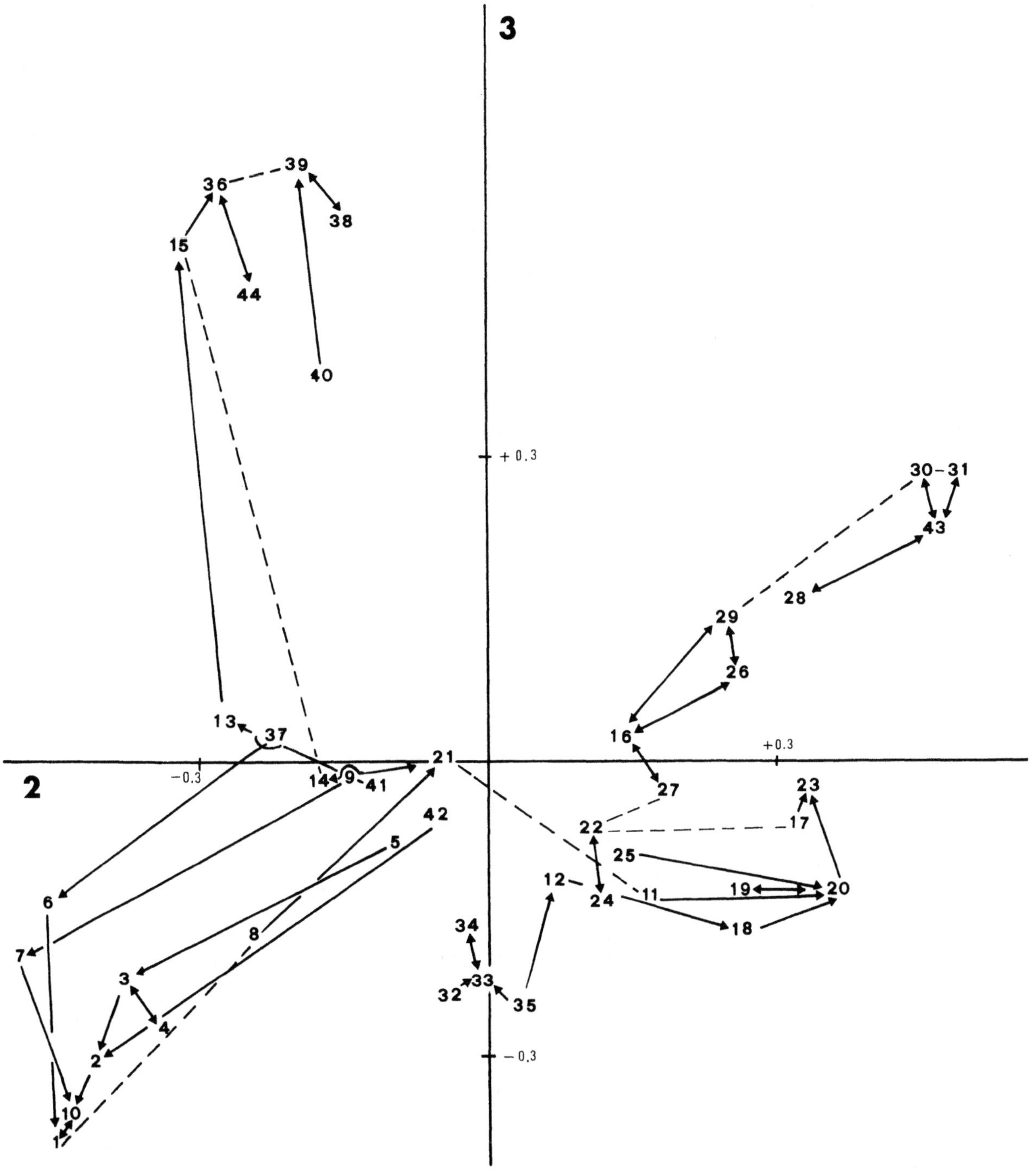

Fig. 3. Ordination of the 44 species in Table 2 on the second and third components extracted from a similarity matrix of species. The arrows trace a minimum spanning tree. The dashed lines show links between the species groups (in Fig. 4).

Table 2. In this table, 58 relevés are ordered according to ordination of species by PCA and the results of MST. The numbers to species are the codes in Fig. 3, symbols a to g identify species groups.

No.	Species	Group	Relevés
			11133121113323 222223334445444155555244523 11234545 1
			6788957340681466942323701802790278655453041?1215643135 8979
43	Scirpus rufus	a	+++
30	Eleocharis pauciflora	a	+ +
31	Carex oederi	a	+ +
28	Eleocharis uniglumis	a	++
29	Centaurium vulgare	b	+ ++
26	Sagina nodosa	b	+++
16	Odontites litoralis	b	++ +
27	Centaurium pulchellum	b	+++
22	Taraxacum sp.	c	++
24	Plantago major	c	+ +
11	Festuca rubra	d	+++++++++ ++++++
20	Poa pratensis	d	++ + ++
23	Sagina procumbens	d	++ + + +
17	Leontodon autumnalis	d	+ + ++++ +
19	Trifolium repens	d	+ + ++
18	Potentilla anserina	d	++ + ++++
25	Agropyron repens	d	++
12	Atriplex litoralis	d	+ ++
35	Tripleurospermum maritimum	d	+ + + +
34	Juncus bufonius	d	++
32	Spergularia salina	d	+ ++ +++
33	Puccinellia distans	d	+++
39	Samolus valerandi	e	+ + ++++
36	Carex extensa	e	++ ++++
38	Carex punctata	e	+++
15	Juncus maritimus	e	++++++++++++
44	Scirpus maritimus	e	+ + ++
40	Oenanthe lachenalii	e	++ ++
13	Parapholis strigosa	e	+ ++ + +
41	Bupleurum tenuissimum	e	++
21	Agrostis stolonifera	f	++++++++++++++ ++++ +++ ++ ++++
8	Plantago maritima	f	+++++++ ++ +++++ + ++++++++ + + + + +++ + ++++ +++
14	Juncus gerardi	f	++++++++ +++ ++ ++++++++ + +++ +++ + ++++++
37	Salicornia radicans	g	+ + + ++
42	Limonium humile	g	+ ++
5	Spartina townsendii	g	+ +
9	Cochlearia anglica	g	+ + +
6	Aster tripolium	g	+ + ++ + +++++ +++ + +++++ + ++++++
7	Spergularia marginata	g	+ + + +++ + ++ + + ++
2	Suaeda maritima	g	+ + ++ + +++++ +++ ++ ++
10	Salicornia europaea	g	+ + + ++++ + + +++ + + + + ++ + ++
4	Halimione portulacoides	g	++ +++ + ++ + + + + +++ ++
3	Limonium vulgare	g	++ + + + + + + + +++ ++
1	Puccinellia maritima	g	+ + ++ +++ + + ++++++++++ +++++ ++ +

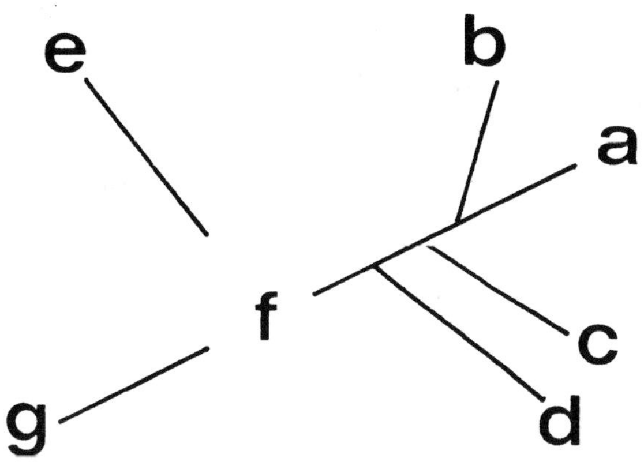

Fig. 4. Graph showing links between species groups in Table 2.

order (Beeftink 1962, cf. van der Maarel 1971, 1974, van der Maarel, Orlóci & Pignatti 1976).

The total number of species has been reduced by excluding the ones with unique presence and also excluding those having low diagnostic value. The diagnostic value of species accords with the index proposed by Feoli (1973). For a given species, the index is given by the ratio of average similarity between the relevés possessing it and the average similarity that these relevés have with those lacking it. Only the species with an index > 1.60 have been taken into consideration.

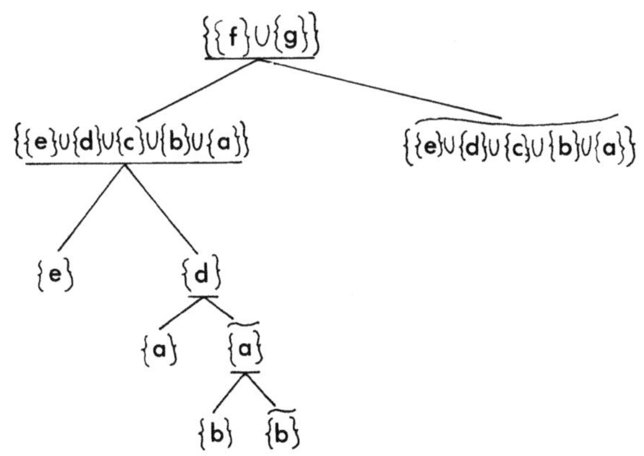

Fig. 5. Hierarchical classification of 58 relevés based on the method here described. Symbols {f}, {g}, {e}, etc. are the sets of relevés defined respectively by the species groups f, g, e, etc.

Fig. 3 includes results from PCA and MST both based on the similarity ratio matrix between the species. The graphic display is based on the second and third components since in this case (non-centered PCA, cf. Noy-Meir 1973) the species ordination reaches the best resolving power (Feoli 1977). MST gives rise to 8 groups of species: a, b, c, d, e, f, g, linked together like in the graph in Fig. 4. The hierarchical classification of the 58 relevés is given in Fig. 5, derived from Table 2.

The whole set of 58 relevés is represented by the reunion of the sets defined by species group g and f. The species of group g are more concentrated on the right of the table, while the species of group f are more concentrated on the left. Associated with this fact there is from right to left of the table an enrichment of relevés with species considered characteristic of syntaxa not included in *Asteretea tripolii*. The relevé ordination relative to the species ordination indicates a gradient of salinity increasing from left to right in Table 2. Along this gradient it is possible to recognise the alliances of the order *Glauco-Puccinellietalia*: *Puccinellion maritimae* (relevés 9 to 55), and *Armerion maritimae* (including, perhaps, also *Puccinellio-Spergularion salinae*: relevés 24, 22, 23, 32, 33, 37) from relevé 56 to relevé 16.

Concluding about the criterion, I suggest that it supports the idea that a set of phytosociological entities can be hierarchically classified on the basis of species presence. However, it may not justify any particular definition of the taxonomic levels. The hierarchy of a set of phytosociological entities is directly dependent on the particular data analysed and may not be extrapolated to situations elsewhere.

Summary

Presented is a criterion for monothetic classification of phytosociological entities on the basis of species presence. The classification criterion springs from the fact that every species, in algebraic terms, gives origin to two equivalence classes. One class includes entities with the species present, and the other class lacking it. The classificatory criterion is embedded in a procedure that establishes the coincidence of two classes according to maximal intersection. The methods employed include principal component analysis, PCA, and the technique of minimum spanning tree, MST, applied to a similarity matrix of species. The method is illustrated by example using data from *Glauco-Puccinellietalia*.

References

Adriani, M. J. 1945. Sur la phytosociologie, la synecologie et le bilan d'eau de halophytes. Wolters, Groningen.

Basagni, M. L., J. Nuvolari & A. Sciacchitano. 1970. Algoritmi di classificazione automatica. Monografia del Centro Zambon, Milano.

Beeftink, W. G. 1962. Conspectus of the phanerogamic salt plant communities in the Netherlands. Biol. Jaarb. Dodonaea 30: 325–362.

Corillon, R. 1953. Les halipédes du nord de la Bretagne. Revue Gén. Bot. 60: 609–658.

Feoli, E. 1973. Un indice che stima il peso dei caratteri per classificazioni monotetiche. Giorn. Bot. Ital. 107: 263–268.

Feoli, E. 1977. On the resolving power of principal component analysis in plant community ordinations. Vegetatio 33: 119–125.

Gillner, V. 1960. Vegetations und Standortsuntersuchungen in den Strandwiesen der schwedischen Westküste. Acta Phytogeogr. Suec. 43: 1–198.

Goodall, D. W. 1954. Objective methods for the classification of vegetation. III. An essay in the use of factor analysis. Aust. J. Bot. 2: 303–324.

Gower, J. C. & C. J. S. Ross. 1969. Minimum spanning tree and single linkage cluster analysis. Appl. Stat. 18: 47–54.

Kessel, S. R. & R. H. Whittaker. 1976. Comparison of three ordination techniques. Vegetatio 32: 21–29.

Maarel, E. van der. 1969. On the use of ordination models in phytosociology. Vegetatio 19: 21–46.

Maarel, E. van der. 1971. Bericht über die Arbeitsgruppe für Datenverarbeitung der Internationalen Vereinigung für Vegetationskunde. Vegetatio 22: 269–273.

Maarel, E. van der. 1974. The Activities of the Working Group for Data-processing of the International Society for Plant Geography and Ecology in 1972–1973. Vegetatio 29: 63–67.

Maarel, E. van der, L. Orlóci & S. Pignatti. 1976. Data-processing in phytosociology, retrospect and anticipation. Vegetatio 32: 65–72.

Moore, J. J., S. J. P. Fitzsimons, E. Lambe & J. White. 1970. A comparison and evaluation of some phytosociological techniques. Vegetatio 20: 1–20.

Noy-Meir, I. 1973. Data transformation in ecological ordination. I. Some advantages of non centering. J. Ecol. 61: 329–341.

Pignatti, S. 1966. La vegetazione alofila della Laguna veneta. Ist. Veneto Sci. Let. Arti 33: 1–174.

Orlóci, L. 1973. Ordination by resemblance matrices. In: Handbook of Vegetation Science. Part V: Ordination and Classification of Communities (ed. R. H. Whittaker), pp. 249–286. Junk, The Hague.

Orlóci, L. 1974. On information transfer in ordination. Vegetatio 29: 11–16.

Whittaker, R. H. & H. G. Gauch. 1973. Evaluation of ordination techniques. In: Handbook of Vegetation Science. Part V: Ordination and Classification of Communities (ed. R. H. Whittaker), pp. 287–321. Junk, The Hague.

Williams, W. T. 1971. Principles of clustering. Ann. Rev. Ecol. Syst. 2: 303–326.

Williams, W. T. & M. B. Dale. 1965. Fundamental problems in numerical taxonomy. Adv. Bot. Res. 2: 35–68.

Westhoff, V. & E. van der Maarel. 1973. The Braun-Blanquet approach. In: Handbook of Vegetation Science. Part V: Ordination and Classification of Communities (ed. R. H. Whittaker), pp. 617–726. Junk, The Hague.

THE CONCEPT OF STRUCTURE IN PHYTOSOCIOLOGY WITH REFERENCES TO CLASSIFICATION OF PLANT COMMUNITIES*

M. REJMÁNEK**
Department of Botany, Charles University, Praha, Czechoslovakia

Keywords: Classification, Plant communities, Structure.

Introduction

In phytosociology the term 'structure' appears to refer prevailingly to the spatial distribution pattern of biomass and representation of life forms which are expressed in different ways. The use of compound terms such as population structure, age structure, ecological structure, competitive structure, production structure, size-class structure, dynamic structure, are more or less common in the literature. In texts describing vegetation structure, we can find paragraphs dealing with physiognomy, species composition, synusiae, consortia, minimal area, species association, etc. In these texts only about one in two authors discuss cyclic changes, i.e. the periodicity of plant communities. In theoretical ecology, on the other hand, sets of differential equations describing dynamics of multipopulation systems are considered as appropriate models for community structure.

The 1966 Rinteln Symposium of the International Society for Vegetation Science (Tüxen 1970), was devoted to structure. However, no clear defintions of structure were given or explicit use made of this concept in the majority of the papers as has been pointed out already in reviews of the volume (e.g. Küchler 1972). This has stimulated the thoughts to be presented here. I hope that my contribution will not be seen as just another terminological exercise, but rather as an attempt to show new ways to understanding vegetation phenomena.

The analysis of the basic phytosociological concept of structure is an inevitable condition for an objective evaluation of structural and organisational complexity (Organisationshöhe, Braun-Blanquet 1964) in plant communities. Furthermore, the approach to the 'biosystematics' of plant communities and the evaluation of a plant community's position in the economy of nature depends in many cases on the knowledge of a plant community's structure.

* Contribution from the Working Group for Data-Processing, International Society for Vegetation Science.

** Thanks are due to Drs. J. Jeník and J. Květ for advice and to the Editors for linguistic help.

The concept of structure

In contemporary science we can observe a process of defining theoretical terms (sensu Carnap 1956) with relevance in several or in many scientific disciplines. Terms which are generally defined but without reference to any specific field of science add new insight into concepts implicit in the terms, thereby promoting the use of the universal methods of system analysis.

Structure of an abstract system has been defined as a set of elements and relationships (e.g. Antomonov 1969). Mathematically, we can conceive structure in the same way as an abstract system: a subset of the Cartesian product of a set of elements and of a set of relationships, i.e. $S \subset E \times R$ (cf. Mesarović 1964). With regard to this definition and to various explicit and implicit definitions of structure in phytosociology and ecology (Bykov 1970, Descoings 1971, Gounot 1969, Knight & Loucks 1969, Mueller-Dombois & Ellenberg 1974, Odum 1962, Pielou 1972, Segal 1971, Shimwell 1971, Suganuma 1967, Vandermeer 1969, Westhoff 1967, Westhoff & van der Maarel 1973, Whittaker 1970), we can propose a relatively broad definition appropriate for phytosociology: the structure of any system is determined by (1) the presence of components which belong to defined classes, (2) the quantitative representation of these components, and (3) the relations among them including their connections and spatial arrangement. However, often the analysis and the description of structure is performed from different points of view. The same subject can thus be regarded as either a component (element), from one point of view, or as a structure, from another. If we accept the above definition, three families of phytosociological structural characteristics should be distinguished: (1) qualitative constituent; (2) quantitative constituent, and (3) relational. To put this division of structural characteristics into proper phytosociological perspective is an immediate objective of the present paper.

(1) *Qualitative constituent characteristics of structure*

Of the components which form plant communities taxa (usually species), life forms, ecoelements, geoelements, size classes, age classes are examples. Classifications based on presence/absence type similarity indices, such as Sørensen's, Kulczynski's and Jaccard's, use only qualitative constituent characteristics (Češka 1966, Cheetham & Hazel 1969). Vegetation analysis based on taxa above the species level has been examined by van der Maarel (1972) and Dale & Clifford (1976). In the last mentioned paper, there was an interesting result in that the use of generic and sub-generic categories instead of species caused only little loss in classification interpretability.

(2) *Quantitative constituent characteristics of structure*
Representation of species with respect to different importance values describes the population structure of a community. Distributions of relative species abundance, biomass, or other importance values (Whittaker 1970, May 1975) and indices of dominance and diversity (DeJong 1975, Pielou 1975, Rejmánek 1975) appear to be most frequently used descriptors of population structure. The sequence of plant communities according to their species diversity values may exhibit a clearly interpretable pattern (Fig. 1). Use of equitability indices in the form of ratios of absolute diversity to maximum possible diversity is very frequent in recent research (e.g. Jacobs 1975). This is not correct, however, except in the rare case when the total number of species in the sampling universe is known (cf. Peet 1974).

Comparisons of population structures with respect to both qualitative and quantitative properties, are the subject of vegetation classifications based on resemblance functions (Orlóci 1975). Distribution of individuals in age classes form the age structure (Namkoong & Roberds 1974). Similarly, quantitative representation of individuals in different categories according to diameter, volume or height forms the dimensional or size class structure (Schmelz & Lindsey 1965). Biological and ecological spectra (or 'structures') can be expressed with respect to importance values of life forms and ecoelements respectively.

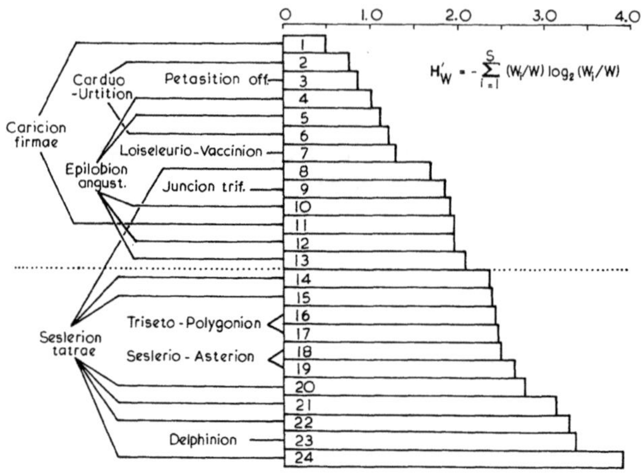

Fig. 1. Species diversity (H'_w) of plant communities in the Dolina Siedmich Prameňov Valley in Belanské Tatry Mts. Primary data by Hadač et al. (1969): W_i – biomass of i-th species, W – total biomass in 1 m^2 quadrat, S – total number of species. Communities above the dotted line belong to extereme habitats (*Caricion firmae*, *Juncion trifidi*, *Loiseleurio-Vaccinion*, *Petasition officinalis*) and initial successional stages (*Carduo-Urticion*, *Epilobion angustifolii*); communities below represent blocked successional stages (*Seslerio-Asterion alpini*), permanent meadows and pastures (*Triseto-Polygonion bistortae*) and mature communities (*Seslerion tatrae*, *Delphinion elati*). 1 – *Dryadetum octopetalae*, 2 – *Carduetum personatae*, 3 – *Petasitetum kablikiani*, 4 and 12 – *Digitalo-Calamagrostetum arundinaceae*, 5 – *Senecio-Rubetum idaei*, 6 – *Geranio-Urticetum dioicae*, 7 – *Vaccinieto-Empetretum nigri*, 8 – *Silenetum prostratae*, 9 – *Juncetum trifidi tatricum*, 10 and 13 – *Rubo-Chamaenerietum angustifolii*, 11 – *Caricetum firmae*, 14 – *Seslerietum tatrae*, 15 – *Hypochoero-Calamagrostetum villosae*, 16 – *Geranio-Alchemilletum crinitae*, 17 – *Alchemillo-Deschampsietum cespitosae*, 18 – *Tortello-Festucetum tatrae*, 19 – *Diantho-Festucetum versicoloris*, 20 and 22 – *Festucetum carpaticae*, 21 – *Carduo glauci-Caricetum tatrorum*, 23 – *Senecio-Adenostyletum alliariae*, 24 – *Aconito-Digitaletum grandiflorae*. Nomenclature follows Hadač et al., op.c.

(3) *Relational characteristics of structure*
Relations among the components of any system can be viewed at different levels with various difficulties, and they can be expressed by all sorts of techniques. Most accessible are the spatial relations, i.e. the spatial structure. Models of spatial structure can be divided into three groups, namely iconical, symbolic and quantitative numerical or graphical. Iconical models (the profile diagrams) are often used in descriptions of tropical vegetation. Symbolic models (e.g. diagrams by Dansereau & Arros 1959) can express selected qualitative constituent characteristics as well as others. Horizontal structure can also be expressed by the use of symbols (this is dominantly done in permanent sample

quadrats). In detailed studies, we may classify plant communities according to their quantitative vertical structure such as the vertical distribution of biomass. The vertical distribution of specific leaf area is another highly complex characteristic of structure. Vertical structure of plant communities is often studied within framework of production ecology (Dykyjová 1971, Květ 1971, Monsi 1968, Tadaki 1970). Some authors speak of production structure. Underground vertical structures do not remain without attention (e.g. Jeník 1971).

Quantitatively expressed horizontal structure represents one of the subjects of quantitative ecology in the sense elaborated by Greig-Smith (1964) and Kershaw (1973). According to them, the first basic feature of horizontal structure appears to be the pattern of populations. Progress has been achieved in this field by the introduction of information statistics (Orlóci 1971). A similar approach is represented by pattern-diversity (Pielou 1966). Other related accounts can be found in Rogers (1974) and Hill (1973), Goodall (1974) and Mead (1974). Models of pattern seem to be very important for understanding of vegetation succession and for the classification of successional stages. Another basic feature of horizontal structure is represented by minimal area. A survey of different approaches and new designs has been recently published by Juhácz-Nagy (1968), Moravec (1973) and Kobayashi (1974). The use of information statistics for an objective evaluation of minimal area appears to be very hopeful. It is necessary to stress that some of the methods used in the analysis of spatial structure are elaborated in great detail in forestry (Leibundgut 1959, Vinš 1961). Spatial structure is often referred to as synmorphology. The first (qualitative) approximation to synmorphology is, as a rule, physiognomy. While the concept of structure in its full implication involves complex latent properties, physiognomy refers to appearance. The association and segregation of co-existing populations in a community (cf. Pielou 1969) may be called a bond structure. Analysis of frequencies represents a broad class of methods used in this field.

The mentioned relational characteristics of structure or partial structures are connected with relations which appear to be formal at first glance. However, biological interpretations of these characteristics are possible. Population, age, dimensional, spatial and bond structures represent the sources of indirect information on the competitive structure of communities, i.e. interference and exploitation between interacting populations (Fig. 2). Competitive structure can be directly studied in isolated mixtures of two or three populations (e.g. Harper 1961, Iwaki

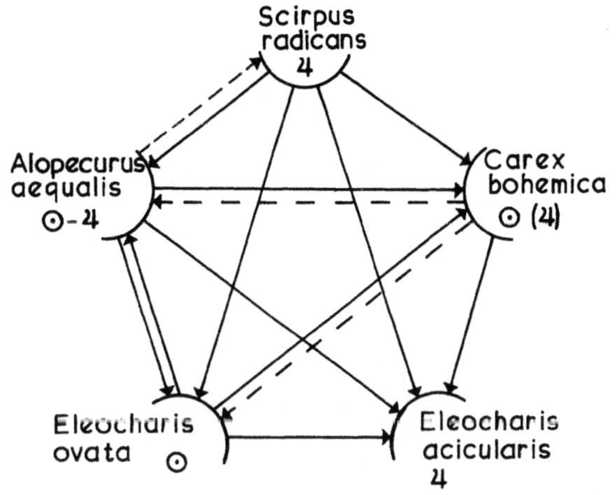

Fig. 2. Simple qualitative model of competitive structure in the aboveground part of *Eleocharito-Caricetum bohemicae* Klika 1935 community. The arrows indicate interference caused by shading. Model has been derived on the basis of vertical structure analysis by point quadrat method (cf. Kubíková & Rejmánek 1973, p. 161).

1959, de Wit 1960, Haizel 1972). Only some first steps have been made in the study of interference between populations in more complicated mixtures (van den Bergh & Elberse 1975). On the theoretical side, matrices of competition coefficients (Levins 1968) serve as models of competitive structures in more complex communities. Symbiotic occurrences and other non-competitive interactions in plant communities also belong to relational structural characteristics. Generally, we can speak of interactive structure. For expressing the complexity of relationships among populations within a community, a highly controversial, and yet sophisticated concept has been used, namely the niche structure of community. Primarily, this concept is useful for theoretical exology (cf. Maguire 1973, Rejmánek & Jeník 1975).

We may express the relationships between the mentioned families of structural characteristics in a scheme (Fig. 3). Thus we obtain the structure of higher order, which we may describe as the structure of structures. Groups of characteristics are the elements. Each element has at least one input and one output; all characteristics are connected partial structures. As a matter of fact, there is only one overall structure, and in this paper, we have mentioned only some of its important facets.

As follows from the foregoing remarks, all changes in plant communities – cyclic and unidirectional – can be considered as changes in the development of structure. It is questionable whether we are justified to designate these

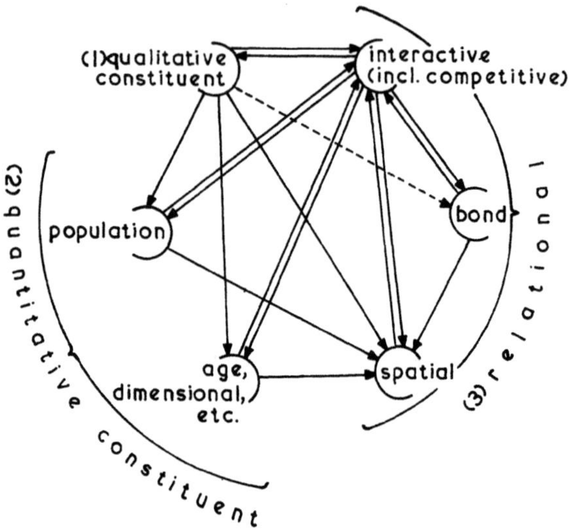

Fig. 3. Determinative relations among important families of structural characteristics.

changes as the temporal or dynamic structure. Kormondy (1969) and Shimwell (1971) state that the description of any change (process) represents a functional approach. Functions always exist in interaction relations of structure. This unity of structure and function is termed *organisation* (cf. Petrusewicz 1965, Teumer 1969). According to this distinction it seems to be correct to regard temporal changes of plant communities as temporal organisation. Cytology already uses the term organisation in this sense.

The role of spatial structure

It follows from Figure 3 that in some way spatial structure integrates various stages of the other structural characteristics. It is clear that structure is conceptually more than just 'spatial relations and pattern', but spatial structure is a highly synthetic feature of the whole. In the major ecosystem processes, such as energy flow, transpiration or soil profile development, the spatial structure of plant communities plays a key role, while the floristic composition may be less important. The results of intensive ecological studies can be extrapolated into the landscape scale only when a method of mapping is used which utilizes spatial parameters of plant communities to an adequate extent. From the biocenotical point of view, there is much evidence showing that composition and diversity of bird communities are rigorously regulated by the vertical structure of vegetation (e.g. MacArthur & MacArthur 1961, Turček 1951). These suggest the potential importance of spatial structure in the classification and ordination of plant communities. Numerical analysis of selected spatial parameters of vegetation has been attempted, e.g., by Knight (1965), Knight & Loucks (1969) and Webb et al. (1970). Walker (1975) pointed out that ordinations based on floristic data only, were in all cases less informative than the ordinations based on combined floristic and physiognomic data. Within a framework of the Braun-Blanquet approach, Dutch phytosociologists pay considerable attention to spatial structure in vegetation studies (Doing 1962; van der Maarel 1966, 1969; Westhoff 1967).

How do we incorporate the parameters of spatial structure in data processing? A simple model for community stratification (Fig. 4) may show this. We consider only two layers; a dominant and a subdominant layer with respect to the degree of their cover. Within the dominant layer, four elements are described. These are trees, shrubs, herbs, and bryophytes or lichens. The same elements are distinguished in the subdominant layer. There is, of course, the special case of communities displaying a single layer which is, necessarily, dominant. In this way we obtain 16 types of spatial structure. These types can be subjected to divisive classification parallel to or combined with an agglomerative or divisive floristic classification.

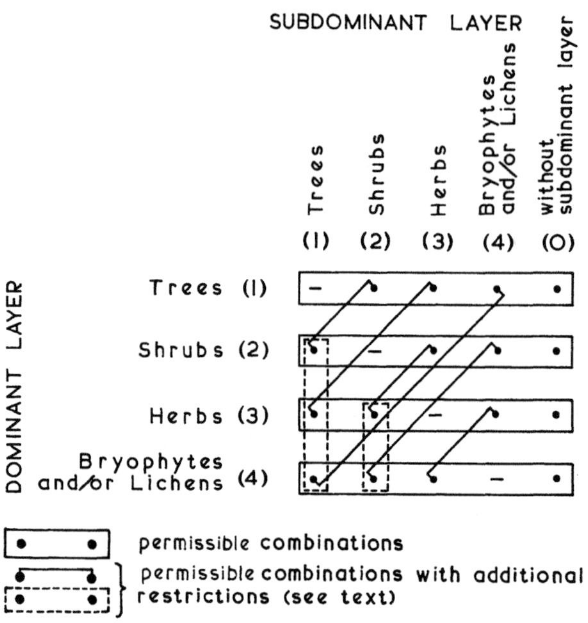

Fig. 4. Division of plant communities into 16 spatial structural types on the basis of dominant and subdominant layers with respect to their cover. Combinations which are 'permissible' are indicated.

Syntaxonomical remarks

Within the Braun-Blanquet approach an arrangement of plant communities has been developed, in which different spatial structures are assigned to different classes (see Westhoff 1967). However, the same class should combine only types of communities which can form permissible combinations in terms of Fig. 4. There are 120 possible combinational pairs within the model, but only 34 combinational pairs are permissible. Further divisions within the pairs in Fig. 4 are not considered but are obviously possible from, for instance, the floristic dissimilarity point of view. All horizontal combinations are generally permissible without additional constraints. For example, all communities with trees in the dominant layer may be grouped in the same class. However, certain communities with a subdominant tree layer can be grouped with communities possessing a dominant tree layer. This appears to be the case in those instances where the tree layer covers more than 50 % in two communities; both of them thus may be regarded as a forest (cf. Jeník & Lokvenc 1962). Generally, cover exceeding 50 % in a subdominant but vertically superposed layer represents a reason for permitting non-horizontal connections (Fig. 4). But we are mainly interested in the impermissible combinations such as combination of a community with trees in the dominant layer and shrubs in the subdominant layer with another community composed of shrubs in the dominant layer and herbs in the subdominant layer.

For spatial-structural reasons, the new class *Rhamno-Prunetea* Rivas Goday et Borja Carbonell 1961 was separated from the *Querco-Fagetea* Br.-Bl. et Vlieger in Vlieger 1937. If we look at the contemporary system of Zurich-Montpellier syntaxa, we may observe sharp contradictions: (1) Shrub communities, i.e. those with shrubs in a dominant and herbs or bryophytes in a subdominant layer, are included in the class *Vaccinio-Piceetea* Br.-Bl. 1939 together with typical forest communities. In partial agreement with the above reflections the new but relatively heterogeneous class *Loiseleurio-Vaccinieta* Eggler 1952, em. Schubert 1960 was established (see Rothmaler 1972, p. 45); (2) Shrub communities with *Alnus viridis*, *Betula carpatica*, *Salix silesiaca*, *S. pentandra*, or *S. grandifolia* in a dominant shrub layer are grouped together with poor herb communities within the single class of *Betulo-Adenostyletea* Br.-Bl. et Tx. 1943 (similarly *Betulo ermanii-Ranunculetea acris japonicae* Ohba 1969), or even within the single alliance *Adenostylion* Br.-Bl. 1926. A new class of *Betulo* (*carpaticae*)-*Alnetea viridis* may be proposed for

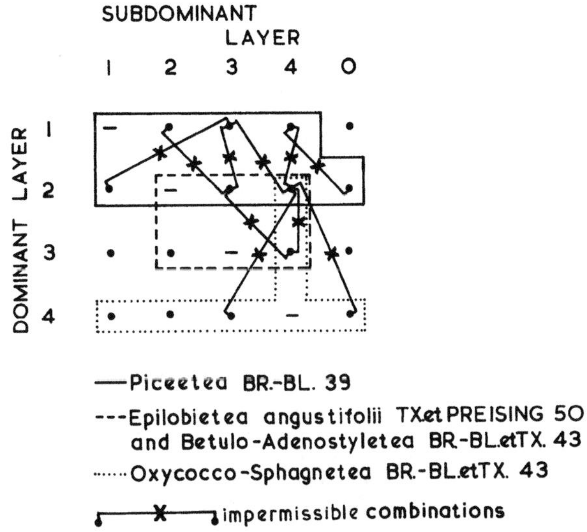

Fig. 5. The same ordering of spatial structural type, of plant communities as in Fig. 4 with an indication of examples of 'impermissible' combinations and examples of syntaxa which do not conform to the suggested rules.'

subalpine deciduous shrub communities. The following alliances would be grouped within this class: *Alnion viridis* Rübel 1933, *Salicion pentandrae* Br.-Bl. 1950, *Salicion silesiacae* Rejmánek, Sýkora et Štursa 1971, and *Salicion grandifoliae* foed. nova (incl. *Salicetum grandifoliae* Horvat 1962, *Salicetum grandifoliae montenegrinum* Lakusić 1966).

Fig. 5 illustrates some examples of so-called impermissible groupings. Such groupings are the reason for separation of the above classes, and for establishment of the class *Urtico-Sambucetea* Doing 1962 em. Passarge 1968 which comprises part of the older *Epilobietea angustifolii* Tx. et Preising in Tx. 1950.

Summary

The concept of structure in phytosociology has been analysed. Three families of structure characteristics have been distinguished: (1) qualitative constituent, (2) quantitative constituent and (3) relational characteristics. Selected structural features are illustrated in some detail and completed by references. The key role of spatial structure has been emphasized and a simple method for the introduction of the parameters of spatial structure in data processing has been proposed. Some remarks to the present syntaxonomy have been added from the spatial structure point of view.

References

Antomonov, Ju. S. 1969. Sistemy, složnost, dinamika. Naukova Dumka, Kiev. 126 pp.

Bergh van den, J. P. & W. Th. Elberse. 1975. Degree of interference between species in complicated mixtures. In: XII International Botanical Congress, Leningrad 1975, Abstracts Vol. 1, p. 138, Leningrad.

Braun-Blanquet, J. 1964. Pflanzensoziologie, Grundzüge der Vegetationskunde, 3rd ed. Springer, Wien-New York. 865 pp.

Bykov, B. A. 1970. Vvedenie v fitocenologiju. Nauka, Alma-Ata. 234 pp.

Carnap, R. 1956. The methodological character of theoretical concepts. Minnesota Stud. Phil. Sci. 1: 38–76.

Češka, A. 1966. Estimation of mean floristic similarity between and within sets of vegetational relevés. Folia Geobot. Phytotax. 1: 93–100.

Cheetham, A. H. & J. E. Hazel. 1969. Binary (presence-absence) similarity coefficients. J. Paleont. 43: 1130–1136.

Dale, M. B. & H. T. Clifford. 1976. On the effectiveness of higher taxonomic ranks for vegetation analysis. Austr. J. Ecol. 1: 37–62.

Dansereau, P. & J. Arros. 1959. Essais d'application de la dimension structurale en phytosociologie. I. Quelques exemples européens. Vegetatio 9: 48–99.

DeJong, T. M. 1975. A comparison of three diversity indices based on their components of richness and eveness. Oikos 26: 222–227.

Descoings, B. 1971. Méthode de description des formations herbeuses intertropicales par la structure de la végétation. Candollea 26: 223–257.

Doing, H. 1962. Systematische Ordnung und floristische Zusammensetzung Niederländischer Wald- und Gebüschgesellschaften. Wentia 8: 1–85.

Dykyjová, D. 1971. Productivity and solar energy conversion in reedswamp stands in comparison with outdoor mass cultures of algae in the temperate climate of Central Europe. Photosynthetica 5: 329–340.

Goodall, D. W. 1974. A new method for the analysis of spatial pattern by random pairing of quadrats. Vegetatio 29: 135–146.

Gounot, M. 1969. Méthodes d'étude quantitative de la végetation. Masson, Paris. 314 pp.

Greig-Smith, P. 1964. Quantitative Plant Ecology. Butterworths, London, 256 pp.

Hadač, E. et al. 1969. Die Pflanzengesellschaften des Tales 'Dolina Siedmich prameňov' in der Belaer Tatra. Vydavatelstvo Slovenskej akadémie vied, Bratislava. 343 pp.

Haizel, K. A. 1972. The productivity of mixtures of two and three species. J. Appl. Ecol. 9: 601–608.

Harper, J. L. 1961. Approaches to the study of plant competition. Symp. Soc. Exp. Biol. 15: 1–39.

Hill, M. O. 1973. The intensity of spatial pattern in plant communities. J. Ecol. 61: 225–235.

Iwaki, H. 1959. Ecological studies on interspecific competition in a plant community I. Jap. J. Bot. 17: 120–138.

Jacobs, J. 1975. Diversity, stability and maturity in ecosystems influenced by human activities. In: W. H. van Dobben & R. H. Lowe-McConnell, eds., Unifying Concepts in Ecology, pp. 187–207. Junk, The Hague/Pudoc Wageningen.

Jeník, J. 1971. Root structure and underground biomass in equatorial forests. In: Productivity of Forest Ecosystems. Proc. Brussels Symp. 1969, pp. 323–331. Unesco, Paris.

Jeník, J. & T. Lokuenç. 1962. Die alpine Waldgrenze in Krkonoše Gebirge. Rozpr. Českosl. Akad. Věd, Ser. Math. - natur 72/1: 1–65.

Juhász-Nagy, P. 1968. On some 'characteristic areas' of plant community stands. In: A. Rényi, ed., Proceedings of the Colloquium on Information Theory, pp. 269–282. Akadémiai Kiadó, Budapest.

Kershaw, K. A. 1973. Quantitative and Dynamic Plant Ecology, 2nd ed. Arnold, London. 308 pp.

Knight, D. H. 1965. A gradient analysis of Wisconsin prairie vegetation on the basis of plant structure and function. Ecology 46: 744–747.

Knight, D. H. & O. L. Loucks. 1969. A quantitative analysis of Wisconsin forest vegetation on the basis of plant function and gross morphology. Ecology 50: 219–234.

Kobayashi, S. 1974. The species-area relation. I. A model for discrete sampling. Res. Popul. Ecol. 15: 223–237.

Kormondy, E. J. 1969. Concepts of Ecology. Prentice Hall, New Jersey. 209 pp.

Kubíková, J. & M. Rejmánek. 1973. Notes on some quantitative methods in the study of plant community structure. Preslia 45: 154–164.

Küchler, A. W. 1972. On the structure of vegetation. Ecology 53: 196–198.

Květ, J. 1972. Growth analysis approaches to the production ecology of reedswamp plant communities. Hidrobiologia 12: 15–40.

Leibundgut, H. 1959. Über Zweck und Methodik der Struktur- und Zuwachsanalysen von Urwäldern. Schw. Zeitschr. Forstw. 110: 111–124.

Levins, R. 1968. Evolution in Chaging Environments. Princeton Univ. Press. 120 pp.

Maarel, E. van der. 1966. On vegetational structures, relations and systems, with special reference to the dune grasslands of Voorne, The Netherlands. Thesis, Utrecht. 170 pp.

Maarel, E. van der. 1969. On the use of ordination models in phytosociology. Vegetatio 19: 21–46.

Maarel, E. van der. 1972. Ordination of plant communities on the basis of their plant genus, family and order relationships. In: E. van der Maarel & R. Tüxen, eds., Basic Problems and Methods in Phytosociology, pp. 183–190. Junk, The Hague.

MacArthur, R. H. & J. M. MacArthur. 1961. On bird species diversity. Ecology 42: 594–598.

Maguire, B. 1973. Niche response structure and the analytical potential of its relationships to the habitat. Amer. Natur. 107: 213–246.

May, R. M. 1975. Patterns of specires abundance and diversity. In: M. L. Cody & J. M. Diamond, eds., Ecology and Evolution of Communities, pp. 81–120. Harvard University Press, Cambridge.

Mead, R. 1974. A test for spatial pattern at several scales using data from a grid contiguous quadrats. Biometries 30: 295–307.

Mesarović, M. D. 1968. System theory and biology – view of a theoretician. In: M. D. Mesarović, ed., System Theory and Biology, pp. 59–87. Springer, New York.

Monsi, M. 1968. Mathematical model of plant communities. In: Functioning of Terrestrial Ecosystems at the Primary Production Level, pp. 391–398. Unesco, Paris.
Moravec, J. 1973. The determination of the minimal area of phytocenoses. Folia Geobot. Phytotax. 8: 23–47.
Mueller-Dombois, D. & H. Ellenberg, 1974. Aims and Methods of Vegetation Ecology. J. Wiley, New York. 547 pp.
Namkoong, G. & J. H. Roberds. 1974. Extinction probabilities and the changing age structure of redwood forests. Amer. Natur. 108: 355–368.
Nicenko, A. A. 1971. Rastitelnaja associacija i rastitelnoe soobščestvo kak pervičnye obekty geobotaničeskogo issledovanija. Nauka, Leningrad. 132 pp.
Odum, E. P. 1962. Relationships between structure and function in the ecosystem. Jap. J. Ecol. 12: 108–118.
Orlóci, L. 1971. An information theory model for pattern analysis. J. Ecol. 59: 343–349.
Orlóci, L. 1975. Multivariate Analysis in Vegetation Research. Junk, The Hague. 276 pp.
Peet, R. K. 1974. The measurement of species diversity. Ann. Rev. Ecol. Syst. 5: 285–307.
Petrusewicz, K. 1965. Dynamics, organisation and ecological structure of population. Ekol. Polska 11: 299–322.
Pielou, E. C. 1966. Species-diversity and pattern diversity in the study of ecological succession. J. Theoret. Biol. 10: 370–383.
Pielou, E. C. 1969. Introduction to Mathematical Ecology. Wiley, New York. 286 pp.
Pielou, E. C. 1972. Measurement of structure in animal communities. In: J. A. Wiens, ed., Ecosystems Structure and Function, pp. 113–135. Oregon State Univ. Press.
Pielou, E. C. 1975. Ecological Diversity. Wiley, New York. 165 pp.
Rejmánek, M. 1975. K ekologické teorii populační struktury společenstev. Thesis, Praha, 139 pp.
Rejmánek, M. & J. Jeník. 1975. Niche, habitat, and related ecological concepts. Acta Biotheoret. 24: 100–107.
Rejmánek, M., Sýkora, T. & J. Štursa. 1971. Fytocenologické poznámky k vegetaci Hrubého Jeseníku. Campanula 2: 31–39.
Rogers, A. 1974. Statistical Analysis of Spatial Dispersion. Pion Limited, London. 164 pp.
Rothmaler, W. 1972. Exkursionsflora für die Gebiete der DDR und der BRD. Volk und Wissen Volkseigener Verlag, Berlin. 612 pp.
Schmelz, D. W. & A. A. Lindsey. 1965. Size-class structure of old-growth forests in Indiana. Forest Sci. 11: 258–264.
Segal, S. 1971. Principles on structure, zonation and succession of a aquatic macrophytes. Hidrobiologia 12: 89–95.
Shimwell, D. W. 1971. The Description and Classification of Vegetation. Sydgwick and Jackson, London. 322 pp.
Suganuma, T. 1967. Phytosociological studies on the semi-natural grassland used for grazing in Japan. II. Ecological strata in pasture vegetation. Bot. Mag. 80: 145–160.
Tadaki, Y. 1970. Studies on the production structure of forest (XVII). Vertical change of specific leaf area in forest canopy. J. Jap. Forest. Soc. 52: 263–268.
Teumer, E. 1969. Philosophische Probleme der Wechselbeziehung von Struktur und Funktion in der Biologie. Fischer, Jena. 130 pp.
Turček, F. J. 1951. O stratifikácii vtačej populácie lesných biocenóz typu Querceto-Carpinetum na južnom Slovensku. Sylvia 13: 71–86.
Tüxen, R., ed. 1970. Gesellschaftsmorphologie (Strukturforschung). Junk, Den Haag. XVI + 356 pp.
Vandermeer, J. H. 1969. The competitive structure of communities: An experimental approach with protozoa. Ecology 50: 362–371.
Vinš, B. 1961. Struktura a vývoj přirozených porostů s jedlí. Práce Výzk.Úst. Les.ČSSR 23: 63–159.
Walker, B. H. 1975. Vegetation site relationships in the Harvard forest. Vegetatio 29: 169–178.
Webb, L. J. et al. 1970. Studies in numerical analysis of complex rain-forest communities. V. A comparison of the properties of floristic and physiognomic-structural data. J. Ecol. 58: 203–232.
Westhoff, V. 1967. Problems and use of structure in the classification of vegetation. Acta Bot. Neerl. 15: 495–511.
Westhoff, V. & E. van der Maarel. 1973. The Braun-Blanquet approach. In: R. H. Whittaker, ed., Handbook of Vegetation Science, Part V: Ordination and Classification of Vegetation, pp. 617–726. Junk, The Hague.
Whittaker, R. H. 1970. The population structure of vegetation. In: R. Tüxen, ed., Gesellschaftsmorphologie, pp. 39–62. Junk, Den Haag.
Wit, C. T. de. 1960. On competition. Versl. Landbouwk. Onderz. 66(8): 1–82.

DELIMITATION AND RANKING OF FLORISTIC-SOCIOLOGICAL UNITS ON THE BASIS OF RELEVÉ SIMILARITY*

Robert NEUHÄUSL**

Botanical Institute, Czechoslovak Academy of Sciences, 25243 Průhonice near Praha, Czechoslovakia

Keywords:
Braun-Blanquet approach, Classification, Floristic homogeneity, Floristic similarity, Numerical syntaxonomy, Oak-hornbeam forest.

Introduction

The present information deals with the possibility of using coefficients of relevé similarity for the delimitation of units in floristic-sociological synsystematics in the sense as used by the Braun-Blanquet approach. The classical method uses for the determination of syntaxonomic units a tabular comparison of phytocoenological relevés with regard to six fundamental criteria:

1. Evaluation of the total floristic composition of phytocoenological relevés.
2. Consideration of the sociological value of all species. The sociological value of a species denotes the degree of linkage between a species and a certain plant community, i.e. its diagnostic value expressed by the degree of fidelity.
3. Evaluation of the quantitative representation, and also the sociability of all species. Sociability is understood as an analytical character.
4. Evaluation of the significance of the group occurrence of a species. The incidence of a species in a plant community is evaluated from the point of view whether it occurs in a group of sociologically, ecologically or chorologically similar species, or more or less in isolation. The group incidence of species of a certain type carries more weight than their singular occurrence.
5. Evaluation of ecological, chorological and functional properties of species represented in the relevés.
6. Utilisation of subjective and objective data on the vegetation with regard to their structure, ecology, dynamics, distribution and function in the landscape.

It is evident that table sorting requires complicated, complex evaluations of phytocoenological relevés, and hence, it may be impeded (a) by insufficient objective criteria (cf. criteria 2, 4, 5, 6); (b) by unavailability of a uniform model (e.g. formula); (c) by the complications in the problem, especially when the problem is on the limit of sience and art. Under these circumstances – that may be even further complicated by insufficient knowledge of the vegetation of a large area and by other formal and methodological impositions – it is understandable that no satisfactory unification of the classificatory approach of vegetation could be achieved.

Although increased objectivity may be attained with now available mathematical-statistical methods in several operations of synthetic phytocoenology, these cannot replace completely the entire synthetic work of the phytocoenologist. For example, numerical methods can be fitted successfully with criterion 1 or 3. The fact that contemporary mathematical procedures do not seem to cover completely or even partially in some cases the remaining criteria (2, 4, 5, 6) notwithstanding the mathematical-statistical methods can disclose aspects and unobvious weaknesses in subjective comparisons, offer clues for classification, and facilitate the procedure of table sorting.

In addition, mathematical-statistical methods may be used in a verification of the results of classical syntaxonomic classifications. However, it should be kept in mind that discrepancies between the results of the mathematical analysis and those of subjective table sorting need not necessarily indicate incorrectness in the latter, since all criteria (1, 2, 3, 4, 5, 6) are involved in subjective table

* Contribution from the Working Group for Data-Processing in Phytosociology, International Society for Vegetation Science.
** The author thanks L. Orlóci and E. van der Maarel for their help in preparing the final form of the manuscript.

sorting and only two (1, 3) in the numerical process. The numerical results should nevertheless initiate revisions in classical tables. In line with these, a verification of syntaxa of the alliance *Carpinion* and other groups of communities, established in classical table sorting, has been performed on data in several studies (see section on Materials and Methods) by means of coefficients of floristic similarity. A generalisation of the results is the object of the present paper.

Materials and methods

A verification of syntaxonomic solutions deriving from the classical method of table sorting has been performed for the following communities:

1. Syntaxa of the alliance *Carpinion* in S-Moravia (Neuhäusl & Neuhäuslová 1968).
2. Syntaxa of the alliance *Carpinion* in central and N-Moravia (Neuhäusl & Neuhäuslová 1972).
3. Associations of *Quercus* and *Quercus-Carpinus* forests in central Bohemia (Neuhäusl & Neuhäuslová 1968a).
4. *Quercus-Carpinus* forests in central Europe (Neuhäusl & Neuhäuslová 1973).
5. A selection of central and easteuropean *Carpinion* associations (Neuhäusl & Neuhäuslová 1972a).
6. Associations of subcontinental peat bogs in Europe (Neuhäusl 1972).
7. Associations of the alliance *Filipendulo-Petasition* in the Czech Lands (Neuhäusl & Neuhäuslová 1975).

In comparison, we used Sørensen's coefficient, as modified by Češka (1966), for calculating mean floristic similarities of relevés. Each syntaxon is represented by a certain number of relevés. The value of the mean similarity of all relevés representing a certain syntaxon is designated as the floristic homogeneity of a syntaxon and calculated by means of formula 1 or 2. On the same principle, using formula 3, we calculated the similarity between the sets of all relevés representing different syntaxa in pairs. This value is referred to as the floristic similarity of syntaxa. We found this method to be advantageous in that the same principle was applied both for the determination of floristic homogeneity in a syntaxon and for that of floristic similarity between two syntaxa. In addition it allowed for a direct comparison of the resulting values.

Delimitation of the problem

The objectives of a numerical evaluation of syntaxa which had been differentiated by the method of classical table sorting may be these:

1. To estimate (a) the total floristic composition of the syntaxa based on a sample of relevés and (b) the floristic similarity between syntaxa.
2. To assess whether values of floristic homogeneity and floristic similarity (see Methods) could serve as suitable criteria for determination of basic syntaxonomic units (associations) and subordinate units (subassociations, variants, etc.).
3. To estimate the degree of change in values of floristic homogeneity in certain associations and of floristic similarity of the most similar (related) associations in various areas together with changes in these values in case of enlargement of the area in which the syntaxa had been studied.
4. To determine the values of floristic homogeneity and floristic similarity particularly for widely distributed associations and those inside and among segments of areas of these associations.
5. To assess changes in the values of floristic homogeneity and mutual similarity among the most closely related associations in different types of vegetation.

Results

Problems regarding objective 1 have been discussed in various studies referred to in the section on Materials and Methods. On the one hand, in all cases, the contents of the group of character- and association-differential species was in agreement with the high degree of mean similarity of relevés within associations. On the other hand, in a majority of cases a high degree of floristic dissimilarity was observed among groups of relevés representing different associations as separated on the basis of different groups of character- and association-differential species. High floristic similarity among associations, distinct in few sociologically and ecologically important species, was found only in exceptional cases. For example, in central Bohemia, high floristic similarity was recorded between the associations *Tilio-Quercetum* and *Potentillo albae-Quercetum*, although they were unrelated in character- and differential species of their particular associations, alliances and orders of any ecological and sociological value. The floristic similarity between these two associations was a consequence of a disturbance of the stands, responsible for a decrease in character- and differential species of the alliance and order to the advantage of common character species of the class and companion species, particularly those indicating disturbance.

Syntaxa determined on smaller and floristically more

uniform area displayed mostly marked floristic differences and could, therefore, be distinguished readily on the bases of larger number of differential or locally characteristic species. However, the value of differential species may often be overestimated and local communities may be considered to be associations, but in a regional comparison they show little syntaxonomic value.

The degree to which values of floristic homogeneity and floristic similarity of syntaxa (objective 2) can be used for an estimation of the rank of a community has been illustrated on the alliance *Carpinion* from S-Moravia (Material 1). All syntaxa of this alliance, from associations to variants, had been distinguished with the Braun-Blanquet method of table sorting and the value of floristic homogeneity of all syntaxa as well as that of floristic similarity among systematically most closely related syntaxa had been calculated. A schematic illustration of the system of associations with the pertinent values of homogeneity and similarity are given in Figure 1. Each row of framed syntaxa refers to one rank. Inside each box, there is with the name of the syntaxon either symbol + or – indicating the value of floristic homogeneity (+ = high homogeneity, roughly 60 or more, – = low homogeneity, roughly 40 or less). Arrows indicate the calculated values of floristic similarity. Similar symbols between the arrows indicate a high (+, approximately 50 or more) or low (–, 40 or less) floristic similarity. This schematic illustration indicates that the associations are characterised by generally high values of floristic homogeneity and by low average values of floristic similarity. A deviation from this situation was observed only in the association *Galio-Carpinetum* and *Carici pilosae-Carpinetum* situated in a transitional area of S-Moravia in which, for chorological reasons, elements of both associations were abundant (in the centres of the areas of both associations there were only elements of one or the other association present); here, the value of floristic similarity was increased to 49. Syntaxa subordinate to associations displayed either analogous or higher values of floristic homogeneity (the average for subasociations was 63, that for variants 60); values of floristic similarity were higher than those among the associations. By contrast, both the values of floristic homogeneity and floristic similarity among syntaxa of a higher rank were lower than those of associations.

Regarding objective 3 we tried to clarify whether the floristic homogeneity and the mutual floristic similarity of certain associations were stable values depending on the area in which they were studied. For this purpose we used the materials 1–3. Table 1 indicates that values of floristic homogeneity of associations in the alliance *Carpinion* change, depending on geographical occurrence, from 50 to 60 for the averages, and 45 to 67 for extremes. Average values of floristic similarity were similar in all regions, but absolute values varied considerably in all

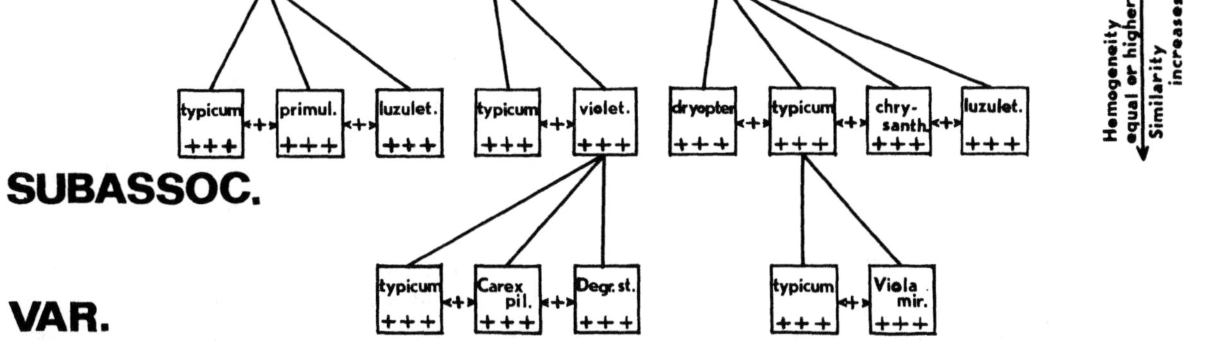

Fig. 1. Syntaxa of the alliance *Carpinion* in South Moravia and numerical characteristics. Explanations are given in the main text.

Table 1. Values of floristic homogeneity and coefficients of similarity for associations of the alliance *Carpinion* in various areas

Area investigated	Size of territory in km^2	Floristic homogeneity average minimum and maximum (in parentheses)	Values of floristic similarity
S. Moravia	4700	60 (55-67)	39 (28-49)
Central and N. Moravia	12000	50 (ca. 50)	39 (32-42)
Central Bohemia	4000	54 (45-66)	39 (35-44)

areas investigated. There was no correlation between the size of the area studied and values of floristic homogeneity. The latter appeared to be independent of the size of the segment of area in which the association was studied (Figure 2). Decrease in homogeneity with increasing size of the area studied could not be confirmed reliably due to considerable variation in these values. It is however concluded that the floristic homogeneity of a certain association displayes different values in the various segments of areas of its distribution, decreasing in proportion to the degree of disturbance and the number of disturbing factors. Values of floristic similarity among the associations varied considerably even over a small area, but the average values were similar.

For a solution of objective 4 we used the results of a syntaxonomic revision of the central and eastern European *Quercus-Carpinus* forests (Materials 4 and 5). By means of the method of table sorting in the alliance *Carpinion*, three widely distributed associations were described for central Europe, and four associations for eastern Europe on climax soils. The central European associations occurred in several geographically isolated local areas. The values of homogeneity of the relevés oscillated from 28 to 48 in the individual local areas; the average values in a large

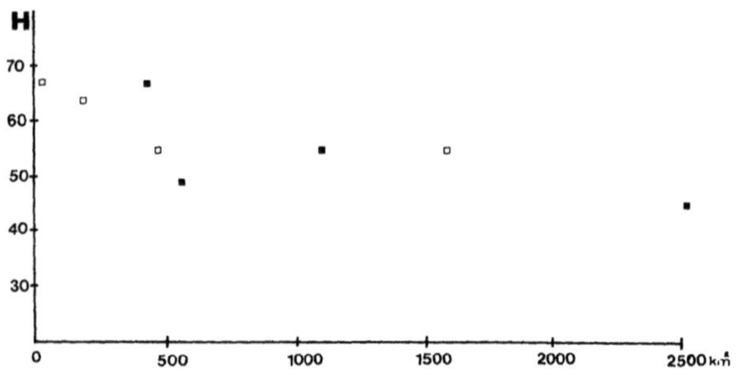

Fig. 2. Correlation between the size of the local area and the values of homogeneity of *Carpinion* associations in South Moravia (□) and central Bohemia (■).

PA NR H	1 173 39	2 434 34	3 12 33	4 702 34	5 11 35	6 61 38	7 92 28	8 121 44	9 36 48	10 830 28	11 146 41	12 445 27	13 176 33
1	+	+	+	+	–	–	+	–	–	+	+	+	+
2	+	+	+	+	+	+	+	+	+	+	+	–	–
3	+	+	+	+	–	+	–	–	–	+	–	+	–
4	+	+	+	+	–	–	+	–	–	+	+	+	+
5	–	+	–	–	+	+	+	+	+	+	–	–	–
6	–	+	+	–	+	+	+	+	+	+	–	–	–
7	+	+	–	+	+	+	+	+	+	+	–	–	–
8	–	+	–	–	+	+	+	+	+	+	–	–	+
9	–	+	–	–	+	+	+	+	+	+	–	+	+
10	+	+	+	+	+	+	+	+	+	+	+	+	+
11	+	+	–	+	–	–	–	–	–	+	+	+	+
12	+	–	+	+	–	–	–	–	+	+	+	+	+
13	+	–	–	+	–	–	–	+	+	+	+	+	+

Fig. 3. Floristic similarity of relevés in the alliance *Carpinion* in the local areas in central Europe.

Explanations: PA – local area of associations – 1 – W-Poland, 2 – Bohemia, western and central Moravia, 3 – Thuringian Slate Mts. in the G.D.R., 4 – SW part of the G.D.R., 5 – Steiermark, Burgenland, 6 – Lower Austria, eastern border of the Alps, 7 – Upper Austria, Bavaria, 8 – Swiss Midlands, N-Switzerland, 9 – SW-Switzerland and Swiss-French borderland, 10 – Baden, Württemberg, NW-Bavaria, Rhineland-Palatinate, Hesse, Vosges borderland, 11 – SW-Mecklenburg, 12 – Lower Saxonian Plain, Münsterland and NW Weser-Uplands, 13 – Rhineland; NR = number of relevés; H = homogeneity; + = similarity > 20 %; – = similarity < 20 %; ——— = border of associations.

number of relevés were roughly 30. The floristic similarity between sets of relevés from different local areas displayed considerable variation.

In Figure 3, floristic similarities are shown between sets of relevés from different segments of area in which the three central European associations occurred. Values of floristic similarity were always higher than 20 (maximum values 39) in sets of relevés of a certain association from different local areas. Similar values were found also in several sets of relevés belonging to different associations. Analysis of associations from the whole area disclosed a decrease in values of floristic homogeneity to about 30; although these values varied considerably in local areas, they never decreased below 20. Similarities among sets of relevés from different local areas varied considerably and did not offer a reliable basis for differentiation of the associations. The determination of the values of floristic homogeneity and similarity among the associations in the whole area of distribution was impeded by a scarcity of basic data and in sufficient pure material. Therefore, the results convey only provisional information.

Better results have been obtained when representative sets of a small number of pure relevés or tables were used instead of all relevés obtained for associations distributed over a large area. With this modification for material 5, the floristic homogeneity of the associations was always greater than 35, mostly around 40, and the values of floristic similarity were always less than 29, mostly around 25.

For a solution of objective 5, concerned with changes in the values of floristic homogeneity of the associations and floristic similarity among the most closely related associations from different types of vegetation, the material available was very small. Widely distributed subcontinental peat bog associations (Material 6) displayed higher values of floristic homogeneity (35 to 53) than the central European associations of the alliance *Carpinion*. The same applied to the floristic similarities; these attained values of up to 35 for the most closely related associations and 28 for less similar associations. The difference between the floristic homogeneity and mutual similarity of the most closely related associations inside the alliance was roughly 15.

Also, the regionally recognized associations in the alliance *Filipendulo-Petasition* in the Czech Lands displayed slightly higher values of floristic homogeneity (about 50) than those recorded for forest associations. The homogeneity of the lower syntaxa remained practically unchanged; also values of similarity among the associations were higher (about 40) than those of the regionally treated forest associations.

In view of the scarcity of material used for the determination of floristic homogeneity and similarity among associations of three different formations it was not appropriate to draw general conclusions. As indicated by the preliminary results, each group of communities keeps its homogeneity to a certain level around which the sample values fluctuate. Also the difference between the average floristic homogeneity and floristic similarity of the most closely related associations was specific of a certain group of associations, but the variability of these values was so high that the significance of the differences could not as yet be proved.

Discussion and survey of results

Westhoff & van der Maarel (1973) summarize results on similarity values on various syntaxon levels in different vegetation types. Their conclusion on a reasonable parallel between syntaxonomical and similarity levels in related communities are only partly supported by our results. Their statement that similarity values for a given syntaxonomic level in different vegetation types are on the same level could not be generally confirmed either.

To this author, the most satisfactory method to be used for a classification of phytocoenoses, which are an organic component of the ecosystem (incapable of independent existence), is still that of classical table sorting. It appeals to users, because, contrary to available numerical methods, it is capable of considering all the criteria employed by phytosociologists in syntaxonomic work. These criteria were discussed. Methods of what became known as objective classification offer only a partial picture of the cluster structure in a relevé sample. Numerical methods may nevertheless contribute considerably to the objectivity of tabular operations, accelerate synthetic work and make it numerically exact. They are also convenient for the verification of the results in conventional tabular synthetic procedures. The application of Sørensen's coefficient of floristic similarity to the relevés (the form of Češka's modification) enables an easy evaluation of a large number of relevés, even without the use of a computer. The mean floristic similarity of all relevés representing a syntaxon, called the floristic homogeneity of the syntaxon, and the mean similarity of all relevés representing two different syntaxa, called the floristic similarity of these syntaxa, are considered as phytosociological descriptors. The employment of these two values enables the identification of the rank of most syntaxa with either local or regional distribution. Syntaxa of the alliance *Carpinion* in a florogenetically uniform area display the following characters:

a) Associations are characterised by a relatively high degree of floristic homogeneity and by relatively low values of mutual floristic similarity. Syntaxa at ranks below associations such as subassociations, variants, etc., display similar or higher values of floristic homogeneity and, simultaneously, higher values of mutual floristic similarity than associations. Syntaxa of a higher rank than association such as alliances, orders, etc., exhibit lower values of floristic homogeneity and also lower values of mutual floristic similarity than these of associations. An occasional exception from this rule was found in several associations in that these displayed a relatively high degree of mutual similarity close to that determined for subassociations. In these cases it had to be decided whether the criterion of floristic similarity was to be considered superior or inferior to the remaining criteria. Other criteria than those of floristic similarity were considered to be decissive even in cases in which associations from transitional areas were analysed, i.e. from areas in which different associations meet, or in cases in which differentiation of these associations were made on the basis of a smaller number of species with high sociological and ecological value. A high floristic similarity may be attributed to the influence of various factors such as the presence of a large number of species indicating a certain degree of disturbance of the phytocoenosis. Floristic differences increase in a natural structure free of disturbance.

The value of floristic homogeneity in an association is usually unstable. If an association is represented by relevés from different segments of area of distribution, the values obtained for this association vary between local areas. The same applied to values of floristic similarity obtained in different regions which also fluctuate considerably around a relatively stable average. The floristic homogeneity of an association appears to be independent of the size of the local area.

Our material was not extensive enough and not pure to allow an evaluation of floristic homogeneity and mutual floristic similarities of associations within the entire area of their distribution. However, it is indicated by the preliminary results that values of floristic homogeneity may decrease to roughly 30 with considerable variation in the different local areas; mutual similarities of the sets of relevés from different local areas display considerable variability and do not provide a reliable basis for further differentiation of the associations. Whenever a representative selection of tables is used, instead of an indiscriminate use of all available tables and relevés characterising the associations, floristic similarity attains values higher than about 35, and the values of floristic similarity roughly 25.

These results pertain exclusively to associations of the alliance *Carpinion* in which the author was particularly interested. For other groups of communities there is still a scarcity of data on floristic homogeneity and floristic similarity. The more widely distributed associations of subcontinental peat bogs have been characterised by higher values of floristic homogeneity (35 to 53) and floristic similarity (28 to 35) than those recorded for central and eastern European plant communities of the alliance *Carpinion*. For grassland and tall-herb communities, the corresponding values have been assessed for associations limited in occurrence locally or regionally. The associations of this group display a relatively high degree of floristic homogeneity and higher values of floristic similarity than those recorded for forest associations.

For the individual types of plant communities it has been impossible as yet to determine on the basis of our preliminary results exact levels and borders for values of floristic homogeneity typical of the basic syntaxa (associations). Although a certain relationship between types of communities and values of homogeneity have been indicated by average results, but variation in these values was too high and the number of data available too small to provide reliable estimates for the significance of the differences. A similar relationship has been shown to exist in the values of floristic similarity and differences between the values of floristic homogeneity and floristic similarity in the most closely related associations.

The application of the coefficient of similarity in a syntaxonomic study of relevés and/or synthetic tables has been found satisfactory as it may verify the correctness of the results. In several instances, mainly in those concerning locally and regionally distributed syntaxa, the method supplied additional criteria for the determination of basic phytocoenological units (associations) and their differentiation from sub- or superordinate units.

Summary

The present paper discusses the use of coefficients of floristic similarity in differentiation of phytocoenological units as an adjunct technique to supplement the tabular sorting method of the Braun-Blanquet approach. The criteria to be considered in tabular synthesis are surveyed and those partial operations which may profit from numerical evaluations in the synthesis have been pointed out. Numerical indices were used to verify that the established syntaxa in table sorting are correct. For syntaxa of the alliance *Carpinion*, conceived either locally or regionally, numerical criteria are established to distinguish units at the basic rank (associations) from those on the subordinate ranks. Associations displayed relatively high floristic homogeneity, the values of the mean floristic similarity of all relevés representing a certain syntaxon, and a relatively lower floristic similarity with the most closely related associations. In syntaxa of a rank below that of an association, the floristic homogeneity was either the same or moderately increased, while the value of floristic similarity among the most closely related units was increased. In syntaxa of a rank higher than that of an association, there should in the author's opinion, occur a considerable decrease in floristic homogeneity and also a low floristic similarity. In the different parts in the distribution area of a certain unit, considerable variation was found in the values of floristic homogeneity, and also, in floristic similarity of the syntaxa; the value of floristic homogeneity decreased particularly in disturbed areas being dependent on the number and intensity of disturbing factors. In a small area, floristic similarity among the associations varied considerably although average values obtained from the various segments of area were similar. An analysis of associations from the whole distribution area indicated a decrease in the values of floristic homogeneity and similarity. The determination of values typical of the associations throughout their area of distribution is still impeded by a scarcity of suitable materials. In addition to associations of the alliance *Carpinion*, the method was used on a small scale for other types of plant communities. It has been postulated by the preliminary results that a certain level around which the values of floristic homogeneity fluctuate, is typical of each type. Also, the difference between the mean floristic homogeneity and floristic similarity of the most closely related associations appears to be specific of the individual types of plant communities.

Zusammenfassung

Der vorliegende Beitrag behandelt die Anwendung von Ähnlichkeitskoeffizienten bei der Abgrenzung der Assoziationen und bei der Rangstufebestimmung von Syntaxa in der floristisch-soziologischen Phytocoenologie (Schule von Braun-Blanquet). Zuerst werden Kriteria erwähnt, die die floristisch-soziologische Systematik bei der syntaxonomischen Auswertung von Aufnahmen und Tabellen im allgemeinen benutzt. Bei durch das klassische Verfahren gefassten Syntaxa des *Carpinion*-Verbandes wurden die Werte der floristischen Homogenität (mittlere Artenähnlichkeit aller ein bestimmtes Syntaxon repräsentierenden Aufnahmen) und der floristischen Ähnlichkeit (auf demselben Prinzip für Syntaxa gleicher Rangstufe festgestellt) bestimmt; die numerischen Daten dienten zur Überprüfung der syntaxonomischen Konzeption. Die Assoziationen sind durch eine relativ hohe floristische Homogenität bei einer relativ niedrigen floristischen Ähnlichkeit mit den nächstverwandten Assoziationen gekennzeichnet. Die der Assoziation untergeordneten Syntaxa wiesen etwa die gleiche oder eine höhere floristische Homogenität als die Assoziationen auf, die floristische Ähnlichkeit zwischen den nächstverwandten Untereinheiten gleicher Rangstufe war jedoch hoch. In den höheren Syntaxa (Verband, Ordnung usw.) sinkt die floristische Homogenität in beträchtlicher Weise sowie auch die floristische Ähnlichkeit mit den nächstverwandten Einheiten gleicher Rangstufe. In verschiedenen Teilarealen von *Carpinion*-Assoziationen wurden grössere Unterschiede in Homogenitäts- und Ähnlichkeitswerten festgestellt; die floristische Homogenität der Assoziationen sinkt meistens mit steigender Anzahl von Störungsfaktoren und mit steigender Störungsintensität. Die floristische Ähnlichkeit zwischen Assoziationen schwankt in beträchtlicher Weise schon in einem kleinen Gebiet, die Durchschnittswerte von verschiedenen Teilarealen sind jedoch ähnlich. Bei der numerischen Analyse der Assoziationen im Rahmen ihrer Gesamtareale sinken die Werte der floristischen Homogenität sowie auch der gegenseitigen Ähnlichkeiten im Vergleich mit lokalen Verhältnissen. Wir entbehren jedoch eine genügende Anzahl von Aufnahmen und 'reinen' Tabellen, um die numerische Charakteristik für Assoziationen im Rahmen ihres Gesamtareals zu gewinnen. Ausser den *Carpinion*-Gesellschaften wurde die numerische Wertung eines geringen Materials auch anderer Vegetationstypen durchgeführt; die vorläufigen Resultate deuten darauf hin, dass die mittleren Homogenitätswerte in verschiedenen Vegetationstypen etwas abweichen, die Einzelwerte jedoch stark schwanken. Ebenso scheinen die Unterschiede zwischen den Werten der floristischen Homogenität und der Ähnlichkeit nächstverwandter Assoziationen für bestimmte Vegetationstypen im Durchschnitt kennzeichnend zu sein.

Formulae:

$$Hom. = \frac{\sum_{i=1}^{k} Pi}{\sum_{i=1}^{k} Ci} \cdot 100 \qquad (1)$$

Pi – partial similarity (s. Češka 1966, p. 96, Formula 9 and values of Pi for different number of relevés in set [n = 1 – 20], p. 98–99).

Ci – percentage constancy of species.

$$Hom. = \frac{nM' - 100}{n-1} \text{ where } M' = \frac{\sum_{i=1}^{k} Ci^2}{\sum_{i=1}^{k} Ci} \qquad (2)$$

n = number of relevés > 20

$$M = \frac{2 \sum_{i=1}^{k_{A+B}} Ci_A \cdot Ci_B}{\sum_{i=1}^{k_A} Ci_A + \sum_{i=1}^{k_B} Ci_B} \qquad (3)$$

M – mean floristic similarity.
Ci_A – percentage constancy of species i in set A.
Ci_B – percentage constancy of species i in set B.
k_A and k_B – number of species in set A and B.

References

Češka, A. 1966. Estimation of the mean floristic similarity between and within sets of vegetational relevés. Folia Geobot. Phytotax. 1: 93–100.

Neuhäusl, R. 1972. Subkontinentale Hochmoore und ihre Vegetation. Studie ČSAV, Praha, 13: 1–121.

Neuhäusl, R. & Neuhäuslová, Zdenka 1968. Mesophile Waldgesellschaften in Südmähren. Rozpr. Čs. Akad. Věd, Ser. Math.-Natur., Praha, 78 (11): 1–84.

Neuhäusl, R. & Neuhäuslová-Novotná, Zdenka. 1968a. Mesophile und subxerophile Waldgesellschaften Mittelböhmens. Folia Geobot. Phytotax. 3: 225–273.

Neuhäusl, R. & Neuhäuslová-Novotná, Zdenka. 1972. Carpinion Gesellschaften in Mittel- und Nordmähren. Folia Geobot. Phytotax. 7: 225–258.

Neuhäusl, R. & Neuhäuslová-Novotná, Zdenka 1972a. Eine einfache Orientierungsmethode zur Beurteilung des Assoziationsranges. In E. v. d. Maarel & R. Tüxen (ed.): Grundfragen und Methoden in der Pflanzensoziologie, p. 211–223. Dr. W. Junk, Den Haag.

Neuhäusl, R. & Neuhäuslová-Novotná, Zdenka. 1973. Eichen-Hainbuchenwälder Mitteleuropas. (Unpublished Ms.)

Neuhäusl, R. & Neuhäuslová-Novotná, Zdenka. 1975. Ein Beitrag zur systematischen Gliederung des Verbandes Filipendulo-Petasition Br.-Bl. 1949. Phytocoenologia 2 (1/2): 183–207.

Westhoff, V. & Maarel, E. van der. 1973. The Braun-Blanquet Approach. In: R. H. Whittaker (ed.): Ordination and Classification of Communities. Handb. Veget. Sci. 5: 617–726.

APPLICATION OF CONSTANCY-SPECIES GROUPS FOR NUMERICAL ORDERING OF PHYTOSOCIOLOGICAL TABLES – THE SYNOPTIC TABLE VERSION*

Jaroslav MORAVEC**, ***

Botanical Institute, Czechoslovak Academy of Sciences, CS 252 43 Průhonice near Praha, C.S.S.R.

Keywords:
Constancy-species groups, Differentiating-species groups, Numerical ordering, Phytosociological tables

Introduction

Present practice in ordering phytosociological tables for the extraction of groups of relevés representing individual syntaxa and species groups differentiating them uses mostly subjective regrouping of relevés and species in the table and the rewriting of the table. The use of computers for the generation of ordered phytosociological tables is so far out of reach for most phytosociologists. The use of even simple numerical techniques without the use of a computer is not practicable in many large sets of relevés (cf. Moravec 1975: 43) for the geometrically increasing number of necessary calculations. Therefore, a procedure has been sought to place the tabular ordering of larger sets of phytosociological data (relevés and synoptic tables) on objective grounds. For sets of some hundreds of relevés of a large region, a successive synthesis of relevés of smaller areas into constancy tables seems to be the best first step. It is noted that many phytosociological data have been published only in the form of constancy tables.

Westhoff & van der Maarel (1973: 652) propose to use the term 'synoptic table' (Übersichtstabelle) for phytosociological tables where constancy (either as a percentage or a class value) for individual synthetized sets of relevés is given. In this paper I use this term in their sense, and the term 'constancy table' (Stetigkeitstabelle) for a table containing constancy values for one synthetized set of relevés only.

*Nomenclature of species follows Ehrendorfer et al. (1973).

**Contribution from the Working Group for Data Processing in Phytosociology, Int. Society for Vegetation Science.
***The author thanks L. Orlóci and E. van der Maarel for linguistic assistance in preparing the manuscript.

In species-rich plant communities the mere presence or absence of species is likely to be more important in syntaxonomic treatments than their quantity. Based on this, the revealing of the deviations in the representation of a certain species group in individual relevés from the mean representation for the given set of relevés has been accepted as the principle objective of a simple numerical ordering of relevés (cf. Moravec 1975). Species are grouped according to their constancy values into groups of a certain range of constancy percentage (*C*) to form *constancy-species groups* (CSGs). Constancy thus serves as an objective measure for grouping the species of a set of relevés. However, the delimination of groups according to certain constancy values is in fact arbitrary in that there is no real basis to prefer CSGs with 1–10, 11–20, 21–30. . . % of constancy range to those of 1 – 5, 6 – 15, 16 – 25. . . % of constancy or to any other.

The term 'Konstanz' seems to have been introduced for the first time by Brockmann-Jerosch (1907) to express the synthetic value of the presence of species in sets (tables) of vegetation relevés. The meaning was later narrowed by Du Rietz et al. (1920) to the presence percentage of species on plots of equal size only, and then used ambiguously both in the broader sense (c.f., e.g. Mueller-Dombois & Ellenberg 1974: 181) and in the narrower (c.f., e.g. Westhoff & van der Maarel 1973: 646).

The application of CSGs in a table of phytosociological relevés for the numerical ordering of relevés is shown by Moravec (1975). In this paper the application of the same method for ordering synoptic tables is described and a method to extract *differentiating – species groups* – (DSGs), on the basis of simple numerical manipulations, is given . The present paper illustrates these procedures on a synoptic table of the herb layer of herb-rich beech forests containing 19 constancy tables of 98 relevés from Bohemia and Moravia.

Description of the procedure

Principles

As shown by Moravec (1975: 41–42) species of a certain CSG (except CSG of 100% constancy) may have distributions in two different ways within a set of relevés – regular or clustered. In the first case each relevé contains the species number of the CSG which equals its mean species number per relevé; such CSG cannot serve to divide and order a set of relevés. In the case of a clustered distribution, the species concentrate in certain relevés only, where all species of the CSG occur, whereas they are missing in the remaining relevés thus dividing the set of relevés into two subsets. In the case of a perfectly regular distribution of given CSG within the set of relevés , the differences between the number of species in individual relevés and the mean species number per relevé are zero. However, in a clustered type of species distribution these differences can reach values other than zero. This difference, expressed as 2nd power (d^2) is used to measure the suitability of a certain CSG for ordering the set of relevés. The same considerations apply to synoptic phytosociological tables with constancy values of relevé groups instead of individual relevés.

Table 1 shows models of species distribution in a synoptic table. Model A represents the regular distribution of constancy values of individual species, model B the agglomerative distribution. As in the relevé tables, these two types of species distribution do not occur in synoptic tables. Models C and D in Table 1 approach real synoptic tables and show that the subsets can easyly be detected by visual inspection as usual in the relevé tables. Moreover, model D demonstrates that a certain CSG can contain species with opposite distribution of constancy within the synoptic tables, such as species a – d concentrated in columns 16 – 18 and species e in columns 19 – 20, having different syntaxonomic significance.

Thus the ordering of a synoptic table on the basis of a numerical representation of CSGs in individual columns is only the first step in the syntaxonomic processing of a table. The second step must follow to rearrange the constancy-species groups in differentiating-species groups, DSGs.

The principle value of such a rearrangement of species groups lies in distinguishing subsets of columns within the ordered synoptic table and in revealing the concentration or the absence (or low constancy) of species in certain subsets. For the latter purpose, differences between the constancy values of species in a subset and in the total synoptic table (i.e. total constancy) expressed in percentages of the higher constancy value $D\%$ are suitable. A sign may be added to indicate whether the difference is positive or negative in comparison with the total constancy. The values of $D\%$ range from 0 (for a species with constancy value equal to that of the

Table 1. Types of constancy distribution of a constancy-species group within a synoptic table

Model	A						B						C						D					
Column	1	2	3	4	5	Total	6	7	8	9	10	Total	11	12	13	14	15	Total	16	17	18	19	20	Total
Number of relevés	5	5	5	5	5	25	5	5	5	5	5	25	5	5	5	5	5	25	5	5	5	5	5	25
Species																								
a	60	60	60	60	60	60	100	100	100	.	.	60	80	100	80	20	20	60	100	80	80	20	20	60
b	60	60	60	60	60	60	100	100	100	.	.	60	100	80	80	20	20	60	80	100	80	20	20	60
c	60	60	60	60	60	60	100	100	100	.	.	60	80	80	100	20	20	60	100	80	100	20	.	60
d	60	60	60	60	60	60	100	100	100	.	.	60	100	80	80	20	20	60	20	40	40	100	100	60
e	60	60	60	60	60	60	100	100	100	.	.	60	80	100	100	20	.	60	100	100	80	.	20	60
Mean species number	3.0	3.0	3.0	3.0	3.0	3.0	5.0	5.0	5.0	0.0	0.0	3.0	4.4	4.4	4.4	1.0	0.8	3.0	4.0	4.0	3.8	1.6	1.6	3.0
d^2	0.0	0.0	0.0	0.0	0.0		4.0	4.0	4.0	9.0	9.0		2.0	2.0	2.0	4.0	4.8		1.0	1.0	0.6	2.0	2.0	

Table 2. Constancy percentages (C) of subsets of columns delimited within models of Table 1 and differences between the constancy values in individual subsets and in the total set expressed as percentages of the higher constancy value (D%)

Model	A				B				C				D			
Subset of columns	1-3		4-5		6-8		9-10		11-13		14-15		16-18		19-20	
	C	D%	C	D%	C	D%	C	D%	C	D%	C	D%	C	D%	C	D%
Species																
a	60	0	60	0	100	+40	0	-100	87	+31	20	-67	87	-31	20	-67
b	60	0	60	0	100	+40	0	-100	87	+31	20	-67	87	+31	20	-67
c	60	0	60	0	100	+40	0	-100	87	+31	20	-67	93	+35	10	-83
d	60	0	60	0	100	+40	0	-100	87	+31	20	-67	33	+45	100	+60
e	60	0	60	0	100	+40	0	-100	93	+35	10	-83	93	+35	10	-83

total table) to 100 (for a species absent from a subset).

The principle for rearrangement of CSGs in differentiating-species groups is illustrated on the above models in Table 2. There is no real basis for delimiting subsets within model A and every division results in subsets in which each species has the constancy of 60% – a value which does not differ from the value for the total set; thus the $D\%$ for each species = 0. In model B, two subsets can easily be distinguished (columns 6–8 and 9–10). The constancy of each species in the first subset equals 100%, in the second one 0%; $D\%$ reaches +40 for the first subset and −100 for the second, indicating concentration of species in the first subset. Model C shows a similar situation, C and $D\%$ ranging from 87 to 93 and +31 to +35 for the first subset (columns 11–13) and from 10 to 20 and −67 to −83 for the second subset (columns 14–15) respectively. In model D while species a, b, c and e concentrate in the first subset (columns 16–18) with values of C and $D\%$ ranging from 87 to 93 and +31 to +35 respectively, species d shows the opposite having $C = 33$ and $D\,\% = -45$. This species is concentrated in the second subset (columns 19–20) with $C = 100$ and $D\% = +40$ whereas the other species have $C = 10$ to 20 and $D\% = -67$ to −83 here.

The rearrangement of CSGs into DSGs proceeds in such a way that starting with CSGs of a high constancy species, having negative $D\%$ values $\leqq -51$ in a single subset are grouped together, then species with these negative values in two subsets are grouped, etc. Species having positive $D\%$ values $\geqq 51$ in a single subset are grouped starting with CSGs of the highest constancy, then species having these $D\%$ values in two subsets are grouped etc. Species with $D\%$ not reaching 51 or −51 are placed at the top of the table in a sequence of descending total constancy.

Illustrating this procedure on model D we see that species a, b, c and e have $D\,\% \geqq -$ in the second subset.

Table 3. Model D (Table 1) ordered both in columns and in species sequence

Column	16	17	18	19	20
Species					
a	100	80	80	20	20
b	80	100	80	20	20
c	100	80	100	20	.
e	100	100	80	.	20
d	20	40	40	100	100

subset. Starting with lower values of $D\%$, we rearrange species of the given CSGs as shown in Table 3 obtaining thus two subsets of columns characterized by two DSGs.

Preparatory calculations

The steps include:

1. Calculation of weighted means of constancy percentages for all species of the total synoptical table (total constancy).
2. Ordering of species according to decreasing total constancy.
3. Grouping of species into CSGs according to these constancy percentages – the width to be adjusted to the total set of relevés and species of the synoptic table. Species with constancy of 100% are not considered further.
4. Calculation of the mean species number ($m_{c(J)}$) for each of the CSG according to the following formula:

$$m_{c(J)} = \frac{1}{100} \sum_{j=1}^{l} C_j$$

where J = constancy-species group, C_j = constancy of the species j (jEJ) in % or in mean percentage of the given constancy class, and l = species number in CSG J. These values are calculated for individual columns and for the total synoptic table.

5. Calculation of squared difference, d^2, between the mean species number of individual columns and of the total synoptic table of each CSG.

The ordering procedure

In this the steps include:

6. Starting with the CSG of the highest constancy (except 100 %) the d^2 values are inspected to check whether they exceed a certain limit (the limit of 2.0 has proved effective – cf. Moravec 1975):

a) If yes, columns with $d^2 \geqq 2.0$ are grouped according to mean species numbers of the given CSG starting with the highest mean species number on the left and ending with the lowest number on the right; columns with d^2 below 2.0 remain in the middle without ordering.

b) If no, the CSG of the nearest lower constancy is treated according to step 6.

7. Subsets delimited in step 6a will be treated according to step 6 using the CSG of the nearest lower constancy.
8. In subsets delimited by iteration of steps 6 and 7,

are there columns without a fixed place in the sequence established by the above procedure?
a) If yes, repeat the ordering procedure within the relevant groups of columns using a lower d^2 value as the limit (e.g. $d^2 \geqq 1.0$).
b) If no, stop the procedure.

Running this procedure by hand, the mean species number of individual CSGs and their d^2 values with descending constancy can be noted on separate cards for each column, as recommended by Moravec (1975).

Ordering of species and extraction of DSGs

The following steps apply:
9. Delimit subsets of columns within the ordered synoptic table.
10. Calculate weighted means of constancy values for each subset.
11. Calculate differences of constancy values between each subset and total table expressed as percentage of the higher value according to the following formula:

$$D\% = 100\ (C_{max} - C_{min}/C_{max}) \qquad (2)$$

12. Search for species not having the $D\%$ values $\geqq 51$ and place them at the top of the table according to descending total constancy.
13. Search for species with the negative $D\%$ value $\leqq -51$ in a single subset having simultaneously the highest total constancy, select further species fulfilling the above conditions for the given subset and order them according to descending total constancy.
14. Apply step 13 to other subsets.
15. Search for species with negative $D\%$ value $\leqq -51$ in two subsets having simultaneously the highest total constancy, select further species fulfilling the above conditions for the given two subsets and order them according to descending total constancy.
16. Apply step 15 for other subsets.
Note: the condition that species selected according to steps 12–16 must not have any positive $D\%$ value $\geqq 51$ included in the procedure.
17. Search for species with positive $D\%$ value $\geqq 51$ in a single subset having simultaneously the highest total constancy, select further species fulfilling the above conditions and order them according to the descending total constancy.
18. Apply step 17 for other subsets.
19. Search for species with positive $D\%$ value $\geqq 51$ in two subsets having simultaneously the highest total constancy, select further species fulfilling the above conditions and order them according to descending total constancy.
20. Apply step 19 for other subsets.
Species groups extracted in steps 17 and 19 can be divided according to the different occurrence of negative $D\%$-values $\leqq -51$ in other subsets.

The above procedure can be modified in steps 6 and 7 in such a way that the sequence of individual CSGs does not follow their descending constancy values but their descending differentiating power expressed as the sum of $d^2 \geqq 1.0$ for the given synoptic table (see later).

Example

Table 4 presents constancy tables of the herb layer of herb-rich beech forests synthetized from 98 relevés from individual sites, or taken from the literature, from Bohemia and Moravia. In the left hand part of the table the columns are not ordered, in the right hand part they are ordered on the numerical basis presented in Table 5 according to the procedure described. The following sequence of columns has been thus established:

3 7 17 15 5 9 14 6 2 19 12 18 1 10 16 4 13 11 8

Using the modification mentioned, the ordering of columns follows the descending values of the sums of $d^2 \geqq 1.0$ of CSGs given in Tab. 5. In this case the CSG of 31–40% constancy initiates the procedure, having the highest sum of d^2 (= 45.5), followed by the CSGs with constancy percentages of 21–30 (sum of d^2 = 23.2), 51–60 (sum of d^2 = 20.0), 61–70 (sum of d^2 = 13.1) and 11–20 (sum of d^2 = 12.6). The sequence of columns is then as follows:

3 9 14 5 15 6 2 7 17 12 1 18 19 8 10 16 4 11 13

This sequence is related to the one given above, however, in that in both sequences nearly the same columns can be found both in the left hand and in the right hand thirds respectively.

The results shown in the right hand part of Table 4 does not give as clear a picture of the table differentiation as in case of a relevé table (Moravec 1975). By inspection and using the numerical values of Table 5 as the an aid four subsets are delimited within the total table:

Table 4. Numerical ordering of a synoptic phytosociological table of beech forests on the basis of the constancy-species groups of the herb layer

		not ordered																			ordered																		
Subsets of columns	Total set																				A				B			C							D				
Columns		1	2	3	4	5	6	7	8	9	10	11	12	13	14	15	16	17	18	19	3	7	17	15	5	9	14	6	2	19	12	18	1	10	16	4	13	11	8
Number of relevés	98	4	2	4	5	6	11	4	3	3	5	3	5	12	5	2	6	3	7	8	4	4	3	2	6	3	5	11	2	8	5	7	4	5	6	5	12	3	3
Mean species number	27.8	29	26	24	29	19	21	25	32	26	28	34	30	39	22	22	27	29	29	29	24	25	29	22	19	26	22	21	26	29	30	29	29	28	27	29	39	34	32
Species																																							
Galium odoratum	98	100	100	75	100	100	100	75	100	100	100	100	100	100	100	100	100	100	100	100	75	75	100	100	100	100	100	100	100	100	100	100	100	100	100	100	100	100	100
Viola reichenbachiana	89	100	100	75	100	83	73	100	100	100	100	100	.	100	100	100	100	100	100	100	75	100	100	100	83	100	100	73	100	100	.	100	100	100	100	100	100	100	100
Mycelis muralis	85	75	100	100	100	17	55	100	67	100	100	67	100	92	100	100	100	100	86	100	100	100	100	100	17	100	100	55	100	100	100	86	75	100	100	100	92	67	67
Poa nemoralis	85	100	100	100	100	50	91	100	67	100	100	67	100	83	60	50	67	100	86	100	100	100	100	50	50	100	60	91	100	100	100	86	100	100	67	100	83	67	67
Oxalis acetosella	83	100	50	25	80	83	100	50	67	100	100	100	60	92	80	100	100	33	86	100	25	50	33	100	83	100	80	100	50	100	60	86	100	100	100	80	92	100	67
Lathyrus vernus	67	75	100	.	80	17	64	100	100	33	40	100	100	83	20	100	33	100	100	75	.	100	100	100	17	33	20	64	100	75	100	100	75	40	33	80	83	100	100
Bromus benekenii	64	50	100	50	.	33	64	75	100	33	100	67	.	83	20	100	83	100	86	88	50	75	100	100	33	33	20	64	100	88	.	86	50	100	83	.	83	67	100
Melica nutans	64	25	100	50	20	33	64	75	100	67	60	100	80	50	40	50	67	100	86	100	50	75	100	50	33	67	40	64	100	100	80	86	25	60	67	20	50	100	100
Dryopteris filix-mas	64	100	50	.	60	100	18	.	100	100	40	100	60	100	100	.	33	67	71	88	.	.	67	.	100	100	100	18	50	88	60	71	100	40	33	60	100	100	100
Urtica dioica	64	50	.	25	60	67	64	50	100	67	100	100	.	75	20	.	100	33	86	88	25	50	33	.	67	67	20	64	.	88	.	86	50	100	100	60	75	100	100
Senecio fuchsii	59	25	50	50	80	67	27	.	100	67	80	100	80	100	80	.	17	33	57	63	50	.	33	.	67	67	80	22	50	63	80	57	25	80	17	80	100	100	100
Scrophularia nodosa	58	75	50	100	80	50	45	75	67	67	80	100	60	75	20	50	17	.	57	50	100	75	.	50	50	67	20	45	50	50	60	57	75	80	17	80	75	100	67
Dentaria bulbifera	56	75	100	.	.	33	64	.	67	67	80	33	60	58	100	50	.	33	100	100	.	.	33	50	33	67	100	64	100	100	60	100	75	80	.	.	58	33	67
Mercurialis perennis	55	75	.	.	.	100	64	.	100	100	80	100	.	42	80	.	50	67	86	63	.	.	67	.	100	100	80	64	.	63	.	86	75	80	50	.	42	100	100
Luzula luzuloides	53	75	100	75	80	.	.	50	.	33	60	67	100	75	20	100	83	100	57	38	75	50	100	100	.	33	20	.	100	38	100	57	75	60	83	80	75	67	.
Lamiastrum galeobdolon	52	75	50	.	40	83	64	.	33	.	80	100	20	58	20	50	67	.	86	63	.	.	.	50	83	.	20	64	50	63	20	86	75	80	67	40	58	100	33
Geranium robertianum	52	100	.	50	80	.	73	.	67	33	60	100	20	100	40	.	67	.	29	38	50	.	.	.	.	33	40	73	.	38	20	29	100	60	67	80	100	100	67
Milium effusum	50	50	50	.	20	67	27	25	100	67	40	33	20	92	40	50	83	33	15	88	.	25	33	50	67	67	40	27	50	88	20	15	50	40	83	20	92	33	100
Hepatica nobilis	50	50	100	.	20	.	36	100	100	.	40	33	100	67	.	100	17	100	100	50	.	100	100	100	.	.	.	36	100	50	100	100	50	40	17	20	67	33	100
Sanicula europaea	47	75	.	75	40	.	36	75	67	.	20	33	40	92	.	50	83	100	15	50	75	75	100	50	.	.	.	36	.	50	40	15	75	20	83	40	92	33	67
Moehringia trinervia	45	75	50	75	40	17	.	.	33	33	80	100	.	83	20	50	67	33	57	50	75	.	33	50	17	33	20	.	50	50	.	57	75	80	67	40	83	100	33
Hieracium sylvaticum	43	75	100	50	80	17	45	50	33	33	.	33	100	58	40	50	.	33	15	38	50	50	33	50	17	33	40	45	100	38	100	15	75	.	.	80	58	33	33
Rubus idaeus	41	50	.	75	60	17	9	25	67	100	60	33	.	58	60	50	33	67	43	25	75	25	67	50	17	100	60	9	.	25	.	43	50	60	33	60	58	33	67
Maianthemum bifolium	40	50	50	.	40	17	27	50	33	67	.	33	40	83	60	100	83	33	.	25	.	50	33	100	17	67	60	27	50	25	40	.	50	.	83	40	83	33	33
Veronica chamaedrys	39	25	.	25	80	.	45	50	.	.	.	33	60	83	.	50	50	100	29	25	25	50	100	50	.	.	.	45	.	25	60	29	25	.	50	80	83	33	.
Athyrium filix-femina	38	50	50	25	100	50	.	.	.	.	100	67	20	33	20	.	67	.	57	50	25	.	.	.	50	.	20	.	50	50	20	57	50	100	67	100	33	67	.
Hordelymus europaeus	36	.	.	.	.	50	45	100	100	33	80	67	.	.	40	.	.	.	57	88	.	100	.	.	50	33	40	45	.	88	.	57	.	80	.	.	.	67	100
Brachypodium sylvaticum	35	.	.	.	40	.	45	50	.	.	100	.	.	.	.	.	100	100	71	88	.	50	100	.	.	.	.	45	.	88	.	71	.	100	100	40	.	.	.
Cardamine impatiens	35	75	100	.	.	.	27	.	.	.	.	67	40	58	20	.	67	33	57	75	.	.	33	.	.	.	20	27	100	75	40	57	75	.	67	.	58	67	.
Prenanthes purpurea	34	50	50	.	80	17	9	.	67	.	20	67	.	75	.	50	50	33	29	50	.	.	33	50	17	.	.	9	50	50	.	29	50	20	50	80	75	67	67
Actaea spicata	32	50	.	25	60	50	.	.	100	33	.	100	20	92	20	.	17	.	.	13	25	.	.	.	50	33	20	.	.	13	20	.	50	.	17	60	92	100	100
Vicia sylvatica	32	25	.	.	60	.	9	.	33	.	40	33	100	58	.	.	17	.	57	63	.	.	.	.	.	.	.	9	.	63	100	57	25	40	17	60	58	33	33
Campanula trachelium	32.	50	50	25	20	.	36	25	33	.	.	67	40	75	.	.	17	67	43	13	25	25	67	.	.	.	.	36	50	13	40	43	50	.	17	20	75	67	33
Impatiens noli-tangere	32	50	.	.	80	17	9	25	67	.	.	100	.	83	20	.	67	.	15	13	.	25	.	.	17	.	20	9	.	13	.	15	50	.	67	80	83	100	67
Alliaria petiolata	29	50	.	.	.	.	64	50	67	33	40	.	20	.	20	.	.	.	100	75	.	50	.	.	.	33	20	64	.	75	20	100	50	40	.	.	.	.	67
Stachys sylvatica	28	25	.	25	.	33	18	25	33	.	60	67	.	67	20	.	17	.	15	38	25	25	.	.	33	.	20	18	.	38	.	15	25	60	17	.	67	67	33
Epilobium montanum	28	.	.	100	40	.	18	25	33	.	40	33	20	42	20	.	50	33	29	25	100	25	33	.	.	.	20	18	.	25	20	29	.	40	50	40	42	33	33
Fragaria vesca	28	.	.	50	60	17	.	.	.	33	20	33	60	67	.	50	17	33	29	25	50	.	33	50	17	33	.	.	.	25	60	29	.	20	17	60	67	33	.
Festuca gigantea	28	.	50	25	40	.	27	.	33	33	40	67	.	33	.	.	83	.	29	38	25	.	.	.	.	33	.	27	50	38	.	29	.	40	83	40	33	67	33
Calamagrostis arundinacea	27	.	.	50	20	.	27	.	.	.	40	33	20	8	.	50	33	33	86	63	50	.	33	50	.	.	.	27	.	63	20	86	.	40	33	20	8	33	.
Carex sylvatica	25	50	.	25	20	17	.	.	.	67	80	33	.	50	40	.	33	.	.	25	25	.	.	.	17	67	40	.	.	25	.	.	50	80	33	20	50	33	.
Vicia sepium	23	.	50	.	40	.	27	50	.	.	.	100	80	17	.	.	.	100	.	38	.	50	100	.	.	.	.	27	50	38	80	.	.	.	.	40	17	100	.
Pulmonaria officinalis	22	25	.	.	20	50	.	.	67	33	.	67	20	75	40	.	.	.	.	.	.	.	.	.	50	33	40	.	.	.	20	.	25	.	.	20	75	67	67
Anemone nemorosa	22	25	.	.	40	.	9	.	33	.	100	.	100	.	20	50	17	33	29	13	.	.	33	50	.	.	20	9	.	13	100	29	25	100	17	40	.	.	33
Paris quadrifolia	22	.	.	.	20	17	.	.	33	67	20	.	.	67	60	100	17	.	.	13	.	.	.	100	17	67	60	.	.	13	.	.	.	20	17	20	67	.	33
Carex pairaei	21	50	.	100	20	.	.	.	100	33	60	.	.	.	20	.	.	.	43	25	100	.	.	.	.	33	20	.	.	25	.	43	50	60	.	20	.	.	100
Myosotis sylvatica	21	25	.	50	40	.	.	25	.	.	.	67	60	50	.	.	.	67	29	.	50	25	67	.	.	.	.	.	.	.	60	29	25	.	.	40	50	67	.
Stellaria holostea	20	.	.	.	.	17	27	100	.	33	40	33	20	17	.	.	17	.	57	.	.	100	.	.	17	33	.	27	.	.	20	57	.	40	17	.	17	33	.
Carex digitata	19	100	50	.	25	.	.	.	.	.	.	.	40	67	.	50	17	.	.	.	25	.	.	50	.	.	.	.	50	.	40	.	100	.	17	.	67	.	.
Galeopsis tetrahit	19	25	.	.	.	33	.	.	.	33	40	67	.	25	20	50	33	.	29	25	.	.	.	50	33	33	20	.	.	25	.	29	25	40	33	.	25	67	.
Dentaria enneaphyllos	18	.	.	.	.	33	27	.	.	.	20	.	.	.	.	.	.	.	57	25	.	.	.	.	33	.	.	27	.	25	.	57	.	20	.	.	.	.	.
Circaea lutetiana	18	50	.	.	20	33	.	.	33	.	20	.	.	67	20	.	33	.	.	.	.	.	.	.	33	.	20	.	.	.	.	.	50	20	33	20	67	.	33
Polygonatum multiflorum	18	50	.	.	.	33	.	.	33	.	.	.	.	17	20	100	50	.	43	13	.	.	.	100	33	.	20	.	.	13	.	43	50	.	50	.	17	.	33
Festuca altissima	17	.	.	.	.	33	18	.	67	67	.	33	.	.	80	.	.	.	15	25	.	.	.	.	33	67	80	18	.	25	.	15	.	.	.	.	.	33	67
Lamiastrum montanum	17	50	50	.	20	33	.	.	33	67	.	.	.	33	60	.	.	.	.	.	.	.	.	.	33	67	60	.	50	.	.	.	50	.	.	20	33	.	33
Gymnocarpium dryopteris	16	25	.	.	40	.	.	.	.	.	80	33	.	.	20	.	33	.	29	25	.	.	.	.	.	.	20	.	.	25	.	29	25	80	33	40	.	33	.
Galeopsis pubescens	16	25	50	50	40	17	36	25	100	.	.	.	.	.	.	.	.	.	15	.	50	25	.	.	17	.	.	36	50	.	.	15	25	.	.	40	.	.	100
Dactylis glomerata	15	.	.	.	.	.	27	100	33	.	.	.	.	8	.	.	17	67	29	.	.	100	67	.	.	.	.	.	.	27	.	29	.	.	17	.	8	.	33
Melica uniflora	14	.	.	.	.	100	.	.	.	100	.	.	.	.	100	.	.	.	.	.	.	.	.	.	100	100	100	.	.	.	.	.	.	.	.	.	.	.	.
Dryopteris carthusiana	13	25	.	.	40	17	9	.	33	.	40	33	.	.	40	.	.	.	15	13	.	.	.	.	17	.	40	9	.	13	.	15	25	40	.	40	.	33	33
Veronica officinalis	12	25	.	50	40	.	.	.	.	.	.	.	20	33	.	.	.	33	.	13	50	.	33	.	.	.	.	.	.	13	20	.	25	.	.	40	33	.	.
Galium rotundifolium	12	50	.	25	60	.	.	.	.	33	.	.	.	17	40	.	17	.	.	.	25	.	.	.	.	33	40	.	.	.	.	.	50	.	17	60	17	.	.
Galeopsis speciosa	12	25	.	.	.	.	.	.	.	.	.	33	.	83	.	.	.	.	.	.	.	.	.	.	.	.	.	.	.	.	.	.	25	.	.	.	83	33	.
Veronica montana	11	.	.	.	.	.	.	.	.	33	20	33	.	33	20	.	50	.	.	.	.	.	.	.	.	33	20	.	.	.	.	.	.	20	50	.	33	33	.
Campanula rapunculoides	11	.	50	50	.	.	27	.	.	33	.	.	40	17	.	.	.	.	.	.	50	.	.	.	.	33	.	27	50	.	40	.	.	.	.	.	17	.	.
Luzula pilosa	11	.	50	25	.	.	.	25	.	.	.	67	40	33	.	.	.	.	.	.	25	25	.	.	.	.	.	.	50	.	40	.	.	.	.	.	33	67	.
Rubus fruticosus	11	.	.	50	60	.	.	.	.	.	.	.	40	17	.	.	.	67	.	.	50	.	67	.	.	.	.	.	.	.	40	.	.	.	.	60	17	.	.

Species with constancy of 10 % and lower are omitted.

A (columns 3, 7, 17, 15), B (3, 9, 14), C (6, 2, 19, 12, 18, 1, 10) and D (16, 4, 13, 11, 8). The homotoneity of these subsets is higher than that of the total synoptic table. Expressed in the basic homotoneity coefficient (cf. Moravec 1971), it is 27.3 for the total table, and 34.5, 39.4, 35.2 and 46.6 for subsets A, B, C and D respectively. Constancy values of species for these subsets as well as differences between these constancy values and those for the total table (*D* % values) are given in Table 6. The *D* % values serve for extraction of DSGs as described below.

According to steps 13 and 14 (respecting the condition that the species selected must not have any positive *D*% values $\geqq 51$) the following DSGs have been extracted: – *Lathyrus vernus, Bromus benekenii, Luzula luzuloides, Hepatica nobilis, Sanicula europaea, Moehringia trinervia, Veronica chamaedrys, Brachypodium sylvaticum, Campanula trachelium, Epilobium montanum, Calamagrostis arundinacea, Vicia sepium, Anemone nemorosa, Myosotis sylvatica, Carex digitata, Galeopsis pubescens, Veronica officinalis* and *Luzula pilosa* characterizing negatively subset B;

Table 5. Numerical basis for ordering of the synoptic phytosociological table (Table 4) of beech forests on the basis of the constancy-species groups

Mean species numbers (above) and $d^2 \geq 1.0$ (below) — not ordered

Column / Constancy (%)	Total set	1	2	3	4	5	6	7	8	9	10	11	12	13	14	15	16	17	18	19	Sum of d^2
91-100	1.0	1.0	1.0	0.8	1.0	1.0	1.0	0.8	1.0	1.0	1.0	1.0	1.0	1.0	1.0	1.0	1.0	1.0	1.0	1.0	0.0
81-90	3.4	3.8	3.5	3.0	3.8	2.3	3.2	3.5	3.0	4.0	4.0	3.3	2.6	3.7	3.4	3.5	3.7	3.3	3.6	4.0	
(d²)						1.2															1.2
61-70	3.2	3.0	3.5	1.3	2.2	2.5	2.7	3.0	5.0	3.0	3.4	4.7	2.4	3.9	2.0	2.5	3.2	4.0	4.3	4.4	
(d²)				3.6					3.2			2.3			1.4				1.2	1.4	13.1
51-60	3.9	5.0	3.5	2.8	3.6	3.3	3.4	1.3	4.3	3.7	5.2	6.0	3.4	5.1	3.6	2.5	3.0	2.3	4.7	4.2	
(d²)		1.2		1.2				6.8			1.7	4.4		1.4		2.0		2.6			20.0
41-50	2.8	3.8	3.0	2.8	2.6	1.2	1.5	2.8	4.0	2.3	2.4	2.7	2.6	4.5	1.6	3.5	2.8	3.7	2.5	3.0	
(d²)						2.6	1.7		1.4					2.9	1.4						10.0
31-40	3.9	4.3	3.0	1.0	5.4	2.0	2.5	3.0	4.3	1.3	3.4	6.3	3.2	6.4	1.8	2.0	5.4	3.7	4.2	5.0	
(d²)				8.4	2.3	3.6	2.0			6.8		5.8		6.3	4.4	3.6	2.3				45.5
21-30	3.2	2.5	1.0	4.3	3.6	1.3	1.9	1.8	4.0	3.0	5.0	5.0	3.8	4.8	2.4	2.5	2.7	3.0	3.9	3.8	
(d²)			4.8	1.2		3.6	1.7	2.0			3.2	3.2		2.6							23.2
11-20	3.0	4.5	2.5	2.8	3.2	3.5	1.7	2.5	3.3	4.0	2.6	3.3	2.0	4.7	4.2	2.0	2.7	1.7	2.9	1.4	
(d²)		2.3					1.7							2.9	1.4			1.7		2.6	12.6

ordered

Column / Constancy (%)	3	7	17	15	5	9	14	6	2	19	12	18	1	10	16	4	13	11	8
91-100	0.8	0.8	1.0	1.0	1.0	1.0	1.0	1.0	1.0	1.0	1.0	1.0	1.0	1.0	1.0	1.0	1.0	1.0	1.0
81-90	3.0	3.5	3.3	3.5	2.3	4.0	3.4	3.2	3.5	4.0	2.6	3.6	3.8	4.0	3.7	3.8	3.7	3.3	3.0
(d²)					1.2														
61-70	1.3	3.0	4.0	2.5	2.5	3.0	2.0	2.7	3.5	4.4	2.4	4.3	3.0	3.4	3.2	2.2	3.9	4.7	5.0
(d²)	3.6						1.4			1.4		1.2						2.3	3.2
51-60	2.8	1.3	2.3	2.5	3.3	3.7	3.6	3.4	3.5	4.2	3.4	4.7	5.0	5.2	3.0	3.6	5.1	6.0	4.3
(d²)	1.2	6.8	2.6	2.0									1.2	1.7			1.4	4.4	
41-50	2.8	2.8	3.7	3.5	1.2	2.3	1.6	1.5	3.0	3.0	2.6	2.5	3.8	2.4	2.8	2.6	4.5	2.7	4.0
(d²)					2.6		1.4	1.7									2.9		1.4
31-40	1.0	3.0	3.7	2.0	2.0	1.3	1.8	2.5	3.0	5.0	3.2	4.2	4.3	3.4	5.4	5.4	6.4	6.3	4.3
(d²)	8.4			3.6	3.6	6.8	4.4	2.0							2.3	2.3	6.3	5.8	
21-30	4.3	1.8	3.0	2.5	1.3	3.0	2.4	1.9	1.0	3.8	3.8	3.9	2.5	5.0	2.7	3.6	4.8	5.0	4.0
(d²)	1.2	2.0			3.6			1.7	4.8					3.2			2.6	3.2	
11-20	2.8	2.5	1.7	2.0	3.5	4.0	4.2	1.7	2.5	1.4	2.0	2.9	4.5	2.6	2.7	3.2	4.7	3.3	3.3
(d²)			1.7				1.4	1.7		2.6			2.3				2.9		

– *Dryopteris filix-mas, Urtica dioica, Senecio fuchsii, Dentaria bulbifera, Mercurialis perennis, Lamiastrum galeobdolon, Milium effusum, Athyrium filix-femina, Carex sylvatica, Galeopsis tetrahit* and *Dryopteris carthusiana* characterizing negatively subset A;

– *Paris quadrifolia* and *Galium rotundifolium* characterizing negatively subset C and

– *Hordelymus europaeus* characterizing negatively subset D.

According to steps 15 and 16 (including also the above additional condition) the following DSGs have been extracted:

– *Geranium robertianum, Cardamine impatiens, Vicia sylvatica, Festuca gigantea* and *Gymnocarpium dryop-*

teris characterizing negatively subsets A and B and
– *Dentaria enneaphyllos* characterizing negatively subsets A and D.

According to steps 17 and 18 the following DSGs have been extracted:

– *Prenanthes purpurea, Actaea spicata, Impatiens noli-tangere, Pulmonaria officinalis, Circaea lutetiana, Galeopsis speciosa* and *Veronica montana* characterizing positively subset D; this group can be divided according to the occurrence of the negative $D\%$ values $\leqq -51$ into two DSGs: *Actaea spicata, Pulmonaria officinalis, Circaea lutetiana,* and *Veronica montana* having these high negative $D\%$ values in subsets A and C and the rest having them in subsets A and B (and C);
– *Alliaria petiolata* characterizing positively subset C;
– *Festuca altissima, Lamiastrum montanum* and *Melica uniflora* characterizing positively subset B and
– *Dactylis glomerata* and *Rubus fruticosus* characterizing positively subset A.

Table 6. Numerical basis for revealing of differentiating-species groups on the basis of differences between the constancy values of species in individual sets and in the total set of columns (difference are expressed in percentages of the higher value of constancy)

	Mean constancy percentages					Differences in %			
Subsets of columns	Total	A	B	C	D	A	B	C	D
Number of relevés	98	13	14	42	29				
Mean species number	27.8	25.2	21.2	26.8	33.6				
Species									
Galium odoratum	98	85	100	100	100	-13	+2	+2	+2
Viola reichenbachiana	89	92	93	81	100	+3	+4	-9	+11
Mycelis muralis	85	100	64	84	90	+15	-25	-1	+6
Poa nemoralis	85	92	64	95	79	-8	-25	+11	-7
Oxalis acetosella	83	46	86	91	90	-45	+3	+9	+8
Lathyrus vernus	67	69	22	76	76	+3	-67	+12	+12
Bromus benekenii	64	77	28	69	69	+17	-56	+7	+7
Melica nutans	64	69	43	74	59	+7	-33	+14	-8
Dryopteris filix-mas	64	15	100	57	79	-77	+36	-11	+19
Urtica dioica	64	31	50	65	83	-52	-22	+2	+23
Senecio fuchsii	59	23	72	51	79	-61	+18	-14	+23
Scrophularia nodosa	58	62	43	57	66	+6	-26	-2	+12
Dentaria bulbifera	56	15	64	81	34	-73	+13	+31	-39
Mercurialis perennis	55	15	93	60	48	-73	+41	+8	-13
Luzula luzuloides	53	77	14	48	69	+31	-74	-9	+23
Lamiastrum galeobdolon	52	8	43	65	59	-85	-17	+20	+12
Geranium robertianum	52	15	21	50	86	-71	-60	-4	+40
Milium effusum	50	23	57	41	72	-54	+12	-18	+31
Hepatica nobilis	50	69	.	62	48	+28	-100	+19	-4
Sanicula europaea	47	77	.	36	72	+39	-100	-23	+35
Moehringia trinervia	45	38	22	38	69	-16	-51	-16	+35
Hieracium sylvaticum	43	46	29	45	45	+7	-33	+4	+4
Rubus idaeus	41	54	50	26	52	+24	+18	-37	+21
Maianthemum bifolium	40	31	43	24	65	-23	+7	-40	+38
Veronica chamaedrys	39	54	.	31	62	+28	-100	-21	+37
Athyrium filix-femina	38	8	29	40	52	-79	-24	+5	+27
Hordelymus europaeus	36	31	43	48	17	-14	+16	+25	-53
Brachypodium sylvaticum	35	38	.	52	28	+8	-100	+33	-20
Cardamine impatiens	35	8	7	48	45	-77	-80	+27	+22
Prenanthes purpurea	34	15	7	26	69	-56	-79	-24	+51
Actaea spicata	32	8	36	10	73	-75	+11	-69	+56
Vicia sylvatica	32	.	.	43	45	-100	-100	+26	+29
Campanula trachelium	32	31	.	31	48	-3	-100	-3	+33
Impatiens noli-tangere	32	8	14	12	79	-75	-56	-63	+59
Alliaria petiolata	29	15	14	60	7	-48	-52	+52	-76
Stachys sylvatica	28	15	21	24	42	-46	-25	-14	+33
Epilobium montanum	28	46	7	21	41	+39	-75	-25	+32
Fragaria vesca	28	31	14	19	45	+10	-50	-32	+38
Festuca gigantea	28	8	7	26	48	-71	-75	-7	+42
Calamagrostis arundinacea	27	31	.	41	17	+13	-100	+34	-37
Carex sylvatica	25	8	36	19	34	-68	+31	-24	+26
Vicia sepium	23	38	.	26	24	+39	-100	+12	+4
Pulmonaria officinalis	22	.	43	5	48	-100	+49	-77	+54
Anemone nemorosa	22	15	7	39	14	-32	-68	+44	-36
Paris quadrifolia	22	15	43	6	38	-32	+49	-73	+42
Carex pairaei	21	31	14	24	14	+32	-33	+13	-33
Myosotis sylvatica	21	39	.	14	35	+46	-100	-33	+40
Stellaria holostea	20	31	14	24	14	+35	-30	+17	-30
Carex digitata	19	15	.	17	31	-21	-100	-11	+39
Galeopsis tetrahit	19	8	28	17	24	-58	+32	-11	+21
Dentaria enneaphyllos	18	.	14	24	.	-100	-22	+25	-100
Circaea lutetiana	18	.	21	7	41	-100	+14	-61	+56
Polygonatum multiflorum	18	13	21	14	21	-17	+14	-22	+14
Festuca altissima	17	.	57	12	10	-100	+70	-29	-41
Lamiastrum montanum	17	.	50	7	21	-100	+66	-59	+19
Gymnocarpium dryopteris	16	.	7	22	17	-100	-56	+27	+6
Galeopsis pubescens	16	23	7	17	17	+30	-56	+6	+6
Dactylis glomerata	15	46	.	10	10	+67	-100	-33	-33
Melica uniflora	14	.	100	.	.	-100	+86	-100	-100
Dryopteris carthusiana	13	.	22	14	14	-100	+41	+7	+7
Veronica officinalis	12	23	.	7	21	+48	-100	-42	+43
Galium rotundifolium	12	8	21	5	21	-33	+43	-58	+43
Galeopsis speciosa	12	.	.	2	38	-100	-100	-83	+68
Veronica montana	11	.	14	2	27	-100	+21	-82	+59
Campanula rapunculoides	11	15	7	14	7	+27	-36	+21	-36
Luzula pilosa	11	15	.	7	21	+27	-100	-36	+48
Rubus fruticosus	11	31	.	5	17	+65	-100	-55	+35

The DSGs extracted by the procedure described are shown in Table 7. The left hand side of this table presents constancy values for individual subsets of columns. From this we can see that the DSG of *Lathyrus vernus,. . . , Luzula pilosa* which differentiates subset B from subsets A, C, and D, is more numerous than the DSG of *Dryopteris filix-mas,. . . , Dryopteris carthusiana* which differentiates subset A from subsets B, C and D. Therefore, the sequence of the subsets is changed in the right hand part of Table 7 showing the first version of the ordered synoptic table both in columns and in species sequence.

Discussion

The above simple example illustrates that the numerical ordering of synoptic phytosociological tables by hand on the basis of constancy-species groups, followed by extraction of differentiating-species groups, and table rearrangement is table rearrangement is suitable for processing data of rather similar vegetation types (belonging to one alliance or at least to one order). The sequence of columns obtained by this procedure suffers however from providing a less suitable basis for the delimitation of subsets of columns than is the case in an ordered relevé table. The DSGs extracted as described contain species of various constancy values. In general, species with a higher constancy value for a subset of columns and with small oscillation of these values within the given subset have a higher syntaxonomic importance than those with low constancy and large oscillation within the subset. Therefore, not every DSG is of syntaxonomic importance. The DSG of *Paris quadrifolia* and *Galium rotundifolium* and of *Dactylis glomerata* and *Rubus fruticosus* are examples having low syntaxonomic importance.

The example presented shows a numerical ordering of a synoptic table of the herb layer of herb-rich beech forests from Bohemia and Moravia which resulted in the delimination of four subsets of constancy tables interpretable as four coena. From these the coenon B seems to be the best delimited one both negatively and positively; it corresponds to the *Melico-Fagetum* Seibert

Table 7. Rearrangement of the synoptic phytosociological table (Table 4) on the basis of differentiating species groups

Subsets of columns	A	B	C	D	B			A				C							D				
Column					5	9	14	3	7	17	15	6	2	19	12	18	1	10	16	4	13	11	8
Number of relevés	13	14	42	29	6	3	5	4	4	3	2	11	2	8	5	7	4	5	6	5	12	3	3
Mean species number	25.2	21.2	26.8	33.6	19	26	22	24	25	29	22	21	26	29	30	29	29	26	27	29	39	34	32
Species																							
Galium odoratum	85	100	100	100	100	100	100	75	75	100	100	100	100	100	100	100	100	100	100	100	100	100	100
Viola reichenbachiana	92	93	81	100	83	100	100	75	100	100	100	73	100	100	.	100	100	100	100	100	100	100	100
Mycelis muralis	100	64	84	90	17	100	100	100	100	100	100	55	100	100	100	86	75	100	100	100	92	67	67
Poa nemoralis	92	64	95	79	50	100	60	100	100	100	50	91	100	100	100	86	100	100	67	100	83	67	67
Oxalis acetosella	46	86	91	90	83	100	80	25	50	33	100	100	50	100	60	86	100	100	100	80	92	100	67
Melica nutans	69	43	74	59	33	67	40	50	75	100	50	64	100	100	80	86	25	60	67	20	50	100	100
Scrophularia nodosa	62	43	57	66	50	67	20	100	75	.	50	45	50	50	60	57	75	80	17	80	75	100	67
Hieracium sylvaticum	46	29	45	45	17	33	40	50	50	33	50	45	100	38	100	15	75	.	.	80	58	33	33
Rubus idaeus	54	50	26	52	17	100	60	75	25	67	50	9	.	25	.	43	50	60	33	60	58	33	67
Maianthemum bifolium	31	43	24	65	17	67	60	.	50	33	100	27	50	25	40	.	50	.	83	40	83	33	33
Stachys sylvatica	15	21	24	42	33	.	20	25	25	.	.	18	.	38	.	15	25	60	17	.	67	67	33
Fragaria vesca	31	14	19	45	17	33	.	50	.	33	50	.	.	25	60	29	.	20	17	60	67	33	.
Carex pairaei	31	14	24	14	.	33	20	100	.	.	.	.	.	25	.	43	50	60	.	20	.	.	100
Stellaria holostea	31	14	24	14	17	33	.	.	100	.	.	27	.	.	20	57	.	40	17	.	17	33	.
Polygonatum multiflorum	15	21	14	21	33	.	20	.	.	.	100	.	.	13	.	43	50	.	50	.	17	.	33
Campanula rapunculoides	5	7	14	7	.	33	.	50	.	.	.	27	50	.	40	.	.	.	.	.	33	67	.
Lathyrus vernus	69	22	76	76	17	33	20	.	100	100	100	64	100	75	100	100	75	40	33	80	83	100	100
Bromus benekenii	77	28	69	69	33	33	20	50	75	100	100	64	100	88	.	86	50	100	83	.	83	67	100
Luzula luzuloides	77	14	48	69	.	33	20	75	50	100	100	.	100	38	100	57	75	60	83	80	75	67	.
Hepatica nobilis	69	.	62	48	.	.	.	.	100	100	100	36	100	50	100	100	50	40	17	20	67	33	100
Sanicula europaea	77	.	36	72	.	.	.	75	75	100	50	36	.	50	40	15	75	20	83	40	92	33	67
Moehringia trinervia	38	22	38	69	17	33	20	75	.	33	50	.	50	50	.	57	75	80	67	40	83	100	33
Veronica chamaedrys	54	.	31	62	.	.	.	25	50	100	50	45	.	25	60	29	25	.	50	80	83	33	.
Brachypodium sylvaticum	38	.	52	28	.	.	.	.	50	100	.	45	.	88	.	71	.	100	100	40	.	.	.
Campanula trachelium	31	.	31	48	.	.	.	25	25	67	.	36	50	13	40	43	50	.	17	20	75	67	33
Epilobium montanum	46	7	21	41	.	.	20	100	25	33	.	18	.	25	20	29	.	40	50	40	42	33	33
Calamagrostis arundinacea	31	.	41	17	.	.	.	50	.	33	50	27	.	63	20	86	.	40	33	20	8	33	.
Vicia sepium	38	.	26	24	.	.	.	.	50	100	.	27	50	38	80	.	.	.	.	40	17	100	.
Anemone nemorosa	15	7	39	14	.	.	20	.	.	33	50	9	.	13	100	29	25	100	17	40	.	.	33
Myosotis sylvatica	39	.	14	35	.	.	.	50	25	67	.	.	.	.	60	29	25	.	.	40	50	67	.
Carex digitata	15	.	17	31	.	.	.	25	.	.	50	.	50	.	40	.	100	.	17	.	67	.	.
Galeopsis pubescens	23	7	17	17	17	.	.	50	25	.	.	36	50	.	.	15	25	.	.	40	.	.	100
Veronica officinalis	23	.	7	21	.	.	.	50	.	33	.	.	.	13	20	.	25	.	.	40	33	.	.
Luzula pilosa	15	.	7	21	.	.	.	25	25	.	.	.	50	.	40	.	.	.	.	.	33	67	.
Dryopteris filix-mas	15	100	57	79	100	100	100	.	.	67	.	18	50	88	60	71	100	40	33	60	100	100	100
Urtica dioica	31	50	65	83	67	67	20	25	50	33	.	64	.	88	.	86	50	100	100	60	75	100	100
Senecio fuchsii	23	72	51	79	67	67	80	50	.	33	.	22	50	63	80	57	25	80	17	80	100	100	100
Dentaria bulbifera	15	64	81	34	33	67	100	.	.	33	50	64	100	100	60	100	75	80	.	.	58	33	67
Mercurialis perennis	15	93	60	84	100	100	80	.	.	67	.	64	.	63	.	86	75	80	50	.	42	100	100
Lamiastrum galeobdolon	8	43	65	59	83	.	20	.	.	.	50	64	50	63	20	86	75	80	67	40	58	100	33
Milium effusum	23	57	41	72	67	67	40	.	25	33	50	27	50	88	20	15	50	40	83	20	92	33	100
Athyrium filix-femina	8	29	40	52	50	.	20	25	.	.	.	.	50	50	20	57	50	100	67	100	33	67	.
Carex sylvatica	8	36	19	34	17	67	40	25	.	.	.	.	.	25	.	.	50	80	33	20	50	33	.
Galeopsis tetrahit	8	28	17	24	33	33	20	.	.	.	50	.	.	25	.	29	25	40	33	.	25	67	.
Dryopteris carthusiana	.	22	14	14	17	.	40	.	.	.	.	9	.	13	.	15	25	40	.	40	.	33	33
Paris quadrifolia	1-	43	6	38	17	67	60	.	.	.	100	.	.	13	.	.	.	20	17	20	67	.	33
Galium rotundifolium	8	21	5	21	.	33	40	25	.	.	.	.	.	.	.	.	50	.	17	60	17	.	.
Hordelymus europaeus	31	43	48	17	50	33	40	.	100	.	.	45	.	88	.	57	.	80	.	.	.	67	100
Geranium robertianum	15	21	50	86	.	33	40	50	.	.	.	73	.	38	20	29	100	60	67	80	100	100	67
Cardamine impatiens	8	7	48	45	.	.	20	.	.	33	.	27	100	75	40	57	75	.	67	.	58	67	.
Vicia sylvatica	.	.	43	45	.	.	.	.	.	.	.	9	.	63	100	57	25	40	17	60	58	33	33
Festuca gigantea	8	7	26	48	.	33	.	25	.	.	.	27	50	38	.	29	.	40	83	40	33	67	33
Gymnocarpium dryopteris	.	7	22	17	.	.	20	.	.	.	.	.	.	25	.	29	25	80	33	40	.	33	.
Dentaria enneaphyllos	.	14	24	.	33	.	.	.	.	.	.	27	.	25	.	57	.	20	.	.	.	.	.
Prenanthes purpurea	15	7	26	69	17	.	.	.	.	33	50	9	50	50	.	29	50	20	50	80	75	67	67
Impatiens noli-tangere	8	14	12	79	17	.	20	.	25	.	.	9	.	13	.	15	50	.	67	80	83	100	67
Galeopsis speciosa	.	.	2	38	.	.	.	.	.	.	.	.	.	.	.	.	25	.	.	.	83	33	.
Actaea spicata	8	36	10	73	50	33	20	25	.	.	.	.	.	13	20	.	50	.	17	60	92	100	100
Pulmonaria officinalis	.	43	5	48	50	33	40	.	.	.	.	.	.	.	20	.	25	.	.	20	75	67	67
Circaea lutetiana	.	21	7	41	33	.	20	.	.	.	.	.	.	.	.	.	50	.	33	20	67	.	33
Veronica montana	.	14	2	27	.	33	20	.	.	.	.	.	.	.	.	.	.	20	50	.	33	33	.
Alliaria petiolata	15	14	67	7	.	33	20	.	50	.	.	64	.	75	20	100	50	40	.	.	.	.	67
Festuca altissima	.	57	12	10	33	67	80	.	.	.	.	18	.	25	.	15	.	.	.	.	.	33	67
Lamiastrum montanum	.	50	7	21	33	67	60	.	.	.	.	.	50	.	.	.	50	.	.	20	33	.	33
Melica uniflora	.	100	.	.	100	100	100	.	.	.	.	.	.	.	.	.	.	.	.	.	.	.	.
Dactylis glomerata	46	.	10	10	.	.	.	.	100	67	.	.	.	27	.	29	.	.	17	.	8	.	33
Rubus fruticosus	31	.	5	17	.	.	.	50	.	67	.	.	.	.	40	.	.	.	.	60	17	.	.

1954 (cf. Moravec 1977). The remaining coena can be interpreted as a depauperate, normal and rich type respectively of another association, assigned to the *Tilio cordatae-Fagetum* Mráz 1960 em. Moravec 1977 (cf. Moravec 1977*).

Processing of synoptic phytosociological tables has hitherto seldom been performed based on numerical techniques except for tables generated by digital computers recently (for more details cf. Westhoff & van der Maarel 1973, Mueller-Dombois & Ellenberg 1974). Numerical ordering of constancy tables was proposed early by Kulzcyński (1928) using a similarity coefficient modified for constancy degrees (classes). To establish the sequence and groups of related associations (each represented by one constancy table), the rearrangement of a matrix of similarity coefficients by trial has been applied. This is in fact so in the matrix technique introduced into phytosociology by the Polish author Kulczyński (1928) but applied mainly for ordering the relevé tables. As noted by Moravec (1975) the disadvantage of the aboven technique is the geometrically increasing number of calculations with increasing number of columns (n = relevés or constancy tables) according to the formula $n(n-1)/2$ and the time consuming rearrangement of the similarity matrix by trial and error. For instance, for our synoptic table (Table 4) 171 calculations of similarity coefficient would have to be made.

Extraction of groups of differentiating species has been and is mostly made by visual inspection of tables using partial tables as an aid. This procedure has been clearly described by Ellenberg (1956) and more recently by e.g. Westhoff & van der Maarel (1973), Mueller-Dombois & Ellenberg (1974). Statistical techniques demanding heavy calculations are seldom employed for this purpose.

Summary

Syntaxonomic treatment of larger sets of vegetation relevés proceeds usually by synthetizing sets of relevés in the columns of constancy tables, and by subsequent processing of the synoptic table, constructed from the columns. During this procedure successive rearrangements of the sequence both of columns and of species (rows) are performed until an ordered synoptic table with extracted subsets of columns and groups of species, differentiating these subsets, is obtained. Hand and eye sorting and rewriting the tables according to subjective judgement is the most usual technique for this processing. To aid the process, a simple numerical procedure is proposed.

For ordering constancy tables (columns), within a synoptic table species are grouped, according to constancy percentages calculated for the total synoptic table (total constancy), into constancy-species groups, CSGs. The mean species number of these CSGs in individual columns is calculated. These values are used to establish the sequence of the columns provided that the squared difference between the mean species number of a CSG in the total synoptic table and in the individual columns exceeds a certain value. For this $d^2 \geqq 2.0$ has proved effective. Thus columns are ordered successively starting with the CSG of the highest constancy in the sequence of mean species numbers of individual CSGs having $d^2 \geqq 2.0$ so as to group columns with low mean species numbers on one side of the table, columns with high mean numbers on the opposite side columns with medium mean species number ($d^2 < 2.0$) remaining in between.

Within a synoptic table with an ordered sequence of columns, subsets of columns are delimited by visual inspection with the help of species number and d^2 values. Differences between the total constancy, and that calculated for individual subsets of columns as percentages of the higher constancy value ($D\%$), serve as the numerical basis for extracting differentiating-species groups, DSGs. This extraction proceeds by successive searching for species with the absolute $D\%$ values $\geqq 51$ first in single subsets, then in two (or more) subsets and arranging these species into groups according to the descending total constancy. Thus the sequence of species is established and a synoptic table ordered both in columns and in species can be drown up.

Zusammenfassung

Die syntaxonomische Bearbeitung umfangreicherer Aufnahmematerialien besteht gewöhnlich in einer Synthese der Aufnahmegruppen in einzelne Stetigkeitstabellen (Kolonnen) und in der Umordnung der daraus hergestellten Übersichtstabelle sowie in der Reihen-

*Numbers of columns in Table 4 correspond to those by Moravec (1977; Tab. 1, Beil. 1) as follows: 1 = 3, 2 = 4, 3 = 1, 4 = 2, 5 = 28, 6 = 11, 7 = 12, 8 = 10, 9 = 30, 10 = 15, 11 = 16, 12 = 5, 13 = 6, 14 = 29, 15 = 7, 16 = 8, 17 = 9, 18 = 13, 19 = 14.

folge der Kolonnen alsauch der der Arten, bis eine geordnete Übersichtstabelle mit ausgeschiedenen Gruppen von Stetigkeitskolonnen und mit den sie differenzierenden Artengruppen gewonnen wird. Diese Umordnung wird meistens mittels einer subjektiven Entscheidung nach der Besichtigung der Übersichtstabelle und mittels der handschriftlichen Umschreibung dieser Tabelle erreicht. Um dieses Verfahren objektiver zu machen, ist eine einfache numerische Methode für die Bearbeitung der Vegetationsaufnahmen und – tabellen mittels der Tabellenarbeit vorgeschlagen.

Für die Ordnung der Stetigkeitskolonnen innerhalb einer Übersichtstabelle sind die Arten nach den für die gesamte Tabelle berechneten Stetigkeitsprozenten (Gesamtstetigkeit) in Stetigkeitsartengruppen, SAG (=CSG), gruppiert und die mittleren Artenzahlen dieser SAG in der gesamten Übersichtstabelle und in den einzelnen Kolonnen berechnet. Diese Werte werden für die Festlegung der Reihenfolge der Kolonnen bei denjenigen Kolonnen verwendet, bei denen das Quadrat der Differenz zwischen den mittleren Artenzahlen der betreffenden Kolonne und der Gesamttabelle, d^2, einen bestimmten Wert erreicht oder überschreitet ($d^2 \geqq 2.0$ hat sich als effektiver Wert bewährt). Auf diese Weise werden die Spalten Schritt nach Schritt in der Reihenfolge der mittleren Artenzahlen mit $d^2 \geqq 2{,}0$ einzelner SAG so geordnet, dass die Kolonnen mit einer niedriger mittleren Artenzahl in einer Seite der Tabelle, diejenigen mit einer hoher mittleren Artenzahl in der gegenüberliegenden Seite gruppiert werden, wobei die Kolonnen mittlerer Artenzahl ($d^2 < 2{,}0$) dazwischen bleiben.

Innerhalb der Übersichtstabelle mit geordneter Reihenfolge der Kolonnen muss man Gruppen der Kolonnen auf Grund der Besichtigung der Tabelle und mit Hilfe der angeführten Werte der Artenzahlen und der d^2 festlegen. Die Differenzen zwischen der Gesamtstetigkeit und der Stetigkeit für die Gruppen von Kolonnen berechnet als Prozente des höheren Stetigkeitswertes ($D\%$ – s. Formel 2) dienen als die numerische Grundlage für die Ausscheidung der differenzierenden Artengruppen, DAG (=DSGs). Diese Ausscheidung verläuft schrittweise durch Aussuchen von Arten mit dem $D\%$-Wert $\geqq$ 51 zuerst in einzigen Gruppen von Kolonnen, danach in zwei (oder mehr) Gruppen und durch Anordnung dieser Arten in Gruppen nach ihrer abnehmenden Gesamtstetigkeit. Dadurch ist die Reihenfolge der Arten festgelegt und die sowohl in den Kolonnen als auch in den Arten geordnete Übersichtstabelle kann geschrieben werden.

References

Brockmann-Jerosch, H. 1907. Die Flora des Puschlav (Bezirk Bernia, Kanton Graubünden) und ihre Pflanzengesellschaften. Leipzig.

Du Rietz, G.E., T.C.E. Fries, H. Osvald & T. A. Tengwall. 1920. Gesetze der Konstitution natürlicher Pflanzengesellschaften. Vetenskapliga och Praktiska Undersökningar i Lappland. Flora och Fauna, Uppsala et Stockholm 7: 1–47.

Ehrendorfer, F. et al. 1973. Liste der Gefässpflanzen Mitteleuropas. 2. Aufl. G. Fischer, Stuttgart.

Ellenberg, H. 1956. Aufgaben und Methoden der Vegetationskunde. In: Walter, H. Einführung in die Phytologie, IV. Grundlagen der Vegetationsgliederung. E. Ulmer, Stuttgart.

Kulczyński, S. 1928. Die Pflanzenassoziation der Pienninen. Bull. Acad. Polon. Sci. Lettres, Cl. Math.-Nat. B (1927): 57–203.

Moravec, J. 1971. A simple method for estimating homotoneity of sets of phytosociological relevés. Folia Geobot. Phytotax. 6: 147–170.

Moravec, J. 1975. Die Anwendung von Stetigkeitsartengruppen zur numerischen Ordnung von pflanzensoziogischen Tabellen. Vegetatio 30: 41–47.

Moravec, J. 1977. Die submontanen krautreichen Buchenwälder auf Silikatböden der westlichen Tschechoslowakei. Folia Geobot. Phytotax. 12: 121–166.

Mueller-Dombois, E. & H. Ellenberg. 1974. Aims and Methods of Vegetation Ecology. J. Wiley et Sons, New York, London, Sydney et Totonto.

Westhoff V. & E. van der Maarel. 1973. The Braun-Blanquet approach In: Whittaker, R.H. (ed), Ordination and Classification of Communities, Handboek of Vegetation Science Vol. 5. Tüxen, R. ed. PP. 617–726. Junk, The Hague.

RANKING SPECIES BASED ON THE COMPONENTS OF EQUIVOCATION INFORMATION*

László ORLÓCI**

Department of Plant Sciences, University of Western Ontario, London, Ont., N6A 5B7, Canada

Keywords:
Information, Ranking, Species weights

Ranking methods have been justified on grounds that since they supply weights which measure species importance, they actually supply the user with the information required to produce a short list of species to economize the analysis without unduly distorting the results. But the weights supplied have relevance only in the population for which they were determined. As further limitation, they can be applied as logical input only in certain methods of data analysis with which they happened to be compatible. Weights representing the independent components of the total sum of squares (Orlóci 1973) are examples. These can be used for input in data analysis only under such circumstances of application in which the sum of squares represents a meaningful concept.

The different ranking methods use different formulations to define weights. In some examples, these are specific functions of the arithmetic mean and standard deviation (Grigal & Goldstein 1971, Grigal & Ohmann 1975), the mean square contingency (Williams, Dale & Macnaughton-Smith 1964) or the product moment correlation coefficient (Macnaugton-Smith, Williams, Dale & Mockett 1964). In others, they represent the specific variance (Orlóci 1975, Rohlf 1977) or a multiple correlation (Gower 1967).

While the species weights often represent quantities that can be derived through a partitioning or other manipulations of the sum of squares, they can in fact differ drastically from the sum of squares. The Sørensen index is an example which Feoli (1973) used in a procedure which gives high weight to species most sharply highlighting the cluster structure in a sample. Others applied different logarithmic expressions of information. In this connection, Orlóci (1976) suggested that for a meaningful ordering of species, weights can be defined as the independent components of the interaction information. The term interaction is used here in a sense similar to its previous uses in the analysis of variance. (A substitute term for interaction is mutual information.) The principal advantage of such a measure of information lies in its suitability to render given nominal, multistate and ordinal variables available for analysis when the ordinary notions of sum of squares or covariance do not apply.

The interaction information is derivable as a term in the additive partioning of the joint information of p species as outlined in Table 1. The first three terms in this table measure the information generated by divergence of distribution $\mathbf{F}_h$ from $\bar{\mathbf{F}}_h$. Similarly, the fourth term measures the divergence of a distribution with elements f_{ijk} from an expected distribution with a characteristic frequency $(f_{i..}/f_{...})(f_{.j.}/f_{...})f_{..k}$; $i = 1, \ldots, s_1$; $j = 1, \ldots, s_2$; $k = 1, \ldots, s_3$. The last term in Table 1 is the sum of the first four. The ideas apply irrespective of the number of species in the sample, although the formulations may have to be changed. Should there be p species, there would be p subscripts to identify a joint frequency and p terms for main effect. There would be p summations in term (4) in which sample size (total frequency) would be raised to the p–1 power. Term (5) would also involve p summations.

The method of Orlóci (1976) is formulated for term (4) representing the interaction information $I(\mathbf{F}_1; \ldots; \mathbf{F}_p)$ in a p species sample. Relying on the components of $I(\mathbf{F}_i; \ldots; \mathbf{F}_p)$ as the weights, the method orders species in a sense of decreasing redundancy with the most interacting species

* Contribution from the Working Group for Data-Processing in Phytosociology, International Society for Vegetation Science.

** The author thanks the National Research Council of Canada for support given.

Table 1. Components of information in a 3-species sample. Generalizations to p-species are discussed in the main text. The symbols F_1, F_2, F_3 represent s_1, s_2, s_3-valued frequency distributions into which the original observations are condensed. The variables measured are nominal respectively with s_1, s_2, s_3 discrete states. $f_{i..}$, $f_{.j.}$, $f_{..k}$ are frequencies with which state i of species F_1, state j of species F_2 and state k of species F_3 occur in the sample. $f_{...}$ represents sample size (the number of quadrats in which the species were measured) and $\bar{F}_1$, $\bar{F}_2$, $\bar{F}_3$ are equidistributions with mean frequencies $f_{...}/s_m$, m = 1, 2, 3, respectively. f_{ijk} is the joint frequency of state i of species F_1, state j of species F_2 and state k of species F_3. The summations are for all terms $i = 1, ..., s_1$, $j = 1, ..., s_2$ and $k = 1, ..., s_3$.

	Source	Information
(1)	Main effect F_1	$I(F_1;\bar{F}_1) = \sum_i f_{i..} \ln(f_{i..} s_1/f_{...})$
(2)	Main effect F_2	$I(F_2;\bar{F}_2) = \sum_j f_{.j.} \ln(f_{.j.} s_2/f_{...})$
(3)	Main effect F_3	$I(F_3;\bar{F}_3) = \sum_k f_{..k} \ln(f_{..k} s_3/f_{...})$
(4)	Interaction $F_1 F_2 F_3$	$I(F_1;F_2;F_3) = \sum_i\sum_j\sum_k f_{ijk} \ln(f_{ijk} f^2_{...}/(f_{i..} f_{.j.} f_{..k}))$
(5)	Joint information	$I(F_1, F_2, F_3) = \sum_i\sum_j\sum_k f_{ijk} \ln(f_{ijk} s_1 s_2 s_3/f_{...})$

Table 2. Sample data from a salt marsh community (from E. van der Maarel). Numerical values in table indicate class symbols of species cover/abundance in relevés.

	Species	Relevé
(1)	*Aster tripolium* L.	00320003320002032232323035553230330332030025200200
(2)	*Spartina patens* Muhl.	00000000000000000000000000000000000000000000099999
(3)	*Atriplex hastata* L.	00000000000000000000000000000000000879970000000000
(4)	*Suaeda maritima* (L.) Dum.	00002552002020023222322228855578955200205000000000
(5)	*Salicornia radicans* Sm.	00000778980000000000000000000000050000000000000000
(6)	*Spartina alterniflora* Lois.	00000000000000000000000000000000000000009899900000
(7)	*Salicornia europaea* L.	00002502020000055537887702333523353000050000000000
(8)	*Puccinellia maritima* (Huds.) Parl.	00000872350000087755003220000230008300050023000000
(9)	*Spartina maritima* (Curt.) Fernald	79898533330000000000000000000000000000000000000000
(10)	*S. townsendii* H. & J. Groves	00000000007889953333333335255333555557550000000000

given the highest weight. Since the most redundant species carry most of the information about the other species in the sample, common sense suggests that the components of $I(\mathbf{F}_1; \ldots ; \mathbf{F}_p)$ may be appropriately used as weights when a short list of species has to be produced which carry maximal information about species in the entire sample. However, when the objective is to locate species which have good potential to characterize isolated groups of relevés, weights are preferred which are high for the species with low interaction. Weights of this sort can be obtained if the equivocation information $D(\mathbf{F}_1; \ldots ; \mathbf{F}_p) = I(\mathbf{F}_1, \ldots, \mathbf{F}_p) - I(\mathbf{F}_1; \ldots ; \mathbf{F}_p)$ is partitioned into additive components. The method of partitioning is simple since the components are the main effects (Table 1) themselves.

The salt marsh data in Table 2 (Orlóci 1976) are used to illustrate the method. The analysis is performed based on program RANKIN which now incorporates an option for the equivocation information. The new version is available to enquiregs free of charge from the author. The results of the analysis are summarized in Table 3. It is clear from inspection of the original data (Table 2), and from the results in Table 3, that the components of the equivocation information do yield an ordering of species which reflects their value to identify uniform blocks in the sample data. The analysis tends to pick out species which show high variability, in the sense, that they span a wide enough range of the measuring scale, but with one state overwhelmingly more frequent than the others. To explain this, the following are offered for consideration.

It is observed that at given s_m and $f_{...}$ the function $I(\mathbf{F}_m ; \bar{\mathbf{F}}_m)$ has its maximum value if the frequencies in $\mathbf{F}_m$ are maximally concentrated. The distribution $\mathbf{F}_m$ = (1 1 46 1 1) exemplifies maximum concentration with $s_m = 5$ and $f_{...} = 50$. It can be shown that at the maximum $I(\mathbf{F}_m\ \bar{\mathbf{F}}_m)$, s_m

Table 3. Determination of weights for species in the sample based on an analysis (RANKIN) of equivocation information (E) and interaction information (M). The method is described in the present paper for E and in a previous paper (Orlóci, 1976) for M.

Species	Rank order		Information		Percentage of total*	
	E	M	E	M	E	M
5	1	7	54.15	2.31	17.97	0.82
9	2	5	49.86	23.19	16.55	8.22
3	3	9	47.79	0.56	15.86	0.20
6	4	8	36.18	1.18	12.01	0.42
4	5	3	24.36	59.34	8.09	21.03
8	6	4	24.25	39.04	8.05	13.84
10	7	1	21.96	71.17	7.29	25.23
7	8	2	18.67	69.01	6.20	24.46
2	9	10	18.40	--	6.11	--
1	10	6	5.64	16.31	1.87	5.78
Total			301.00*	282.11*	100.00	100.00

is a monotone, increasing function of sample size ($f_{...}$). Experiments with it have shown that, for instance, at the theoretical maximum of $I(\mathbf{F}_m ; \bar{\mathbf{F}}_m)$. the distribution $\mathbf{F}_m$ is 3-valued if $f_{...} = 10$ and 11-valued if $f_{...} = 1000$, the maximum occurs at $s_m = 129$. Thus, the species favoured by rankingbased on $I(\mathbf{F}_m ; \bar{\mathbf{F}}_m)$ are those which are variable, with states approaching the s_m threshold, and at the same time, which have concentrated frequency distributions.

The species in Table 2 are differentially variable with different nominal s_m values and with fixed $f_{...}$ at 50. The species to obtain high weights are those whose s_m comes closest to the threshold 11 and with maximally concentrated frequencies. Species 2 comes nearest to such a state, followed by species 9, and so on. While species 2 highlights a sharp group structure, it is not given high weight because its performance is not sufficiently variable ($s_2 = 2$). Species 7, while considerably more variable ($s_7 = 6$), does not receive a high weight either because its distribution is not sufficiently concentrated. Concentration implies uniform performance in a block no matter at what level. Most such uniform blocks in the sample data are due to absences.

It is reasonable to conclude that the method of ranking which uses the components of the equivocation information as weights can provide a meaningful ordering for species. The rank order is, however, related to the efficiency of species to describe group structure in the sample. Since the equivocation information signifies a possession of specific properties by the species distributions, the method presented complements the ranking procedures already proposed.

Summary

The results presented in this paper suggest that for species to be weighted according to their suitability to characterize isolated groups of relevés in a phytosociological table, the equivocation information may serve as a suitable weight. The appropriate formulations are derived and computed for some data from a salt marsh community.

References

Feoli, E. 1973. An index for weighing characters in monothetic classifications. (Italian with English summary). Giorn. Bot. Ital. 107: 263–268.

Gower, J.C. 1967. A comparison of some methods of cluster analysis. Biometrics 23: 623–637.

Grigal, D.F. & R.A. Goldstein. 1971. An integrated ordination classification analysis of an intensively sampled oak-hickory forest. J. Ecol. 59: 481–492.

Grigal, D.F. & L.F. Ohmann. 1975. Classification, description and dynamics of upland plant communities within a Minesota wilderness area. Ecol. Monogr. 45: 389–407.

Macnaughton-Smith, P., W.T. Williams, M.B. Dale & L.G. Mockett. 1964. Dissimilarity analysis: a new technique of hierarchical sub-division. Nature 202: 1034–1035.

Orlóci, L. 1973. Ranking characters by a dispersion criterion. Nature 244: 371–373.

Orlóci, L. 1975. Measurement of redundancy in species collections. Vegetatio 31: 65–67.

Orlóci, L. 1976. Ranking species by an information criterion. J. Ecol. 64: 417–419.

Rohlf, F. J. 1977. A note on the measurement of redundancy. Vegetatio 34: 63–64.

Williams, W.T., M.B. Dale & P. Macnaughton-Smith. 1964. An objective method of weighting in similarity analysis. Nature 201: 426..

A PHYTOSOCIOLOGICAL SURVEY OF THE IRISH MOLINIO-ARRHENATHERETEA USING COMPUTER TECHNIQUES*

J.J. MOORE, S.J.[1] & A.M. O'SULLIVAN[2]

[1] Department of Botany, University College, Dublin 4, Ireland
[2] Agricultural Institute, Johnstown Castle, Wexford, Ireland

Keywords: Grassland survey, Numerical phytosociology, Phyto programme

Introduction

Braun-Blanquet & Tüxen (1952) described a number of grassland types in Ireland, the most widespread of which they labelled *Centaureo-Cynosuretum* (Br.-Bl. & Tx. 1952). No further grassland surveys were done until 1962 when the second author commenced a detailed grassland survey as a Ph. D. thesis. On the completion of this thesis in 1965, in which about 600 relevés had been incorporated, the phytosociology of all the common grassland types was reasonably clear. The work also showed that about 250 species – or one fifth of the total Irish flora – were to be found in the lowland grasslands.

In analysing the 600 grassland relevés it became clear to the authors that much of the tedious table-making procedures might be accomplished by computer. Unfortunately, the user core of the computer available in Dublin up to 1965 was quite inadequate for all but the simplest phytosociological rearrangements.

In 1966 the University computing facilities were greatly enlarged with the acquisition of an I.B.M. 1620 computer having 40K storage and ancillary equipment. Moore (1966, 1973) produced a suite of programs for re-arranging a phytosociological array according to the principles of Braun-Blanquet. It was now possible to analyse 60 relevés and to print an association table containing this number. The University computing facilities were further improved in 1971 by the addition of an I.B.M. 360/50 which was augmented by extra core in early 1973. It is now possible to analyse and print a table of up to 650 relevés and 260 species at one time.

* Contribution from the Working Group for Data-Processing in Phytosociology, International Society for Vegetation Science.

The field survey

In the field, relevés were made at strict 5 km intervals along the first and second-class roadways in the Irish Republic. The first field on the left-hand side after the car speedometer had registered 5 km was entered, unless 1) it was in tillage; 2) it adjoined a farmyard, or 3) it had a bull! A 5 m by 5 m quadrat area in the central part of the field was normally described, a visual check having first been made as to the homogeneity of the stand. The homogeneity of the stand was checked on the basis of the visual uniformity of the floristic, physiognomic and habitat features.

The field card

Bearing the new computer facilities in mind a special field card had been devised on which the range of species likely to be found were pre-printed and coded. Analysis of the original 600 relevés indicated what the species list should be. Two sets of 73 species each were thus printed on the front of the card and a space was left on the back for the long-hand recording of additional species. The card was made from Manilla paper, buff brown in colour to minimise sun dazzle. It was punched for holding in a ring binder in the field.

Completing the field card

Within the quadrat all species occurring were first ticked off on the field card. Any additional species not on the pre-printed cards were written in and later coded for subsequent transfer to punch cards. When all had been noted, the appropriate Braun-Blanquet + to 5 cover-abundance rating was given to each. To facilitate the later card punching stage a 7 was used instead of the + value. Also, the sociability ratings were dropped after 1966 since it was not considered desirable to print these out in the computer produced tables.

A soil profile description was made at each relevé site and surface soil (0–10 cm) samples were collected with a core sampler. These data were recorded on the back of the field card, some of them in a coded form for ease of transfer later onto punch cards.

About 2,000 relevés of Irish grassland stands were made in the period 1965 – 1971 by the second author.

Processing of the field data

As stated already, each field record card has room for 146 of the commoner species. This corresponds to two 80-column punch cards per relevé. The first 73 columns of the punch card are reserved for species, the remaining 7 columns being for details of the broad vegetation type, the author and the relevé number. The Braun-Blanquet weighting for a species is punched in its corresponding column on the punch card.

Processing of the data occurs with the program suite PHYTO. The name 'PHYTO' really covers a series of programs in Fortram IV for the manipulation of phytosociological tables according to the principles of Braun-Blanquet. Using them a series of editions of a table may be rapidly produced until an acceptably finished form of the table appears. The central program 'PHYTO' re-arranges both the rows (species) and the columns (relevés) of a phytosociological table on the basis of one of a number of options. The use of this central program presupposes that the data have been transferred onto disk in a suitable form by means of one or more of the following ancillary programs:

CAPHY transfers data from cards punched in the 'Dublin' format onto random access disk file for immediate processing by PHYTO.

CASTO transfers data from the 'Dublin' card format to permanent storage on magnetic tape.

TRISTO transfers data from the 'Trieste' card format (i.e. each row of a published or manuscript table punched on a single card cf van der Maarel, Orlóci & Pignatti 1976, Pignatti 1976) to permanent storage on tape.

STOPHY transfers selected data from the permanent tape store to disk for processing by PHYTO.

It will also produce suitable input for the Clustan IA package written by Wishart (1969).

A preliminary phytosociological classification is first carried out using the 146 species. The program CAPHY ('card to PHYTO') transfers the data onto disk for rapid access by the program PHYTO by means of which this preliminary classification is carried out. The depth of this classification will depend on the nature of the data and the degree of existing knowledge; in some cases it may be approximately to the level of alliance; in others, where a large number of relevés are available, it may be to the level of sub-association. Table 1 shows the preliminary classification adopted in the case of the grassland data.

Table 1. Preliminary classification of grassland data with corresponding punch card code

Community	Col. 71	Col. 72
Lolio-Cynosuretem	1	
Centaureo-Cynosuretum typicum	2	
dry sub-ass	2	1
wet sub-ass	2	2
Alliance: *Calthion*	3	
Alliance: *Molinion*	4	
Alliance: *Nardo-Galion*	5	

Transfer of data to permanent storage on magnetic tape

By the time this preliminary classification has been achieved, the extra difficult species such as bryophytes will have been identified, and a consistent code for such species determined. One extra punch card was all that was required for the additional species in the grassland data; two or even three extra cards are needed in vegetation types which are more species-rich. Any soil chemical analyses will also have been completed. All of this extra material is punched with the ecological data occupying a card with the label '9' in column 80. On this card is also punched the preliminary classification determined in the previous stage.

The program CASTO ('card to store') now transfers all the data on magnetic tape where the relevés are arranged in sequential order.

Final processing

The program STOPHY ('store to PHYTO') enables one to select a number of relevés from the permanent storage on tape and transfer the data onto a temporary disk file. A separate file is used for the floristic and the ecological data.

A number of selection criteria may be employed. All the

relevés belonging to any of the provisional classificatory units already decided on in stage 2 may be transferred and a final 'clean' table is obtained by means of a few runs of the PHYTO program. Alternatively only those relevés containing one or more specified species may be selected, or else a random selection may be made.

The production of a satisfactory table by PHYTO will usually entail dictating the final order of the relevés by means of a deck of relevé ordering cards. This deck may then be used in conjunction with the programs COMSTAT and PLOTMAP to obtain basic statistics of the quantitative ecological data (e.g. mean level of available soil nutrients in each community) together with their standard error and a distribution map of the relevés of the different subsidiary communities.

Capability of the 'PHYTO' program

In the classical Braun-Blanquet method one starts with a raw table and proceeds by means of differential species, tested in a number of partial re-writings of the table ('Teil-Tabellen'), to a first draft of a differentiated table. By using the 'Autodivide' option of PHYTO one can proceed immediately to this latter stage.

The program first seeks out the best pair of differential species: it ignores the rarer species and the relatively constant species and seeks the two most highly correlated species among the remainder. The degree of exclusion is at the discretion of the user, but a default option is provided which excludes species present in less than 10% of the relevés and in more than 50% of them. That species pair is chosen which has the maximum number of joint occurrences and the minimum number of non-joint occurrences in the table being analysed. The main emphasis is thus on presence/absence, but the abundance of a species in a relevé will affect the outcome of the test. The exact procedure is described in the program manual (Moore 1973).

The data matrix is scanned to detect any species with similar distribution among the relevés to this pair of differential species and the whole table is sub-divided into two on the basis of the presence or absence of the complete set, those relevés with most of the species group being to the left of the table. The relevés which contain none of this species group are then re-analysed in exactly the same way.

The usual result of this process is that the table is divided into three units, two of them characterised by two sets of differential species and the third (middle) unit without any of the differentials as in the diagram below.

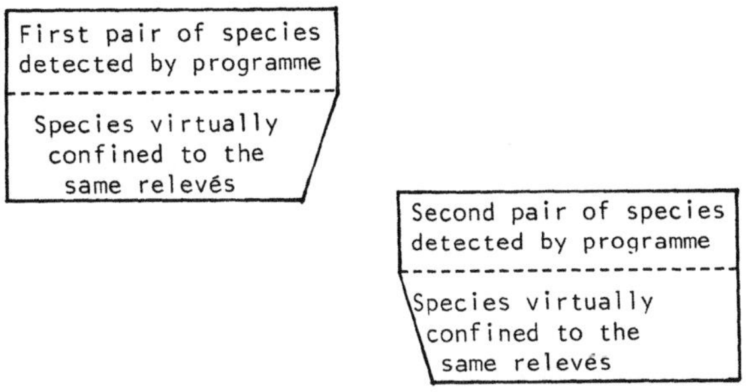

At this stage one can have the program select out sets of character species known from previous work or from the literature and print them after the differential species under appropriate headings. After these have been printed, the remainder of the species are printed out in descending order of their presence in the table.

The frequencies of the species in absolute numbers and in percentages are printed on the left of the species' names for each of the three sub-divisions and for the whole table. This feature provides the basic data for the construction of constancy tables. A separate page gives the relevé number for each column of the table as well as the position occupied by the relevé on the disk store and also the number of species recorded in the relevé.

After examining this table, one may wish to try out other differential species' combinations than those selected in the previous run. Instead of relying on the species pair selected by the program, one may instruct the program to test two opposing sets of differential species and, if desired, further sub-division of each differential block may be carried out. The remainder of the table will appear as described in the previous paragraph.

The final ordering of the table is best done by eye. The order of the relevés and of the species may be dictated in the input to the program. The 'disk position' of each relevé is punched on a separate card and the order of these input cards (which is readily altered by re-arranging the cards) determines the order of the relevés on the printed table. The species may be grouped in sets of character and differential species as desired; sometimes the user may wish to have the species printed in a definite order rather than allow the program to sort them into descending order of presence and this is easily arranged.

Discussion

Time and cost

One of the main points in favour of the PHYTO program is its effectiveness in handling data sets of up to 650 relevés. A single run of the PHYTO program will seldom exceed five minutes unless one asks for a further sub-division of the first three communities distinguished. Table 2 gives the time and commercial cost of running the different programs on the University College, Dublin computer in September 1973.

One of the factors which considerably increases time with large data sets is the necessity of using disk file for back-up storage and the consequent loss of time in transferring data between disk and computer core. Careful thought should be given to minimising the number of such transfers. In PHYTO the transfer is only made twice: at the beginning and at the final print-out. However, a major flaw in the program at present is that the user is left to estimate the degree of rarity (variable MINS) of those species which will not be transferred to core. If his estimation is incorrect and more than 110 species are eligible for transfer, the whole transfer process must begin again. It is intended to have this variable calculated in CAPHY and STOPHY and store its value on the disk file, thus eliminating a frequent source of error which increases the time used.

With large data sets, it is also desirable that the time used should increase approximately in direct proportion to the linear dimensions of the data matrix. This is so with PHYTO. However, in any program which needs to calculate similarity coefficients between relevés, the time increases as the square of the number of relevés and once one reaches 300 relevés the amount of core needed to store the similarity matrix becomes exorbitant for the type of computer available (90K bytes).

Table 2. Approximate budget for running a data set of about 350 relevés with 180 species. The times include that needed for compiling except in the case of PHYTO

Program		CPU Time (in seconds)	Costs in pounds sterling Time	Ancilliaries	Total
CASTO*		498	17.60	0.60	18.20
CAPHY		105	4.41	0.40	4.81
STOPHY		130	5.46	0.40	5.86
PHYTO	Auto	180	7.56	0.40	7.96
	Diff.**	170	7.14	0.40	7.54
	Dict.	103	4.32	0.50	4.82
PLOTMAP		60	2.16	2.50	4.66

* This is the time taken for the total grassland data set of 1804 relevés.
** This includes a sub-division.

Ecological effectiveness

There is nothing radically different in this set of programs from the familiar Braun-Blanquet technique of tabular comparison of relevés. One's final assessment of the effectiveness of the programs will therefore depend on one's opinion on the usefulness of the Braun-Blanquet techniques for the analysis of phytosociological data.

An ecologist intuitively uses several criteria in arriving at a satisfactory classification: overall floristic similarity between relevés of any community; physiognomical similarity – much influenced by the dominant species; presence of a group of character species; similar ecological parameters. A computer algorithm, on the other hand, necessarily uses a single criterion for either sub-divisive or agglomerative classification. The various re-allocation routines that have been developed (Wishart 1969) do admittedly try to overcome this limitation.

The PHYTO program was developed with the conviction that a satisfactory phytosociological classification can only be developed by a series of successive approximations with more and more ecological insights being incorporated towards the final stages. An effort was therefore made to develop a relatively simple algorithm which would initially give ecologically acceptable results (despite any flaws from the point of view of theoretical statistics) and which would not be unduly time-consuming with large data sets. Routines should also be incorporated which would permit easy subsequent manipulation of the table with varying degrees of user intervention.

The algorithm that has been used (and is described in section 1.5.2 of the manual), has in fact proved more satisfactory than maximising a statistical measure of correlation or association between all possible pairs of species in order to find a division pair.

Using a correlated pair of division species has proved more successful than using a single division species as in the 'association analysis' of Williams & Lambert (1959). The two methods often give similar results, but at times the single species chosen in 'association analysis' can give an ecologically meaningless sub-division. On the other hand the fact that at least two species have correlated occurrences in several relevés must have some ecological significance.

The program has been tested with a modification to choose the three most highly correlated species for sub-division. The results seldom differ from the standard version where two division species are chosen; any improvement obtained is not worth the extra computer time needed for the calculation.

The following criticism, based on experience, may be levelled against the programs in their present form: the ordering of the relevés in the table is not sufficiently automated. The ordering of the relevés at present is based on the number of differential species possessed by each relevé. This means that the section of the table without differentials is not subjected to any ordering. It might be desirable to incorporate a feature which would order the relevés within each classificatory category on the basis of their overall similarity to an artificial 'average relevé' for the category.

In conclusion, it should be emphasised that no computer technique will substitute for the ecological insight of the experienced field worker. It would be completely Utopian to aim at the production of a program which would eliminate the long and painful poring over data needed before a satisfactory arrangement is achieved. The computer is a useful tool – hopefully it will never become master of any ecologist.

Summary

The Irish approach to the computer processing of grassland relevés is introduced. A suite of programs in Fortran IV has been developed for the purpose. The preparation and analysis of the relevés using these programs is explained. The cost and ecological effectiveness is discussed. It is concluded that the approach is not radically different to the familiar Braun-Blanquet technique of tabular comparison of relevés.

Zusammenfassung

Es wird über ein irisches Verfahren betreffs der Datenverarbeitung von Wiesenaufnahmen berichtet. Für diesen Zweck wurde eine Reihe von Fortran-IV Programmen entwickelt. Die Vorbereitung und Analyse der Aufnahmen mittels dieser Programme wird erläutert. Die Kosten und die ökologische Effektivität werden besprochen. Die Schlussfolgerung lautet dass das Verfahren sich nicht grundsätzlich von der üblichen Braun-Blanquet-Technik zum tabellarischen Aufnahmenvergleich unterscheidet.

References

Braun-Blanquet, J. & R. Tüxen. 1952. Irische Pflanzengesellschaften. Veröff. Geobot. Inst. Rübel 25: 224–415.

Maarel, E. van der, L. Orlóci & S. Pignatti. 1976. Data-Processing in phytosociology, retrospect and anticipation. Vegetatio 32: 65–72.

Moore, J. J. 1973. Phyto. A suite of programs in Fortran IV for the manipulation of phytosociological tables according to the principles of Braun-Blanquet. Ms Dept. Botany Univ. Coll. Dublin. 15 pp.

Pignatti, S. 1976. A system for coding plant species for data-processing in phytosociology. Vegetatio 33: 23–32.

Williams, W. T. & Lambert, J. M. 1959. Multivariate methods in plant ecology. 1: Association analysis in plant communities. J. Ecol. 47: 83–101.

Wishart, D. 1969. A suite of programs for cluster analysis and other multivariate procedures. Ms. St. Andrews.

TABORD: A PROGRAM FOR STRUCTURING PHYTOSOCIOLOGICAL TABLES*,**

E. van der MAAREL, J.G.M. JANSSEN & J.M.W. LOUPPEN***

Division of Geobotany, University of Nijmegen, Toernooiveld, 6525 ED Nijmegen, The Netherlands

Keywords:
Arrhenatheretum, Classification, Clustering, Data-processing, Ordination, Phytosociological table, Salt marsh

Introduction

Since Benninghoff & Southworth (1964) published an abstract of their computer program for ordering phytosociological tables according to the Braun-Blanquet (1964) method along the lines developed by R. Tüxen and collaborators in the 1930's (which were first described in detail by Ellenberg 1956, see also Mueller-Dombois & Ellenberg 1974), many similar programs have been devised. They have been briefly reviewed by Westhoff & van der Maarel (1973), who distinguished two types of program: *a.* for tables with comparatively many relevés and few species, *b*, for tables with comparatively few relevés and many species. The *Type a* programs concentrate on finding groups of species with a similar distribution within the table which can be used in finding species/relevé blocks. Typical examples are the programs by Moore (1973, Moore et al. 1970), Ceska & Roemer (1971) and Stockinger & Holzner (1972, see Holzner, Werger & Ellenbroek 1978).

The *Type b*-programs concentrate on relevé grouping on the basis of mutual (dis) similarity values and produce tables with the relevé groups ordered according to their group (dis) similarities. A typical example is the program by Spatz (1972, Spatz & Siegmund 1973, see also Mueller-Dombois & Ellenberg 1974).

In both types of program the objective is to impose a *diagonal structure* on the table, i.e. to obtain a sequence of relevé groups from left to right and one of species groups (as they differentiate the relevé groups) from top to bottom so that the species relevé blocks occur from left top to bottom right.

Thus most table rearrangement programs are essentially concerned with both relevés and species. This twofold approach is also characteristic for the traditional manual ordering of phytosociological tables according to Braun-Blanquet and Tüxen, with the difference, of course, that usually only small tables are sorted by hand. There would be much value in a computer program approaching a table in two directions. The two-parameter *inosculate* method of Dale (Dale & Anderson 1973, Dale & Webb 1975) may be a good start, but no results on phytosociological tables are available yet.

In most of the programs referred to, a finishing touch by the investigator is needed, which mainly consists of changing the position of species. In some programs, this final step is incorporated in an additional program.

Since the 1973 review, entirely new programs have not become known to us, but most of the existing programs have been refined and further improved (see Westhoff & van der Maarel 1978 and van der Maarel, Orlóci & Pignatti 1976 for references). Dale & Quadraccia (1973) introduced an interactive tabular sorting program with a graphic display. Clearly, the automized final rearrangement of a table will usually both improve its structure and diminish errors.

* Nomenclature of salt marsh species follows Lausi, Kortekaas & Beeftink (1978); for names of Arrhenatheretum species see Oberdorfer (1970): Pflanzensoziologische Exkursionsflora für Süddeutschland, 3. Aufl., Ulmer, Stuttgart.

** Contribution from the Working Group for Data-Processing in Phytosociology, International Society for Vegetation Science.
*** The program has been tried and discussed by students and colleagues in our Department. We thank them for their help, particularly Drs. Wil Kortekaas and Drs. Willem Schenk. The first author introduced the program and prepared a draft of the manuscript at the Department of Plant Ecology, University of Lund, during his stays there from 1975–1977 under a grant of the Swedish Natural Science Research Council (NFR-B252605). The invitation of Prof. Nils Malmer for the visits to his department and the discussions of results from TABORD with fil. kand. Stefan Persson, fil. kand. Sven Jensén and fil. mag. Carin Tyler are gratefully acknowledged. Meanwhile Persson (1977) completed a manual on TABORD in Swedish. Finally we thank Prof. László Orlóci for his comments on the manuscript.

Our program was initially developed in 1970 for the Working Group for Data-Processing. Janssen (1972) published the essentials of the program and a first application, whereafter Janssen and van der Maarel made the program available for external use under the name TABORD in 1972. This version has been supplied to many users and regularly improved on in various details, mainly by Louppen.

TABORD structures comparatively large phytosociological tables, i.e. with many relevés and many species, which is considered a major advantage in phytosociological data-processing (van der Maarel 1974, van der Maarel, Orlóci & Pignatti 1976). TABORD differs in approach from most of the existing programs which are not adapted to such large tables, either because they require too much computer storage and time, or because the finishing manual work is too complicated. TABORD is written in FORTRAN IV. A listing on punchcards is available on request.

Description of TABORD

TABORD is essentially a clustering procedure based on relevé similarity, combined with a procedure for obtaining a diagonal structure of clusters in the table. The program includes the following steps:

Establishment of initial clusters

The procedure starts with the establishment of a number of initial clusters. Usually the investigator has at least some idea about the floristic structure of the relevé set to be clustered and tabularized. Therefore the procedure normally starts with the formation of initial clusters by the investigator. If there is no such preconception or if one wishes to avoid any personal influence on the clustering process by subjective choices at the beginning, the program generates initial clusters automatically, by assigning relevé 1 to cluster 1, rel. 2 to cl. 2, etc., up to the k-th relevé, if the number of initial clusters is chosen k; rel. $k+1$ is assigned to cl. 1, etc. up to the last relevé.

Clearly such a start means a considerable saving of both computer storage and calculation time as compared with clustering methods building up a hierarchy from a single-relevé (dis) similarity matrix. As will be discussed below the possible drawbacks of automatical initial clustering can be coped with in some subroutines incorporated by the program.

Another initiation would be the application of some numerical classification procedure prior to TABORD to arrive at the initial clusters. This would only fit into the TABORD strategy if the numerical procedure would be a very rapid one. In Lund (Persson 1977), the monothetic binary divisive program DIVINF, based on maximum decrease in the value of the information measure I (Lance & Williams 1968) is used.

Relocation

The initial clusters are subjected to a relocation procedure, which we adopted from the RELOC subroutine in the CLUSTAN package of clustering programs (cf. Wishart 1969, 1975). The position of each relevé is compared to the centroid of each initial cluster and placed in the cluster with which it shows the closest link. After the position of all relevés is checked, a second iteration starts in which again all relevés are compared with the centroids of the clusters formed during the first iteration. Usually after 4 to 6 iterations the clusters are stable. To avoid endless iterations in the exceptional case of relevés having equal (dis) similarity to two or more clusters, a maximum number of iterations is specified.

Measurement of (dis)similarity

To measure (dis)similarity we adopted 26 criteria from the total set of 38 in CLUSTAN, i.e. 13 coefficients for either numerical or binary data, including the information measure, product moment correlation, Canberra metric, Euclidean distance and similarity ratio. We have mostly used the *similarity ratio*, name and formula introduced by Wishart (1969), and found it an attractive measure of similarity. It is related to the well-known Jaccard formula, but it handles quantitative scores of species more efficiently:

$$S_{x,y} = \frac{\sum x_i y_i}{\sum x_i^2 + \sum y_i^2 - \sum x_i y_i} \qquad (i = 1, \ldots, n) \qquad (1)$$

where x_i and y_i are the scoresof species i in relevés x and y and n is the number of species.

Comparisons of values obtained with this formula to those of the Jaccard and Sørensen formulae (cf. Goodall 1973) show that they are intermediate between the somewhat lower Jaccard and the somewhat higher Sørensen values.

The similarity ratio can also be used for the comparison

of one single relevé with a cluster in the following manner (cf. Janssen 1975):

$$S_{x,z} = \frac{N\sum x_i z_i}{N^2 \sum x_i^2 + \sum z_i^2 - N\sum x_i z_i} \qquad (i = 1, \ldots, n) \qquad (2)$$

where x_i is the score of species i in relevé x, z_i is the sum of scores of species i in the relevés of cluster z, N is the number of relevés in cluster z and n the number of species occurring in either relevé or cluster.

The derivation of (2) follows from:

$$S_{x,\bar{z}} = \frac{\sum x_i \bar{z}_i}{\sum x_i^2 + \sum \bar{z}_i^2 - \sum x_i \bar{z}_i} \qquad (3)$$

where $\bar{z}_i$ is the centroid of cluster z and $\bar{z}_i$ the average score of species i in cluster z.

Since $\bar{z}_i = (1/N)\, z_i$ we can write (3) as:

$$S_{x,z} = \frac{\frac{1}{N}\sum x_i z_i}{\sum x_i + \frac{1}{N^2}\sum z_i^2 - \frac{1}{N}\sum x_i z_i} \qquad i = 1, \ldots, n. \qquad (4)$$

After multiplying by N^2 we obtain (2) from (4).

The results of the relocation procedure are printed iteration by iteration. For each cluster formed an average within-cluster similarity is defined as the average similarity of the individual relevés to the centroid of the cluster. The similarity between all pairs of clusters is also printed. The similarity between clusters a and b is calculated for the centroids of a and b as follows:

$$S_{a,b} = \frac{N_a N_b \sum a_i b_i}{Nb^2 \sum a_i^2 + Na^2 \sum b_i^2 - N_a N_b \sum a_i b_i} \qquad i = 1, \ldots, n \qquad (5)$$

where N_a and N_b are numbers of relevés in clusters a and b, a_i and b_i are the sums of scores of species i in clusters a and b, and n is the number of species occurring in either cluster.

In the computation procedure prior to eventual relocation, self comparisons may be included when the similarity between a relevé and its own cluster is calculated. If this is not desirable it is possible to have the relevé removed from its own cluster when calculating the similarity.

Naturally, species scores are expected as numerical values. In the Nijmegen versions we use a 1–9 code, which covers the Braun-Blanquet scale of combined estimates (cf. Westhoff & van der Maarel 1973, van der Maarel 1979). In the printed table the original Braun-Blanquet, or any other symbols may be used. The species scores are expected in a matrix, either with the relevés in columns, as is usual with phytosociological tables, or with the relevés in rows.

Homogenization of clusters

If the investigator wishes to avoid the inclusion of deviating relevés with a low similarity to all clusters, including its own cluster, it is possible to homogenize the clusters by applying a threshold value. During relocation, relevés are removed to a residual group whenever the similarity values with the cluster centroids are less than the threshold value. One can also run the program first without a threshold value, obtain an idea on the average similarity of the clusters obtained and then run a second round with a threshold value adapted to the cluster similarity level.

Fusion of clusters

The program allows clusters to be fused. Logically, the first fusion to be performed is between the two clusters with the closest link. The cluster fusion is treated as a single cluster and the relocation procedure is repeated for the cluster fusion and all other clusters. This procedure goes on until either a minimal number of clusters or a fusion limit i.e. a minimal similarity level, is reached. Both limits are arbitrarily established. A minimal number of final clusters may be desirable for survey or mapping. A minimal similarity (i.e. homogeneity) level may result from considerations concerning the homogeneity of types or syntaxonomical levels (see Westhoff & van der Maarel 1973, 1978 for a review of numerical-syntaxonomical studies, and Kortekaas, van der Maarel & Beeftink 1976 for an example based on the similarity ratio).

Further considerations related to cluster formation and fusion include the minimum size of clusters, the homogeneity threshold value and the handling of the residual relevés. A minimum cluster size may be desired where the relevé set under consideration is to be generalized in terms of common types, or, in a table output, to avoid an irregular diagonal structure. Relevés of clusters which decrease in size below the adopted minimum during relocation are placed in the residual group. On the other hand a minimum cluster size of one may be useful in order to avoid the run-off of relevés into the residue, Although the residual relevés are processed in subsequent relocation rounds, the small but phytosociologically acceptable clusters which were formed, are lost. It is thus recommended to avoid residuals and consequently admit the single-relevé clusters in a first pass of the rearrangement.

A high threshold value should not be applied when the initial clusters are automatically formed by the program: many relevés may be placed in the residual group and may not be replaced in a cluster during the following cycles, because of the relatively low homogeneity of the initial clusters and the threshold value.

Construction of the table

The main structure of the table is determined by the sequence of the clusters determined after the fusions. The program provides several options to obtain a sequence.

The first initiates the table with the cluster which shows the lowest value of the ratio: number of constant species/number of relevés. A species is considered constant for a cluster if its frequencey in the table exceeds a level specified by the user. One may choose here the rather high value of 80%, thus obtaining eu-constant species in the Braun-Blanquet terminology, or the more effective value of 60%, thus obtaining constants in the normal Braun-Blanquet sense. The rationale for the suggested values is that in this way a relatively high number of constant species is retained for further clusters. If two clusters have the same value for this ratio, the cluster with the highest within-similarity is chosen.

The second cluster is chosen according to the highest number of constant species in common with cluster 1. If no cluster shares any constant species in common with cluster 1, this cluster is placed in the table and a next starting cluster is selected following the same procedure.

The phytosociological table is then printed as follows: The relevés of the starting cluster are placed to the left, in order of their current registration numbers, followed by the relevés of cluster 2, also in order of their registration numbers. The species are then listed: in the top part of the table are the species that are constant in cluster 1 but not in cluster 2, arranged in order of their registration numbers. In the salt marsh studies this number follows from the Trieste code (Pignatti 1976). Any ad hoc or other systematical numbering will be treated likewise. An advantage of using the Trieste code here is that large series of constant species occurring in one cluster are ordered according to genus, family and order. Species are printed with their registration number, name and any other specification the user has stored (up to 24 alphanumerical characters). The species which are both constant in cluster 1 and 2 follow in the sequence, again in order of their registration numbers, then the constant species of cluster 2, those which are also constant in cluster 3, etc. The remaining species, not reaching constancy in any cluster, are printed last.

The second way of obtaining a cluster sequence is to ordinate clusters and take the end cluster on an axis as starting cluster. In the present version of TABORD one may chose between the 1st and 2nd axis of a principal components analysis of a matrix of standardized Euclidean distances, following Orlóci (1966) with extraction of components according to Householder as programmed by Roskam (1971), known as program ORDINA. In addition one may adopt a sequence according to position principal components obtained iteratively in the so-called Jacobi-type solution (Harman 1967). The listing of the species is as described above.

Repetition of relocation and fusion procedure with other options. It is possible to repeat the whole relocation and fusion procedure with other options concerning homogeneity threshold and/or fusion limit. Each round takes the final classification of the former round as the initial classification. One may choose various objectives, which are then achieved through a process which is essentially a successive approximation (Poore 1962). An obvious objective is to start at a high fusion limit in order to obtain a table with many homogeneous clusters and to gradually decrease the fusion limit. Another logical procedure is to gradually increase the threshold value in order to obtain clusters with an increasing level of homogeneity but with a growing residual group.

In cases where one is in doubt about an effective number of initial clusters, one may start with a high number, allow fusions to occur and simultaneously decrease the fusion limit and increase the threshold value. Another motive for a repetition of the procedure may be uncertainty about the optimal constancy level to arrive at a comprehensible diagonal table structure.

Additional information presented in TABORD

In addition to the results of the calculations described above, the following information is produced by TABORD:

- The occurrences of species in the final clusters as absolute frequency (presence).
- The occurrences of constant species in the clusters.
- The number of constant species common to pairs of clusters.

– In addition to these, the significance of occurrence or absence of species in the clusters is calculated based on a chi-square test and the results are produced as chi-square values per species per clusters and as significance signs levels with signs for positive or negative associations at P = 0.05, P = 0.025 and P = 0.01. Since many frequency values in the cells of the two x two tables involved are very low, the results of this test are best considered in a less rigorous sense, as deterministic measures of the diagnostic value of the species. Such a measure is of particular use for species which are not constant. Species with low frequency but high positive association value resemble the average faithful species in Braun-Blanquet syntaxonomy. – the sequence of clusters is determined by the result of an ordination, distances between clusters, loadings on four dimensions and an ordination diagram with dimensions 1 and 2 are produced. Although the cluster arrangement option based on constant species may be preferred (see Discussion) it is useful to have one round with an ordination included because of the additional insight in the structure of the data.

Finishing touch by computer assisted visual sorting.

The phytosociological table produced by TABORD is usually liable to some improvement, both in the sequence of relevés and in the sequence of species. The efficiency of the cluster arrangement is often partly determined by the size of the cluster and the constancy level required. This could result in suboptimal sequences. Furthermore, the relevés within one cluster are arranged according to their current numbers, which are of course arbitrary and this might not be optimal for visual interpretation.

Changes may be preferred in species sequence in the case of species with a low presence in a cluster but a high positive association value. Species with an overall low table frequency might be deleted alltogether and placed in an addendum.

It is very difficult, if not impossible, to automize procedures for such additional changes. Visual sorting seems a more appropriate approach and, of course, the simple changes of columns and rows can be done with computer assistance, including display (cf Dale & Quadraccia 1973). We developed a simple 'shuffle program' for replacing relevés and species.

Capacity

The capacity of the present TABORD version run at Nijmegen on an IBM 370–158 system is adapted to a computer storage of 512 K bytes to which we have an instant access. The requirement is determined roughly by:

$$R = 0{,}004(S(N+3C+7)+5N+C(C+7)+100)+150 \text{ K bytes} \quad (6)$$

where S = number of species, N = number of relevés, C = number of initial clusters.

Taking a ratio of 1 : 10 for initial clusters: relevé number as a reasonable average, this means that 300 relevés containing 250 species (and 30 initial clusters) could be handled, or 500 relevés containing 150 species (and 50 initial clusters). The previous version of TABORD had fixed dimensions, i.e. 200 relevés, 250 species and 20 initial clusters, requiring 384 K bytes.

Examples

Three examples will be discussed dealing with a. a simply structured table with few species; b. a relatively simple table with many species; c. a rather complicated table.

Spartinetea

Table 1 presents 50 relevés of *Spartinetea* communities selected for testing multivariate methods for the Working Group for Data-Processing (Van der Maarel, Orlóci & Pignatti 1976). The selection contains relevés of the *Spartinetum maritimae*, *Spartinetum townsendii*, *Spartinetum alterniflorae* and *Spartinetum patentis* (5 relevés each) and 30 relevés of variants and transitions, towards *Salicornietum* communities. Obviously the pure types should be reproduced in any TABORD output, while some freedom may exist with respect to the transitional relevés. Various options were tried, the results of which may be summarized as follows:

The structure of the selection allowed a visual interpretation of an initial classification array of 11 clusters, including the five *pure* types. Relocation only led to some rearrangement of three clusters within the *Spartinetum townsendii* variants. All resulting clusters showed a high internal homogeneity, with average similarity values of over 0.80, but between cluster similarities were considerably lower with one possible fusion at the level of 0.69 and one at 0.66 and further fusions at still lower levels. Inspection of the resulting table revealed that all 11 clusters could be recognized by their own characteristic species combinations, so no further fusions were performed. In this example

		00000	00000	00000	00000	00000	00000	00000	00000	00000	00000
		00000	00001	11111	11112	22222	22223	33333	33334	44444	44445
		12345	67890	12345	67890	12345	67890	12345	67890	12345	67890
TRIGLOCHIN MARITIMUM	66003	-----	-2---	-----	-----	-----	-----	-----	-----	---2-	-----
AGROSTIS STOLONIFERA	240120	-----	-----	----2	-----	-----	-----	-----	-----	-----	-----
SPARTINA ALTERNIFLORA	283001	-----	-----	98999	-----	-----	-----	-----	-----	-----	-----
SPARTINA MARITIMA	283003	79889	53333	-----	-----	-----	-----	-----	-----	-----	-----
SPARTINA TOWNSENDII	283004	-----	-----	-----	-----	78998	55333	33333	35333	55755	25555
SPARTINA PATENS	283005	-----	-----	-----	99999	-----	-----	-----	-----	-----	-----
PHRAGMITES COMMUNIS	333001	-----	-----	-----	2-252	-----	-----	-----	-----	-----	-----
PUCCINELLIA MARITIMA	384008	-----	87253	---23	-----	-----	88775	--352	2-23-	3--5-	-----
SCIRPUS MARITIMUS	468016	-----	-----	--5--	-----	-----	-----	-----	-----	----5	-----
JUNCUS ACUTUS	936002	-----	-----	-----	---2-	-----	-----	-----	-----	-----	-----
JUNCUS MARITIMUS	936029	-----	-----	-----	-552-	-----	-----	-----	-----	-----	-----
ATRIPLEX HASTATA	2229017	-----	-----	-----	-----	-----	-----	-----	-----	87979	-----
HALIMIONE PORTULACOIDES	2229101	-----	23322	-----	-----	22---	-22-2	2--2-	-2253	--3--	355--
SALICORNIA PERENNIS	2255001	-----	77889	-----	-----	-----	5----	-----	-----	-----	-----
SALICORNIA FRUTICOSA	2255002	-----	-----	-----	----2	-----	-----	-----	-----	-----	-----
ARTHROCNEMUM GLAUCUM	2255003	-----	-----	-----	-2---	-----	-----	-----	-----	-----	-----
SALICORNIA EUROPAEA	2257001	---2-	5-22-	-----	-----	-----	35553	88777	-2523	---5-	33533
SUAEDA MARITIMA	2261007	---2-	552--	-----	-----	2---2	22322	32222	88789	--25-	55555
SPERGULARIA MARGINATA	2450006	-----	-----	-----	-----	-----	3----	-----	-----	-----	-----
OENANTHE LACHENALII	6046009	-----	-----	-----	--3-2	-----	-----	-----	-----	-----	-----
LIMONIUM VULGARE	6351017	---3-	-----	---2-	-3522	-----	-----	---2-	-----	-----	-----
PLANTAGO CRASSIFOLIA	8116009	-----	-----	-----	----2	-----	-----	-----	-----	-----	-----
PLANTAGO MARITIMA	8116021	-----	-----	-----	-----	-----	-----	-----	-----	---2-	-----
ASTER SQUAMATUS	8900023	-----	-----	-----	--2--	-----	-----	-----	-----	-----	-----
ASTER TRIPOLIUM	8900025	--3-2	--323	--225	----2	--2--	-3223	3232-	3523-	3323-	53353
INULA CRITHMOIDES	9061006	-----	-----	-----	-22-3	-----	-----	-----	-----	-----	-----
SONCHUS ARVENSIS	9595001	-----	-----	-----	22---	-----	-----	-----	-----	-----	-----

Table 1. 50 *Spartinetea* relevés (selection f, Working Group for Data-Processing in Phytosociology, see van der Maarel, Orlóci & Pignatti 1976). Species are listed according to the Trieste code number, four digits for the genus code, three digits for the species code.

	00000	000	00	00000	0000000000	0000	0	00000	00000	00000	00000
	00000	001	00	22233	3333444445	3333	2	22222	44444	11111	11112
	12345	890	67	78904	6789067890	1235	6	12345	12345	12345	67890
	00000	000	00	00000	0000000000	0000	0	00000	00000	00000	00000
	00000	000	00	11111	0000000000	1111	0	00000	00000	00000	00000
	11111	222	88	00000	7777777777	1111	9	55555	66666	33333	44444
SPARTINA MARITIMA	79889	333	53	-----	----------	----	-	-----	-----	-----	-----
ASTER TRIPOLIUM	--3-2	323	--	32232	3523-53353	323-	-	--2--	3323-	--225	----2
PUCCINELLIA MARITIMA	-----	253	87	87755	2-23------	--32	8	-----	3--5-	---23	-----
HALIMIONE PORTULACOIDES	-----	322	23	22-22	-2253355--	2---	-	22---	--3--	-----	-----
SALICORNIA PERENNIS	-----	889	77	-----	----------	----	-	-----	-----	-----	-----
SUAEDA MARITIMA	---2-	2--	55	23222	8878955555	3222	2	2---2	--25-	-----	-----
SPARTINA TOWNSENDII	-----	---	--	53333	3533325555	3333	5	78998	55755	-----	-----
SALICORNIA EUROPAEA	---2-	22-	5-	55537	-252333533	8877	3	-----	---5-	-----	-----
SPERGULARIA MARGINATA	-----	---	--	-----	----------	----	3	-----	-----	-----	-----
ATRIPLEX HASTATA	-----	---	--	-----	----------	----	-	-----	87979	-----	-----
SPARTINA ALTERNIFLORA	-----	---	--	-----	----------	----	-	-----	-----	98999	-----
SPARTINA PATENS	-----	---	--	-----	----------	----	-	-----	-----	-----	99999
PHRAGMITES COMMUNIS	-----	---	--	-----	----------	----	-	-----	-----	-----	2-252
LIMONIUM VULGARE	---3-	---	--	----2	----------	----	-	-----	-----	---2-	-3522
SONCHUS ARVENSIS	-----	---	--	-----	----------	----	-	-----	-----	-----	22---
INULA CRITHMOIDES	-----	---	--	-----	----------	----	-	-----	-----	-----	-22-3
ASTER SQUAMATUS	-----	---	--	-----	----------	----	-	-----	-----	-----	--2--
PLANTAGO MARITIMA	-----	---	--	-----	----------	----	-	-----	---2-	-----	-----
PLANTAGO CRASSIFOLIA	-----	---	--	-----	----------	----	-	-----	-----	-----	----2
OENANTHE LACHENALII	-----	---	--	-----	----------	----	-	-----	-----	-----	--3-2
ARTHROCNEMUM GLAUCUM	-----	---	--	-----	----------	----	-	-----	-----	-----	-2---
SALICORNIA FRUTICOSA	-----	---	--	-----	----------	----	-	-----	-----	-----	----2
JUNCUS MARITIMUS	-----	---	--	-----	----------	----	-	-----	-----	-----	-552-
JUNCUS ACUTUS	-----	---	--	-----	----------	----	-	-----	-----	-----	---2-
SCIRPUS MARITIMUS	-----	---	--	-----	----------	----	-	-----	----5	--5--	-----
AGROSTIS STOLONIFERA	-----	---	--	-----	----------	----	-	-----	-----	----2	-----
TRIGLOCHIN MARITIMUM	-----	---	-2	-----	----------	----	-	-----	---2-	-----	-----

Table 2. Structured table of the *Spartinetea* relevés of table 1, obtained with TABORD, with options threshold value 0.55, fusion limit 0.75, and constancy limit 0.70. Relevé numbers are given in the top three rows of the tables, cluster numbers is rows 4–6.

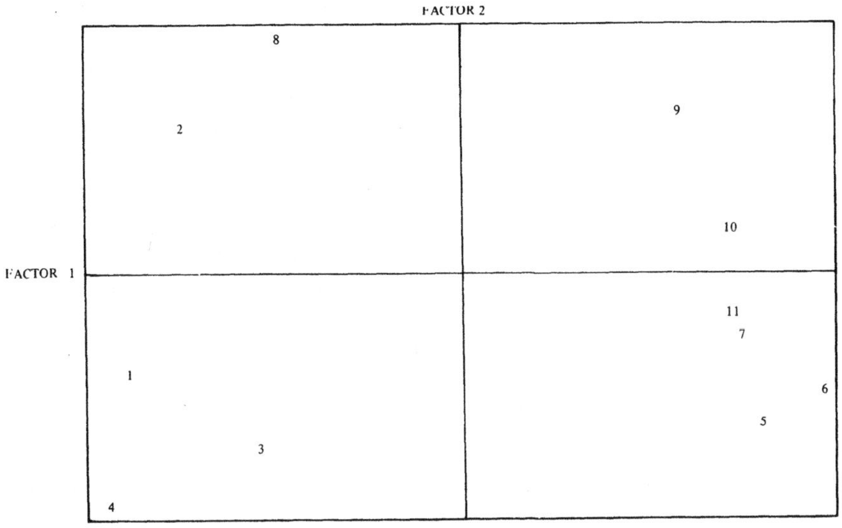

Fig. 1. PCA-ordination of 11 clusters of *Spartinetea*-relevés as presented in table 2.

	00000	000	00	0	0000	0000	00000000000	00000	00000	00000	00000
	00000	001	00	2	2233	3333	44444533332	22222	44444	11111	11112
	12345	890	67	7	8904	6789	06789012356	12345	12345	12345	67890
	00000	000	00	0	0000	0000	00000000000	00000	00000	00000	00000
	00000	000	00	0	1111	1111	10000000000	00000	00000	00000	00000
	11111	222	88	9	0000	1111	77777777777	55555	66666	33333	44444
SPARTINA MARITIMA	79889	333	53	-	----	----	-----------	-----	-----	-----	-----
SALICORNIA PERENNIS	-----	889	77	5	----	----	-----------	-----	-----	-----	-----
SPERGULARIA MARGINATA	-----	---	--	3	----	----	-----------	-----	-----	-----	-----
PUCCINELLIA MARITIMA	-----	253	87	8	8775	5--3	22-23------	-----	3--5-	---23	-----
HALIMIONE PORTULACOIDES	-----	322	23	-	22-2	22--	--2253355--	22---	--3--	-----	-----
ASTER TRIPOLIUM	--3-2	323	--	-	3223	2323	-3523-53353	--2--	3323-	--225	----2
SUAEDA MARITIMA	---2-	2--	55	2	2322	2322	28878955555	2---2	--25-	-----	-----
SALICORNIA EUROPAEA	---2-	22-	5-	3	5553	7887	7-252333533	-----	---5-	-----	-----
SPARTINA TOWNSENDII	-----	---	--	5	5333	3333	33533325555	78998	55755	-----	-----
ATRIPLEX HASTATA	-----	---	--	-	----	----	-----------	-----	87979	-----	-----
SPARTINA ALTERNIFLORA	-----	---	--	-	----	----	-----------	-----	-----	98999	-----
SPARTINA PATENS	-----	---	--	-	----	----	-----------	-----	-----	-----	99999
PHRAGMITES COMMUNIS	-----	---	--	-	----	----	-----------	-----	-----	-----	2-252
LIMONIUM VULGARE	---3-	---	--	-	----	2---	-----------	-----	-----	---2-	-3522
INULA CRITHMOIDES	-----	---	--	-	----	----	-----------	-----	-----	-----	-22-3
JUNCUS MARITIMUS	-----	---	--	-	----	----	-----------	-----	-----	-----	-552-

Table 3. Structured table of *Spartinetea* relevés as obtained in table 2 with additional changes in the sequence of relevés and species. Species lacking a high constancy in any cluster have been left out.

a threshold value (usually we use values between 0 and 0.50) did not make any difference, because of the high homogeneity level of the clusters involved.

An automatic initial array of 10 clusters with 5 relevés each led to an almost identical end classification, which would be completely identical after fusions down to the 0.60 level. However, this occurred only when no threshold value was applied. The same attempt with threshold value 0.30 led to a large residual group, obviously because of the very low homogeneity of the random clusters at the beginning.

Table 2 shows the resulting phytosociological table based on the constant-species-efficiency criterion (In our examples only data on species occurrences in the 1–9 ordinal transform scale and no data on site and vegetation structure are presented, Evidently all additional information on the stands and relevés involved can be incorporated in the structured table.). The *Spartina maritima* cluster with only one constant species comes first and clusters 2, 8, 9, 10, 11, 17, 5 and 6 with *Spartina townsendii* follow logically. All these clusters correspond to subassociations and variants described by Kortekaas et al (1976). Cluster 3, with *Spartina alterniflora*, is placed independently; cluster 4, with *Spartina patens*, joins cluster 3. Each cluster can be characterized with its own combination of constant species. (In this and other examples we used 0.70 as constancy level). The chi-square test confirms the visual observations that the *Spartina patens* cluster has two species with a constancy below the level chosen, but with a significant association to that cluster.

On closer inspection this table can be improved. Cluster 9 (one relevé) should join cl. 10, cl. 7 and 11 could change position. Also some species may be replaced. To consider possible changes the result of the (ORDINA) ordination of the 11 clusters is used (Fig. 1). The table output of this TABORD run as such was also considered as an alternative but found less satisfactory, but the main sequence of table 1 could be altered to conform to the ordination pattern by indeed placing cl. 9 between cl. 8 and 10, and cl. 11 between cl. 10 and 7. In this way a spiral is followed in which both dimensions are involved and moreover cluster 1 instead of the more extreme cl. 4 is taken as a starting cluster. The species sequence is adapted to the new cluster sequence and through visual inspection the diagonal structure is improved. This finishing touch has been done with the shuffle program. Table 3 shows the end result.

Arrhenatheretum

Table 4 presents a version of the *Arrhenatheretum* table used by Ellenberg (1956) and Mueller-Dombois & Ellenberg (1974) to demonstrate the process of tabular phytosociological synthesis according to Braun-Blanquet (cf Westhoff & van der Maarel 1973). Table 4 gives the so-called differentiated table, based on table 9.7 in Mueller-Dombois & Ellenberg with the deviating relevé 19, already struck by the authors, placed to the end of the table. For this example the original % biomass estimations were converted into intervals of the ordinal transform scale. Only those species are listed which are either frequent in the table as a whole or characteristic for one of the community types involved. This list corresponds to table 9.7, left in Mueller-Dombois & Ellenberg (op. cit.), whilst the species not listed are those of table 9.7. right, called remaining species by the authors (op. cit.).

The 25 relevés in this example have been put in a random order (table 9.2. op. cit.), which we do not reproduce here. Three types of vegetation were distinguished in the differentiated table, viz. *Bromus-Arrhenatherum* (columns 1–8), *Geum-Arrhenatherum* (col. 9–14), and *Cirsium-Arrhenatherum* (col. 15–24, with deviant relevé 19 in col. 25).

Treatments were based on various initial arrays with 8–12 clusters. These numbers of intial clusters seemed effective in view of the expected number of types on one hand and the total number of relevés on the other hand. Treatments based on an initial array derived from visual inspection of table 4, and on two automatic arrays, with a fusion limit of 0.70, led to similar results. A cluster of *Bromus erectus* relevés was recognized, the same as in table 4, but without rel. 2 (col. 8), but we miss many of the differential species as indicated by Mueller-Dombois & Ellenberg (from *Bromus erectus* to *Linum catharticum*). In one case rel. 1 and 4 (table 4, col. 3 and 1), which both miss *Salvia pratensis* and *Viola hirta*, form a separate cluster. Rel. 19, the deviant one, always forms a cluster by itself.

The *Geum-Arrhenatherum* type (for which, incidentally, *Geum urbanum* is not typical) is less clearly traceable. In most outputs a cluster is formed which contains the six relevés of the type, but includes other relevés: 13, 23, also 2 in some cases. This result is confirmed by a Q-type factor analysis shown by Mueller-Dombois & Ellenberg (1974, fig. 10.12) themselves, which they have not used in the tabular sorting process.

The *Cirsium-Arrhenatherum* type is still weaker. In most attempts the relevés assigned to this type are distributed over several clusters, including single-relevé clusters (e.g. rel. 14 and 25). In one automatic initial array attempt, which produced only 5 final clusters, the relevés of this type were at once fused with those of the *Geum-Arrhenatherum* type. Again, Mueller-Dombois & Ellenberg (op. cit.) produce evidence supporting this similarity, viz. in the dendrogram of fig. 10.15.

The final TABORD table as seen in Table 5, is based on an initial classification array of 13 clusters, with 0.65 as fusion level and 0.45 as threshold value. Only three clusters are left, one of which is a single-relevé cluster with rel. 23. The two large cluster clearly show the existence of two rather than three community types in the material.

In order to enable a more detailed interpretation of the relevé set the program was run with a higher fusion limit. Fig. 2 shows the result of the ordination (ORDINA) in this TABORD run, with the same initial array but a fusion

	00000000	000000	0000000000	0
	01001020	111202	2110102213	1
	40195342	682052	3326781547	9
BROMUS ERECTUS	8677775-	------	----------	-
SCABIOSA CULUMBARIA	211-1---	------	--------3-	-
THYMUS SERPYLLUM	132--1--	------	--------1-	-
SALVIA PRATENSIS	-4-3244-	------	----------	-
KOELERIA PYRAMIDATA	3333----	------	----------	-
FESTUCA OVINA	-2--33-2	------	----------	-
CAMPANULA ROTUNDIFOLIA	2221211-	1-2---	2----1-12-	-
VIOLA HIRTA	-3-3111-	-1----	----------	-
BRIZA MEDIA	-223---2	3-----	----------	-
LINUM CATHARTICUM	1-11-1--	1-----	----------	-
GEUM RIVALE	--------	11111-	-2133421-2	2
HOLCUS LANATUS	--------	----2-	1232333632	1
MELANDRIUM DIURNUM	--------	2-1--1	-113--1422	1
ALOPECURUS PRATENSIS	--------	-3--35	-4-54562--	-
LYSIMACHIA NUMMULARIA	--------	--11--	--111211--	-
LYCHNIS FLOS-CUCULI	--------	2-----	--1-1111--	1
GLECHOMA HEDERACEA	--------	--1-11	----11---1	-
CIRSIUM OLERACEUM	--------	---111	2135363366	1
DESCHAMPSIA CAESPITOSA	--------	------	--5-734-25	3
ANGELICA SILVESTRIS	--------	------	21-2---123	-
CAREX ACUTIFORMIS	--------	------	----4-23-5	3
FILIPENDULA ULMARIA	--------	------	-1-3-3--1-	-
PIMPINELLA MAGNA	--------	------	-11---2--1	-
POLYGONUM BISTORTA	--------	--1---	-------4-2	2
ARRHENATHERUM ELATIUS	35146372	665646	7564546754	6
DACTYLIS GLOMERATA	44434654	667455	5665556654	2
GALIUM MOLLUGO	25333533	444433	5536442364	3
POA PRATENSIS	45456568	556445	6343442543	2
PLANTAGO LANCEOLATA	22222212	322233	5412142134	1
FESTUCA PRATENSIS	-3-44334	455366	3373556333	5
CHRYSANTHEMUM LEUCANTH	14232341	2-3312	2224121-33	1
RANUNCULUS ACER	-3111111	1-1223	1313211123	1
VERONICA CHAMAEDRYS	1211111-	2-2211	31121122-2	1
ACHILLEA MILLFOLIUM	25433331	452434	141---1162	-
DAUCUS CAROTA	122212-2	121211	3312-41--2	-
RUMEX ACETOSA	1-1-1---	2-2212	1213331232	2

Table 4 25 *Arrhenatheretum* relevés taken from table 9.7 in Mueller-Dombois & Ellenberg (1974). The remaining species as listed in this table have been left out.

	00000000000000000	0000000	0
	00001111111122222	0000112	0
	56781234678901235	1349054	2
	00000000000000000	0000000	0
	00000000000000000	0000000	0
	44444444444444444	1111111	2
GEUM RIVALE	1324312-111212--1	-------	-
HOLCUS LANATUS	22233-23-3-1-3-16	-------	-
CIRSIUM OLERACEUM	15663-16-3-113123	-------	-
RUMEX ACETOSA	1323322321-221212	1-1--1-	-
ARRHENATHERUM ELATIUS	44445555666666677	1334567	2
DACTYLIS GLOMERATA	55455765666246556	4643445	4
GALIUM MOLLUGO	36444456434342353	3523533	3
POA PRATENSIS	43344634545242565	4545566	8
PLANTAGO LANCEOLATA	32441243312122351	2222221	2
FESTUCA PRATENSIS	63355533475536633	-3-4343	4
CHRYSANTHEMUM LEUCANTH	1432132322-13122-	2313424	1
RANCUNCULUS ACER	2331213211-121311	11-1311	1
VERONICA CHAMAEDRYS	1221121-21-122132	1111211	-
ACHILLEA MILLFOLIUM	3-2--246415-41411	4323533	1
DAUCUS CAROTA	1224-13-112-2113-	221221-	2
BROMUS ERECTUS	-----------------	7787675	-
SALVIA PRATENSIS	-----------------	-4-3424	-
CAMPANULA ROTUNDIFOLIA	---1-2-21------21	2121221	-
VIOLA HIRTA	----------1------	-1-3311	-
FESTUCA OVINA	-----------------	-3--23-	2
BRIZA MEDIA	--------3--------	2--32--	2
POLYGONUM BISTORTA	--2--1-----2----4	-------	-
PIMPINELLA MAGNA	--1---1--1---2---	-------	-
FILIPENDULA ULMARIA	-3-3--11---------	-------	-
CAREX ACUTIFORMIS	--5-4------3-2--3	-------	-
ANGELICA SILVESTRIS	-23---12-------21	-------	-
DESCHAMPSIA CAESPITOSA	--537--2-5-3-4---	-------	-
GLECHOMA HEDERACEA	1-1111--------1--	-------	-
LYCHNIS FLOS-CUCULI	---11---21-1-1--1	-------	-
LYSIMACHIA NUMMULARIA	-1-211---1--11--1	-------	-
ALOPECURUS PRATENSIS	35-54-4---3--65-2	-------	-
MELANDRIUM DIURNUM	-32--11221-1-11-4	-------	-
LINUM CATHARTICUM	--------1--------	1111---	-
KOELERIA PYRAMIDATA	-----------------	3-333--	-
THYMUS SERPYLLUM	-------1---------	211-3--	-
SCABIOSA COLUMBARIA	-------3---------	1-2-11-	-

Table 5. Structured table of the *Arrhenatheretum* relevés of table 4 obtained with TABORD, with threshold value 0.45, fusion limit 0.65 and constancy limit 0.70.

limit of 0.80. Fig. 2 resembles the diagram presented by Mueller-Dombois & Ellenberg (op. cit. fig. 10.12) in showing the main environmental gradient from dry to moist underlying the variation in the table. This gradient can be followed along the horseshoe configuration from clusters 1 and 3 via cl. 2, 12 and 4, to cl. 5, 13 and 7. Finally Table 6 presents the floristic composition of these 8 clusters in the described sequence using species with a differentiating value for one or two clusters. This table can be used for the interpretation of Fig. 2 and also for the inspection of the orginal division into three community types. Note that some of these species are not used by Mueller-Dombois & Ellenberg (op. cit. table 9.7.) to differentiate their types. *Helictotrichon pubescens* and *Heracleum sphondylium* are even usable as differentiators for the *Geum-Arrhenatherum* type.

Thus, in one combined TABORD-analysis we obtain most of the numerical results produced in separate analyses by Mueller-Dombois & Ellenberg and moreover structured directly in a phytosociological table.

Glauco-Puccinellietalia

Table 7 presents the 58 relevés of another selection of the Working Group for Data-Processing. The relevés are placed in order of the collection from 17 tables of communities from the order *Glauco-Puccinellietalia*. Communities 1–8 are represented with 3 relevés (col. 1–24), comm. 9 with 4 rel. (col. 25–28), comm. 10 with 3 rel. (col. 29–31), comm. 11 and 12 with 4 rel. (col. 32–39), comm. 13 with 3 rel. (col. 40–42), comm. 14–17 with 4 rel. (rol. 43–58). The various community types involved can be

FACTOR 1 PLOTTED AGAINST FACTOR 2

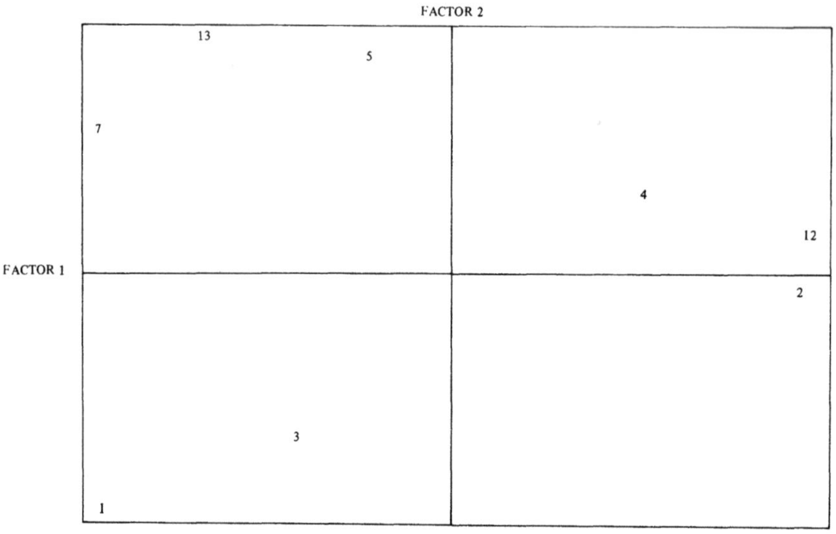

Fig. 2. PCA-ordination of 8 clusters of *Arrhenatheretum* relevés as obtained with TABORD with threshold value 0.55 and fusion limit 0.80. Cluster 1: relevés 1, 4; cl. 2: rel. 2; cl. 3: rel. 3, 9, 10, 15, 24; cl. 4: rel. 5, 12, 13, 16, 18, 20, 22; cl. 5: rel. 6, 7, 8, 14; cl. 7: rel. 19; cl. 12: rel. 23; cl. 13: rel. 11, 17, 21, 25. See also table 6.

	00	00000	00	0000000	0000	0000	0
	00	00112	02	0111122	0001	1122	1
	14	39054	23	5236802	6784	1715	9
	00	00000	00	0000000	0000	0000	0
	00	00000	01	0000000	0000	1111	0
	11	33333	22	4444444	5555	3333	7
	BB	BBBBB	BC	GGCGGGG	CCCC	CCCC	C
LINUM CATHARTICUM	11	11---	--	---1---	----	----	-
CAREX FLACCA	33	---2-	--	---3---	----	-3--	3
KOELERIA PYRAMIDATA	33	-33--	--	-------	----	----	-
BROMUS ERECTUS	78	77675	--	-------	----	----	-
CAMPANULA ROTUNDIFOLIA	22	11221	-2	-2-2---	--1-	-1-1	-
VIOLA HIRTA	--	13111	--	----1--	----	----	-
SALVIA PRATENSIS	--	43424	--	-------	----	----	-
SCABIOSA COLUMBARIA	12	--11-	3-	-------	----	----	-
FESTUCA OVINA	--	3-23-	2-	-------	----	----	-
SILENE INFLATA	--	---1-	3-	-------	1---	----	-
TRISETUM FLAVESCENS	--	43-43	34	5-45456	----	-45-	4
CREPIS BIENNIS	--	1---1	14	32--222	2-51	111-	-
HERACLEUM SPHONDYLIUM	1-	2---1	-2	3-71121	14--	-2-1	-
HELICTOTRICHON PUBESCENS	2-	-41-2	2-	6246473	53--	-41-	-
AJUGA REPTANS	--	-2---	1-	121-111	1231	2121	-
GEUM RIVALE	--	-----	--	-112111	324-	3121	2
CIRSIUM OLERACEUM	--	-----	-2	1-1--11	5666	3333	1
FILIPENDULA ULMARIA	--	-----	--	---1---	3-31	----	-
ANGELICA SILVESTRIS	--	-----	-2	--1----	23-2	---1	-
HOLCUS LANATUS	--	-----	-1	2-2----	2333	3336	1
LYCHNIS FLOS CUCULI	--	-----	--	---2---	--1-	1111	1
DESCHAMPSIA CAESPITOSA	--	-----	--	-------	-53-	754-	3
CAREX ACUTIFORMIS	--	-----	--	-------	-5--	4-23	3
GLYCERIA FLUITANS	--	-----	--	-------	----	----	6
PHALARIS ARUNDINACEA	--	-----	--	-------	----	----	7

Table 6. Structured table of the *Arrhenatheretum* relevés of table 4 according to a Tabord-run with threshold value 0.55 and fusion limit 0.80. The resulting 8 clusters are arranged according to the ORDINA-configuration presented in Fig. 2.

(table 6 cont'd) Relevé and cluster numbers are indicated in rows 1–3 and 4–6 respectively. The community type according to Mueller-Dombois & Ellenberg (1974, table 9.7) is indicated in row 7: B = ***Bromus-Arrhenatherum*** type, G = *Geum-Arrhenatherum* type, C = *Cirsium-Arrhenatherum* type.

derived from the dominant species characterizing them (see further Feoli 1977a).

After various trials with varying threshold values and fusion levels, three runs were compared with a given fusion level, 0.70, and threshold, 0.20, all with 17 initial clusters, one according to the community representation as described in Table 7, one automatic array, and one similarly deviced regular array. The similarity between the three classifications was very high: 39 relevés were placed in the same multi-relevé clusters and 8 relevés became single-relevé clusters in all attempts.

A new run was applied to an array of 18 initial clusters on the basis of these results. Only 6 relocations were needed to achieve stable clusters. The resulting clusters were considered reasonably clear and homogeneous, but still some clusters incorporated deviant relevés. Thus a further run was applied with 24 initial clusters, based on fusion limit 0.65 and threshold 0.45 (Table 8). This table may be further improved manually with the help of the ORDINA cluster ordination as shown in Fig. 3. This is not demonstrated here. It appears that on fusion limit 0.65 the 24 clusters remain unfused and that, indeed, deviant relevés remain on their own.

	000	000	000	000	000	000	000	000	0000	000	0000	0000	000	0000	0000	0000	0000
	000	000	000	111	111	111	122	222	2222	233	3333	3333	444	4444	4445	5555	5555
	123	456	789	012	345	678	901	234	5678	901	2345	6789	012	3456	7890	1234	5678
TRIGLOCHIN MARITIMUM	5--	-5-	-3-	---	3--	232	53-	2--	332-	2--	-3--	---2	22-	-7-2	2-23	25-5	5---
PARAPHOLIS STRIGOSA	---	---	---	3--	---	---	---	---	----	---	----	----	---	---7	----	-3-7	8---
AGROSTIS STOLONIFERA	---	---	---	---	232	223	---	232	2-33	335	----	3-22	-2-	-2--	2533	--7-	----
POA PRATENSIS	---	---	---	---	-35	-22	---	---	----	-3-	----	----	---	----	----	----	----
PUCCINELLIA DISTANS	---	---	---	---	---	---	---	555	----	---	----	----	---	----	----	----	----
PUCCINELLIA MARITIMA	798	57-	---	---	---	---	999	-2-	22--	2--	9989	3---	2-5	9598	----	--72	53--
FESTUCA RUBRA	---	2-8	---	---	233	333	---	---	-288	777	----	--55	---	----	----	----	----
ELYTRIGIA PUNGENS	---	7-5	---	2--	---	---	---	---	----	---	----	----	---	----	-5--	3---	--73
SCIRPUS RUFUS	---	---	---	---	---	575	---	---	----	---	----	----	---	----	----	----	----
SCIRPUS MARITIMUS	---	---	---	--2	---	---	---	---	----	---	----	----	-2-	----	3-58	----	----
CAREX EXTENSA	---	---	---	---	---	---	---	---	----	---	----	----	-22	----	5573	----	----
JUNCUS GERARDII	---	---	---	998	837	335	---	3-2	9933	2-2	----	9998	75-	----	83-2	9388	----
JUNCUS MARITIMUS	---	---	---	2--	---	---	---	---	----	---	----	----	588	----	3823	-7--	--23
ATRIPLEX HASTATA	---	---	---	---	---	---	---	3-5	----	-33	---3	----	---	----	3-22	2-2-	--3-
HALIMIONE PORTULACOIDES	2-5	752	272	32-	---	---	---	---	----	---	----	----	---	8333	----	-3--	-32-
HALIMIONE PEDUNCULATA	---	---	---	---	---	---	---	---	----	---	2833	-2--	---	----	----	----	----
SALICORNIA RADICANS	---	---	---	---	---	---	---	---	----	---	----	----	--2	3--3	----	25--	----
SALICORNIA EUROPAEA	-3-	22-	3-2	---	---	---	---	---	23--	---	3557	57--	2--	5--2	----	-223	52--
SUAEDA MARITIMA	22-	27-	975	-2-	---	---	-2-	---	----	---	2-2-	----	---	5---	2-2-	5--3	5---
SPERGULARIA MEDIA	-22	2--	2-3	---	---	---	--3	---	-2--	2--	3-3-	----	-23	33-3	2---	-3-3	22--
SPERGULARIA MARINA	---	---	---	---	---	---	---	555	----	---	88--	53--	---	----	----	----	---2
COCHLEARIA ANGLICA	-2-	---	---	---	---	---	---	---	----	---	----	----	--3	-5--	--2-	5---	----
POTENTILLA ANSERINA	---	---	---	---	--3	-23	---	---	---3	-52	----	3---	---	----	----	----	----
DAUCUS CAROTA	---	---	---	---	---	---	---	---	----	---	----	----	---	----	----	----	--5-
GLAUX MARITIMA	-22	--2	--3	---	2-2	535	--5	2--	2723	312	3111	1335	721	1111	3772	-78-	5--2
ARMERIA MARITIMA	--2	---	---	---	-3-	---	---	---	---3	--2	32--	533-	225	35-7	----	-385	72-2
LIMONIUM VULGARE	227	352	387	--2	---	---	---	---	----	3--	----	----	2--	----	2---	-7-3	----
PLANTAGO MARITIMA	--5	325	-2-	---	253	323	--3	22-	7233	5-5	23-5	5537	225	-5-7	-32-	-723	5--2
ASTER TRIPOLIUM	722	25-	-2-	222	---	---	723	---	33-2	3--	----	----	-23	2-35	2-3-	25--	35--
ARTEMISIA MARITIMA	--2	225	---	---	---	---	---	---	----	777	----	----	---	----	----	----	--7-
LEONTODON AUTUMNALIS	---	---	---	---	253	2-2	---	---	--2-	---	----	8---	---	----	----	----	----

Table 7. 58 relevés of 17 *Glauco-Puccinellietalia* communities (selection e, see further table 1).

The main structure of the table reflects the variation present in the relevé selection comprising the *Puccinellietum maritimae* of the low salt marsh, the *Plantagini-Limonietum* of the upper low marsh, various forms of the *Juncetum gerardii* of the wet upper salt marsh, and the *Artemisietum maritimae* of the dry upper marsh. This picture is confirmed in Fig. 3, where the position of the major dominants is indicated. Thus the main division within the data set, marked by the alliances *Puccinellion maritimae* and *Armerion maritimae*, as shown by Feoli (1977, 1977a) for the same selection, is confirmed. Due to the relatively high fusion limit chosen, many of the clusters have a distinct species combination, e.g. clusters with *Armeria maritima*, *Scirpus rufus*, *Puccinellia distans* and *Carex extensa*. It should be realized that these clusters were built into the selection (see selection procedure mentioned above). Thus it is no surprise to find these clusters back in the final configuration. Still many of the single relevés ended up in quite different positions as compared with the original position in the author's tables.

Discussion

According to the general phytosociological approach as developed in our Department we attempted to combine the theoretical and technical advantages of selected numerical methods with the conceptual framework and the common sense of the Braun-Blanquet approach (see van der Maarel 1975). TABORD is an outcome of this combined approach.

The essentials of TABORD incorporate elements of the intuitive table rearrangement method of Braun-Blanquet and Tüxen. The first step, i.e. initiating clusters and homogenizing them, resembles the polythetic subdivisive element in the Braun-Blanquet approach as pointed out by Moore et al. 1970, Moore & O'Sullivan 1972). The structuring of the table as realized in TABORD is another link to the classical approach.

The procedure of finding a diagonal structure of relevés combines two ideas: the choice of the most efficient starting cluster as the one with relatively many relevés and few constant species and the subsequent arrangements of

	0000000000000000000000000000000000000000000000000000000000
	0012240004554554444111221153351223340233445500011152223333
	1290153453426564218678785436783568906901790778901212342345
	0000000000000000000000000000000000000000000000000000000000
	0000000000221121122000110012211111110000111100000000001111
	8888881111103325944777115663380000002222444733344449992222
PUCCINELLIA MARITIMA	79999985792-85355---------73---22--2-2------------------2-9989
ASTER TRIPOLIUM	7272332252-5535-32-----2-------33----3--23---2-2222-------
HALIMIONE PORTULACOIDES	2----35758-33-33--------------------2------2272 32--------
ARTEMISIA MARITIMA	------222---------------------------5777---7--------------
SALICORNIA EUROPAEA	-3-----22532252----------257---23--2-----------3-2-------3557
SUAEDA MARITIMA	22-2---2753--5------------------------22--975-2-5---2-2-
SPERGULARIA MEDIA	-2--3-22-333322332------------------2----2--2---2-3-------3-3-
LIMONIUM VULGARE	22----735-37------------------------223--2---387--2-------
PLANTAGO MARITIMA	----3-532-3775-55233233335255227237255-5-2---2-----22-23-5
TRIGLOCHIN MARITIMUM	5-53----5-5525-7-2-2322-------333-22-2--223--2----22--3--
PARAPHOLIS STRIGOSA	----------7378-----------------------------------3--------
JUNCUS GERARDII	----------83-----533353373899-899987-2-28-2----99893-2----
ARMERIA MARITIMA	------2--353772552-----3-88532---3-2---2--------------32--
JUNCUS MARITIMUS	-----------7----388----------3-----5----3232---2----------
SALICORNIA RADICANS	---------3-53---2-------------------------------2---------
COCHLEARIA ANGLICA	-2---------5---53--------------------2--------------------
GLAUX MARITIMA	-2--5-2----7-5---27535232-8-322273572 3-2372---3----2--3---
AGROSTIS STOLONIFERA	------------2-2522333237 3--22-22--335233---------232----
CAREX EXTENSA	---------------225--------------------573-----------------
SCIRPUS RUFUS	-------------------575------------------------------------
FESTUCA RUBRA	------2-------------3338833----2-255-8777-----------------
POTENTILLA ANSERINA	-------------------23-33--3--------52---------------------
POA PRATENSIS	-------------------22--53----------3----------------------
LEONTODON AUTUMNALIS	------------------2-22-35-3--2----------------------------
ATRIPLEX HASTATA	------------------------2----------333223--------23-5---3
SPERGULARIA MARINA	--------------------------532---------------------55588--
ELYTRIGIA PUNGENS	------7-----------5----------3-----5-------7---2--3-------
SCIRPUS MARITIMUS	-----------------2--------------------358------2----------
DAUCUS CAROTA	---------------------------------------5------------------
PUCCINELLIA DISTANS	--555-----
HALIMIONE PEDUNCULATA	---------------------------2-------------------------2833

Table 8. Structured table of *Glauco-Puccinellietalia* relevés of table 7 obtained with TABORD, with threshold value 0.55, fusion limit 0.75, and constancy limit 0.70.

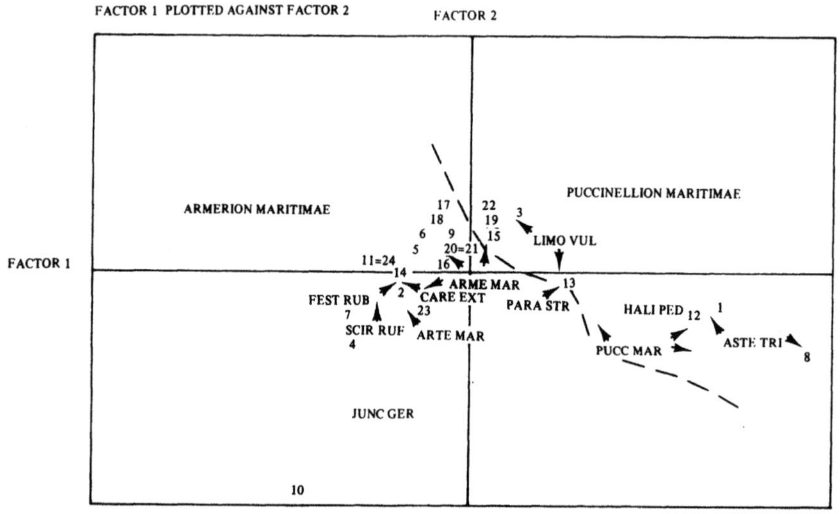

Fig. 3. PCA-ordination of 24 clusters of *Glauco-Puccinellietalia* relevés as obtained with TABORD, with threshold value 0.45 and fusion limit 0.65. See also table 8. The position of the main dominant species and the borderline between *Puccinellion maritimae* and *Armerion maritimae* are indicated.

clusters as well as (joint) constant species, combined with a check on the basis of an ordination of the clusters. In this way, an optimal structure may be expected, also with multidimensional data sets.

Finally we should emphasize the great speed of the program. Simple tables are fully elaborated in seconds, large and complicated tables within tens of seconds.

Some drawbacks we have encountered may now be mentioned: First the final cluster structure is dependent on the structure of the initial array. This would suggest the adoption of an initial array cluster based on a numerical classification, but then some of the practical advantages would be lost. We would prefer a combination of runs based on different initial arrays, in which the number of initial clusters varies. Since nowadays some of the numerical techniques can be done at high speed and with a large capacity, it is becoming feasible to adapt the results of such a classification as input for TABORD.

A second drawback is the distortion brought about by deviant relevés. This drawback may be overcome by removing such relevés from the table. This can be done during the final preparation of the table, on the basis of visual inspection.

Thirdly, users sometimes remark that the tables produced in TABORD are difficult to read. However, if this happens the data-set is probably very complex. It may, for instance, incorporate different plant communities of a region. For this, however, TABORD can hardly be blamed.

This brings us to a final comment. TABORD can be a guide in creating some order in an apparently chaotic ensemble of phytosociological data. In doing so, it does not claim statistical perspective of objectivity. We are essentially dealing with a vegetational polyversum which cannot be approached with statistical methods assuming a universum. We consider an approach based on successive approximation (Poore 1962) as more productive.

Summary

TABORD, a program in FORTRAN IV for structuring phytosociological tables, is described. It is a clustering procedure based on relevé similarity combined with a procedure for obtaining a diagonal structure in the table presenting the clusters arrived at. The following steps are described:

- Establishment of initial clusters, either by the investigator or on the basis of previous knowledge on the data, or automatically by the program.
- Relocation, based on the RELOC-subroutine in the CLUSTAN-package of clustering programs. Various measures of (dis-) similarity are available, of which the similarity ratio is emphasized and descirbed.
- Homogenization of clusters through removal of deviant relevés by adopting a threshold value.
- Fusion of clusters on the basis of the between-cluster similarity pattern by adopting a fusion limit.
- Construction of the table. The first cluster is chosen either as one showing the lowest value of the ratio number of constant species: number of relevés, or as the one placed at the end of an axis of a cluster ordination.
- Repetition of relocation and fusion procedure with other options.
- Finishing touch by computer assisted visual sorting. Various additional information on cluster structure and species occurrence in clusters is presented. The capacity of the program for an IBM 370–158 system, with 512 K bytes instantly available, is roughly 300–500 relevés with 250–150 species and 30–50 initial clusters.

Three examples are discussed. An example with relevés of *Spartinetea* showed that the resulting TABORD table reflects the various communities and subunits as described in earlier work. An example of *Arrhenatheretum* relevés was used to demonstrate various numerical options in TABORD, which confirmed in a concise way what was shown inthe original treatment through many separate analyses. The third example comprises relevés of 17 *Glauco-Puccinellietalia* communities. Again the TABORD table very well reflects the structure and earlier ecological interpretations.

TABORD is considered a numerical elaboration of the intuitive table sorting procedure according to the Braun-Blanquet method.

References

Benninghoff, W.S. & W.C. Southworth. 1964. Ordering of tabular arrays of phytosociological data by digital computer. Abstr. 10. Int. Bot. Congr. Edinburgh: 331–332.

Braun-Blanquet, J. 1964. Pflanzensoziologie, Grundzüge der Vegetationskunde. 3. Aufl. Springer, Wien-New York, XIV + 865 pp.

Češka, A. & H. Roemer. 1971. A computer program for identifying species-relevé groups in vegetation studies. Vegetatio 23: 255–276.

Dale, M.B. & D.J. Anderson. 1973. Inosculate analysis of vegetation data. Austr. J. Bot. 21: 253–276.

Dale, M.B. & L. Quadraccia. 1973. Computer assisted tabular sorting of phytosociological data. Vegetatio 28: 77–73.

Dale, M.B. & L.J. Webb. 1975. Numerical methods for the establishment of associations. Vegetatio 30: 77–87.

Ellenberg, H. 1956. Grundlagen der Vegetationsgliederung. 1. Teil: Aufgaben und Methoden der Vegetationskunde. In: H. Walter. Einführung in die Phytologie 4, (1). Ulmer Stuttgart, 136 pp.
Feoli, E. 1977. On the resolving power of principal component analysis in plant community ordination. Vegetatio 33: 119–125.
Feoli, E. 1977. A criterion for monothetic classification of phytosociological entities on the basis of species ordination. Vegetatio 33: 147–152.
Harman, H.H. 1967. Modern factor analysis. University of Chicago Press, Chicago, London.
Holzner, W., M.J.A. Werger & G.A. Ellenbroek. 1978. Automatic classification of phytosociological data on the basis of species groups. Vegetatio 38: 157–164.
Janssen, J.G.M. 1972. Detection of some micropatterns of winter annuals in pioneer communities of dry sandy soils. Acta Bot. Neerl. 21: 603–610.
Janssen, J.G.M. 1975. A simple clustering procedure for preliminary classification of very large sets of phytosociological relevés. Vegetatio 30: 67–71.
Kortekaas, W.M., E. van der Maarel & W.G. Beeftink. 1976. A numerical classification of European Spartina communities. Vegetatio 33: 51–60.
Lance, G.N. & W.T. Williams. 1968. Note on a new information-statistic classificatory program. Computer J. 11: 195.
Lausi, D., W.G. Beeftink, W.M. Kortekaas. 1978. Survey of salt marsh relevés, selected for the Working-Group for Data-Processing in Phytosociology. (in prep.).
Maarel, E. van der. 1974. The Working Group for Data-Processing of the International Society for Plant Geography and Ecology in 1972–1973. Vegetatio 29: 63–67.
Maarel, E. van der. 1975. The Braun-Blanquet approach in perspective. Vegetatio 30: 213–219.
Maarel, E. van der. 1979. Transformation of cover-abundance values in phytosociology and its effects on community similarity. Vegetatio 39 (in press).
Maarel, E. van der, L. Orlóci & S. Pignatti. 1976. Data-processing in phytosociology, retrospect and anticipation. Vegetatio 32: 65–72.
Moore, S.J., J.J. 1973. PHYTO. A suite of programs in FORTRAN IV for the manipulation of phytosociological tables according to the principles of Braun-Blanquet. Ms Dept. Botany, University College, Dublin, 15 pp.
Moore, S.J., J.J., P. Fitzsimons, E. Lambe & J. White. 1970. A comparison and evaluation of some phytosociological techniques. Vegetatio 20: 1–20.
Moore, J.J. & A. O'Sullivan. 1970. A comparison between the results of the Braun-Blanquet method and those of 'cluster-analysis'. In: R. Tüxen (ed) 'Gesellschaftsmorphologie'. Ber. Int. Symp. Rinteln 1966: p. 26–29. Junk, The Hague.
Mueller-Dombois, D. & H. Ellenberg. 1974. Aims and methods of vegetation ecology. Wiley, New York, XX + 547 pp.
Orlóci, L. 1966. Geometric models in ecology. I The theory and application of some ordination methods. J. Ecol. 54: 193–215.
Orlóci, L. 1976. Ranking species by an information criterion. J. Ecol. 64: 417–419.
Orlóci, L. 1976. TRGRPS – An interactive algorithm for group recognition with an example from Spartinetea. Vegetatio 32: 117–120.
Persson, S. 1977. Datorprogram för bearbetning av vegetationsdata. 1. Klassifikations-program – dokumentation och handhavande. Meddn. Avd. Ekol. Bot. Lunds Univ. 33: 1–68.
Pignatti, S. 1976. A system for coding plant species for data-processing in phytosociology. Vegetatio 33: 23–32.
Poore, M.E.D. 1962. The method of successive approximation in descriptive ecology. Adv. Ecol. Res. 1: 35–68.
Roskam, E. 1971. Programme ORDINA: Multidimensional ordination of observation vectors. Programme Bull, 16, Psychology Lab. Nijmegen, 8 pp.
Spatz, G. 1972. Eine Möglichkeit zum Einsatz der elektronischen Datenverarbeitung bei der pflanzensoziologischen Tabellenarbeit. In: E. van der Maarel & R. Tüxen (eds.). Grundfragen und Methoden in der Pflanzensoziologie. Ber. Int. Symp. Rinteln 1970: 251–258, Den Haag.
Spatz, G. & J. Siegmund. 1973. Eine Methode zur tabellarischen Ordination, Klassifikation und ökologischen Auswertung von pflanzensoziologischen Bestandaufnahmen durch den Computer. Vegetatio 28: 1–17.
Stockinger, J.J. & W.F. Holzner. 1972. Rationelle Methode zur Auswertung pflanzensoziologischer Aufnahmen mittels Elektronen-rechner. In: E. van der Maarel & R. Tüxen (eds.). Grundfragen und Methoden in der Pflanzensoziologie. Ber. Int. Symp. Rinteln 1970: 239–248. Den Haag.
Westhoff, V. & E. van der Maarel. 1973. The Braun-Blanquet approach. In: R.H. Whittaker (ed.). Handbook of Vegetation Science. Part V Ordination and classification of communities. p. 617–726. Junk, The Hague. 737 pp.
Westhoff, V. & E. van der Maarel. 1978. The Braun-Blanquet approach. 2nd ed. In: R.H. Whittaker (ed.). Classification of plant communities, p. 287–399. Junk, The Hague, 408 pp.
Wishart, D. 1969. Clustan 1A. Computing Lab. St. Andrews, 7 pp.
Wishart, D. 1975. CLUSTAN 1C, user manual. Computer Centre University College London, 124 pp.

AUTOMATIC CLASSIFICATION OF PHYTOSOCIOLOGICAL DATA ON THE BASIS OF SPECIES GROUPS*, **

W. HOLZNER[1], M.J.A. WERGER[2] & G.A. ELLENBROEK[2]***

[1]) Botanisches Institut, Universität für Bodenkultur, Gregor Mendel Str. 33, A-1180 Wien, Austria.
[2]) Division of Geobotany, University of Nijmegen, Toernooiveld, Nijmegen, The Netherlands.

Introduction

In phytosociological tables species and relevés are ordered in such a way that the species with a more or less similar distribution pattern over the relevés are grouped together and, similarly, relevés with a more or less similar species content are placed next to one another. As stated previously (Werger 1974a), membership of a species to a species-group is not necessarily determined by association with any individual species in that group, but can also be based on the restriction to the general distribution range in the table of a group of associated species. The same applies mutatis mutandis to the relevés belonging to a relevé group (cf. Hull 1964–65). In this way clusters of relevés characterized by groups of differential species are easily spotted in the table (cf. Van der Maarel et al. 1978).

Obviously the groups of species in a phytosociological table are not all fully differential, but mutual overlaps in occurrences between the species groups occur at various (hierarchical) levels and to various degrees. This makes it difficult to represent the floristic (and sociological) relationships in a phytosociological table: in case all species groups would be mutually exclusive (or fully differential), or if one species group would only overlap with one other species group, a simple diagonal structure of the phytosociological table would adequately represent the pattern in species and relevés in the data set comprised in the table. Since mutual overlap between all species groups commonly occurs, however, not every species group can be represented in a single, coherent cluster in the phytosociological table, and consequently the diagonal structure of the table is distorted. With (automatic) classification this may prevent presentation of the data in an easily perceptible way.

An illustration of this problem is given in Fig. 1 for six differential species groups (a-f) of equal importance, which are mutually overlapping in distribution over the relevés. It is thus impossible to arrange this type of vegetation data objectively in a phytosociological table which closely resembles the classic phytosociological table structure, and it is therefore impossible to computerize the procedures to arrive at such a table. When phytosociologists nevertheless succeed in presenting a phytosociological table with an easily interpretable structure, it is because they subjectively overemphasize the importance of some species groups and present them in the table as coherent species/relevé clusters. This approach is justified and facilitated because of the species groups distinguishable in vegetation data gathered in nature not being all of equal importance. Furthermore, the interpretability, the usual wide extrapolation possibilities, as well as the consistency of the phytosociological table structure when new vegetation data are added to it, indicate the appropriateness of the approach (cf. Whittaker 1962, Westhoff & van der Maarel 1973, Werger 1974b).

It would, however, nevertheless be of significant value to phytosociologists if a comparable table structure could be obtained by automatic procedures.

Possibilities for automatization

When only small segments of a phytosociological table are studied the difficulties outlined in the introduction do not occur. In such sufficiently small segments (e.g. comprising

* Nomenclature of species follows usage in Werger (1973).
** Contribution from the International Working Group on Data Processing in Phytosociology. Part of this paper is based on an unpublished lecture presented at the Prague meeting of the Working Group (Stockinger & Holzner 1973b). Copies of that manuscript and of a description of the mathematical background of the method (Stockinger & Holzner 1973a) are available from the first author. The computer program can be studied at Vienna, Nijmegen, and at the C.S.I.R.O., St. Lucia Qld., Australia (Dr. M. B. Dale).
*** Supported by a grant from the 'Fonds zur Förderung der wissenschaftlichen Forschung', Austria.

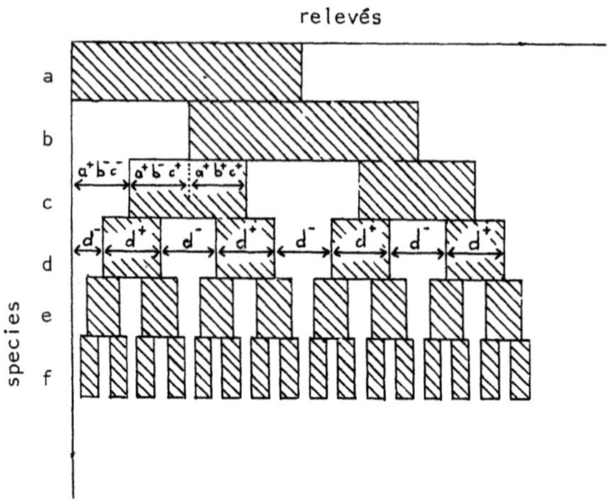

Fig. 1. Table display of phytosociological data with six differential species groups (a–f) of equal importance and mutually overlapping in distribution over the relevés.

Species	Relevés
HELODO	———2–2–22–2223222——
MYRAFR	–333222–343322234236323——
RHUDIV	———2—42—2–224244——
PENSET	————12—22–22–434——
HELNUD	2—2222—2–2322222–233——
HYPAET	————1–2–2–2—2222——
CYMVAL	——————4–5434——
CARSPI	—————3222232——
RHUDEN	———33—22–32–43224——
KOECRI	——2——222—22332——

Fig.2. An *r*-spot defined by species and relevés. The relevés are ordered according to their *r*-value which increases from left to right.

only the species groups a and b in Fig. 1) the differential species groups can be presented in single coherent clusters. It is possible to develop a model which is, based on the selection of such sufficiently small table segments and allows the presentation of floristic relationships in a phytosociological table of a more or less 'classic' structure along generally definable rules. The mathematical background of such a model was discussed in detail by Stockinger & Holzner (1973a).

The r-spot

In this model all sociological relevant species groups in a data file are determined and each group is located in a 'representative spot' (or *r*-spot). All species occurring in the relevés defined by a sociological species group, but not belonging to that group, are omitted from the relevés, and for each relevé a value *r* is calculated which indicates the strength of the species group still present in that relevé;

$$r(x, a) = n \sum_{s \varepsilon X} d(s, a)$$

in which n = the number of scores $\neq 0$ of species of group x in relevé a, and $d\,(s, a)$ = score value of species s in relevé a. The relevés characterized by species group x are ordered within their *r*-spot according to their *r* value (Fig. 2).

The essential species groups

Stockinger & Holzner (1973a) defined a grouping coefficient g for species, which allows for a comparison of species of different sizes with one another. The g value for each group forms a natural threshold for the g values of the smaller partial groups contained in it. Thus, a species group can be disregarded for presentation in a *r*-spot if there is a larger group that contains the smaller group and has a higher g value. Only those species groups which are not contained in another species group with a higher g value, are used for further analyses. Such species groups are called essential species groups.

Thus, the essential species group is sociologically significant just because it is a group of species that is not contained in any other larger and more strongly tied species group. Consequently, all essential species groups together should, in dependence of the choice of the critical value for the grouping coefficient g, adequately and fully describe the sociological structure of the data file.

Often the large sociologicaly relevant groups will contain partial groups which are especially important or interesting for a more detailed sociological interpretation. Such a partition whithin groups can occur at several hierarchical levels. A presentation of all species in disjoint clusters is therefore appropriate for a tabular presentation of the phytosociological pattern only in special cases. The larger the data set and the variation contained in it, the smaller the chances are that all species can be presented in disjoint clusters.

The diagonal table

For each species there is at least one essential species group in which the species is represented, and each essential species group can be located in a *r*-spot. Thus, it is possible

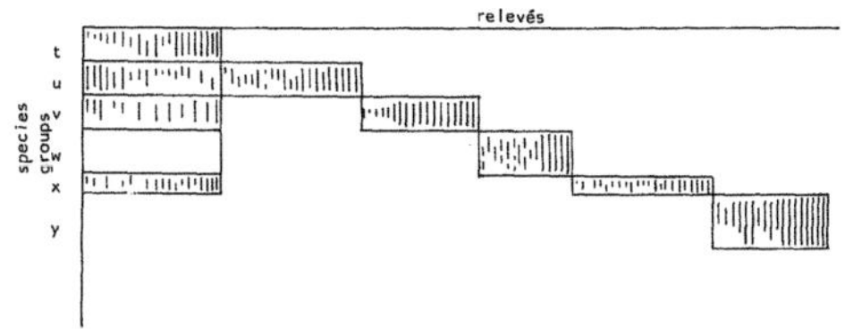

Fig. 3. Diagonal table. The essential species groups are located on the diagonal and sociological relationships between them are indicated by spots away from the diagonal.

to construct a complete phytosociological table: vertically the essential species groups are ordered and horizontally the relevés belonging to the essential species groups. Such a table contains the essential species groups diagonally from upper left to lower right (Fig. 3). The *r*-spots of the essential species groups indicate the extent and importance of the clusters whilst the sociological relationships between the essential species groups are indicated by the spots which have been formed away from the diagonal.

In this way an automatic tabulation based on *g* and *r* functions, is possible for each data file. However, within the essential species groups species and relevés are not repeated, and only those disjoint clusters which are situated on the diagonal are certain. In practice such diagonally arranged tables based on all essential species groups are far too large. Therefore it is useful to choose some (optimated) essential groups for presentation of the sociological structure by means of objective criteria. This choice, based on approximation procedures, leads to an optimalization of the table structure in that the diagonally arranged optimated essential species groups present an overview over the most important sociological relationships present in the data file. By deleting repeated occurrences of species and relevés in this table a structure comparable to a 'classic' phytosociological table is obtained.

Alternatively, the results of the clustering procedure also can be displayed by plexuses (cf. McIntosh 1973) which contain more information on the relations between the species in one group than a table display does (Stockinger & Holzner 1972).

An example

To test the performance of the program in its selection of essential species groups, vegetation data of scrub, dwarf scrub, grassland and riverine forest formations of the Upper Orange River area, South Africa, were used. This data file consists of 560 relevés containing 345 species occurring more than once in the relevés. The data have been collected according to the Braun-Blanquet approach. They have been analysed in a 'classical' phytosociological manner by Werger (1973), who classified the vegetation into seven major groupings. Syntaxonomically the groupings are different from one another at the rank of class. Several vegetation units of subordinate rank were distinguished as well (see Table 1). The results of the computer analyses are compared with the interpretation given by Werger (1973).

All seven major groupings are distinguished by the computer program, but details are recognized with different degrees of success. The species groups are presented in plexuses (cf. McIntosh 1973).

In one large plexus (Fig. 4) with *g* values between 0.745 and 0.494, sixteen species are grouped which together characterize the *Diospyrion lycioidis*, and in particular the *Rhoo-Diospyretum*. In one partial group the two most common species in one subassociation are grouped with four of the five most common species characteristic for the *Rhoo-Diospyretum*. The other partial group mainly consists of all species characteristic for the alliance plus a few *Rhoo-Diospyretum* species which are also comprised in the other partial group.

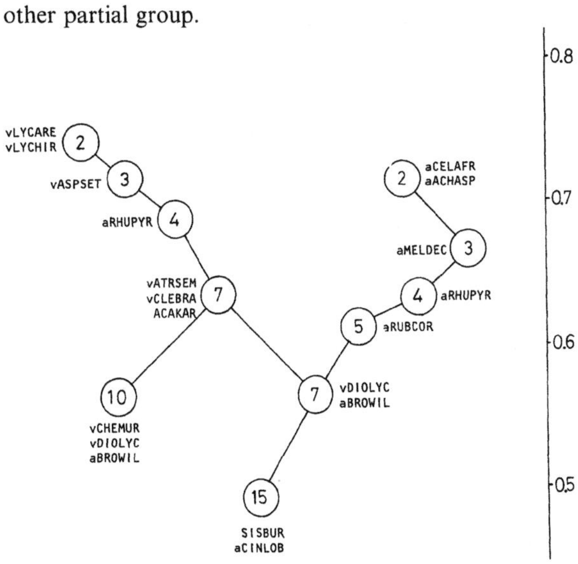

Fig. 4. *Diospyrion lycioidis* plexus. The scale indicates the *g* values at which fusions occur. The horizontal arrangement of the clusters has no meaning. The numbers refer to the number of species contained in each cluster. The names of the species added to each cluster are abbreviated (capitals) and can be retrieved from Appendix 1. The indicator value of the species in the classification by Werger (1973) is shown as follows: a. character or differential species of an association; v. character or differential species of an alliance; o. character species of an order; k. character species of a class.

Table 1. Phytosociological classification according to Werger (1973).

I	*Diospyrion lycioidis*:	1. *Rhoo-Diospyretum* (with 2 subassociations)
		2. *Zizipho-Acacietum karroo*
II	Three grassland communities (not hierarchically classified).	
III	*Pentzio-Chrysocomion*:	1. *Hermannio-Nestleretum* (with 3 subassociations)
IV	*Stipagrostion*:	1. *Enneapogono-Stipagrostietum* (with 2 subassociations)
V	*Rhoetea erosae*	
	Grewio-Rhoetalia	
	Indigofero-Rhoion:	1. *Rhamno-Rhoetum*
		2. *Rhoo-Aloetum*
		Blepharido-Rhoetum
	Rhoetalia ciliato-erosae	
	Hibisco-Rhoion:	1. *Osteospermetum* (with 2 subassociations)
		2. *Stachyo-Rhoetum* (with 2 subassociations)
		3. *Nanantho-Rhoetum*
		Mayteno-Oleetum (with 2 subassociations)
		Setario-Buddleietum
VI	*Pentzietea incanae*	
	Pentzio-Rhigozetalia	
	Enneapogono-Rhigozion:	1. *Zizipho-Rhigozetum* (with 2 subassociations)
		2. *Melhanio-Hermannietum*
	Zygophyllion:	1. *Monechmatetum* (with 2 subassociations)
		2. *Nestlero-Pteronietum* (with 2 subassociations)
		Eriocephalo-Eberlanzietum
VII	*Lycium prunus-spinosa* community	

Another independent species grouping contains the two most important character species of the *Zizipho-Acacietum*, *Ziziphus mucronata* and *Setaria verticillata*, fusing at $g = 0.742$, together with the always abundant species *Acacia karroo* at $g = 0.667$. This group remains separated from the plexus of Fig. 4, but in the diagonal table its relation to the *Diospyrion lycioidis* will be immediately apparent.

Two of the three grassland communities distinguished by Werger (1973) are described by the species group of Fig. 5. The plexus shows that the two communities are clearly separated in different parts of the configuration and connected through a number of shared species. The fusions in the cluster are between $g = 0.760$ and $g = 0.487$. The third community is not distinguished by the program by its own species group, which was to be expected; that community has no distinct species group but consists of a transition from grassland to karroid dwarf scrub with a typical species mixture.

The *Pentzio-Chrysocomion* is less clearly distinguished in four fusions between $g = 0.504$ and $g = 0.434$ in which *Walafrida saxatilis*, *Pentzia globosa*, *Tragus koelerioides*, *Chrysocoma tenuifolia* and *Nenax microphylla* are subsequently grouped. All these species, except the latter, are typical for the alliance but not specific for the association. *Nenax microphylla* occurs more typically on sites with very shallow soils. The recognition of the *Pentzio-Chrysocomion* in the diagonal table on the basis of this species group is rather difficult since these are the species which intrude widely into the contact communities of the *Pentzio-Chrysocomion* (Werger 1973). These species also dominate in the *Pentzio-Chrysocomion*, however, and the characteristic species of the *Hermannio-Nestleretum* are not very abundant and/or constant. It is therefore no surprise that the association species group has not been recognized by the programme.

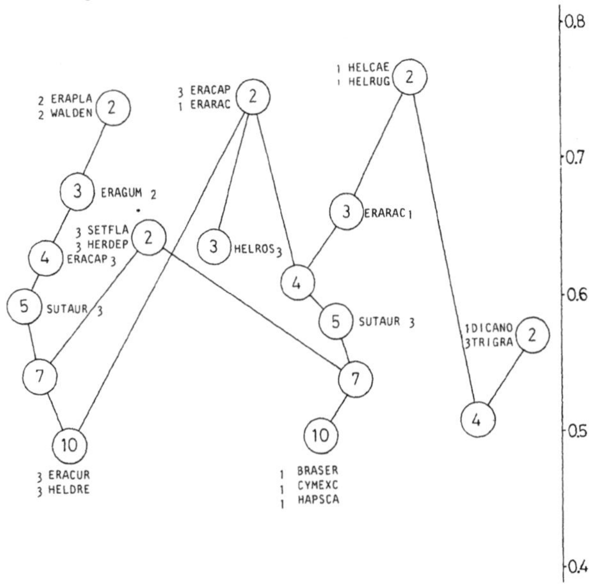

Fig. 5. Plexus of grassland communities. For explanation see Fig. 4. 1. differential species of community 1; 2. differential species of community 2; 3. species common to communities 1 and 2.

The *Stipagrostion* is readily distinguished by the fusion of *Stipagrostis obtusa*, *S. ciliata* and *Pentzia calcarea* at $g = 0.586$ and joined by *Plinthus cryptocarpus* at $g = 0.517$. Another small species group delineates one subassociation of the *Pentzio-Stipagrostietum*. The differences between the two associations in the alliance are expressed by their species combinations and not by exclusive species. Therefore, these communities do not have different *r*-spots on the diagonal.

Thirteen species groups indicate various units of the *Rhoetea erosae*. Some of these species groups are very small, however, and are not always readily interpretable. The *Rhamno-Rhoetum* and with it the order *Grewio-Rhoetalia* (and the alliance) is strongly characterized by a large species group (Fig. 6). Twenty-one species characteristic of the association are combined in this plexus with nine species thought by Werger (1973) to be more typical of the alliance or order. The fusions lie between $g = 0.853$ and $g = 0.546$. Another small group of species clearly characterizes the *Blepharido-Rhoetum*. The *Rhoo-Aloetum* seems only weakly be distinguished in another small species group with *Aloe ferox* as the only characteristic species of the association.

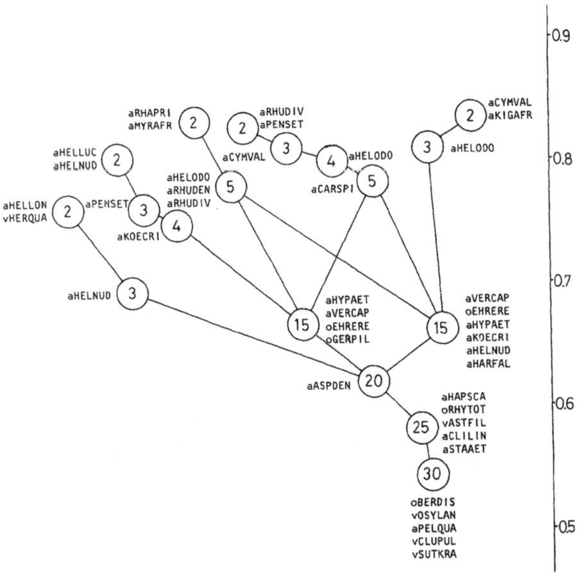

Fig. 6. *Grewio-Rhoetalia* plexus. For explanation see Fig. 4.

Twenty-nine species characteristic of the class, four of the order *Rhoetalia ciliato-erosae* and one other species are combined in a single group with g-values between 0.653 and 0.379 (Fig. 7). Other groups characterize subunits. The group of order species is overwhelmed by the class species and is thus not separately distinguished. The *Hibisco-thoion* is characterized by a small species group at $g = 0.533$ to $g = 0.456$ and contains *Hibiscus marlothianus*, *Nenax microphylla*, *Aptosimum depressum*, *Blepharis villosa* and *Osteospermum leptolobum*. Two other small species groups characterize the *Osteospermetum* (*Eriocephalus spinescens*, *Phymaspermum parvifolium*, *Osteospermum leptolobum* and *Hibiscus marlothianus*; g-values between 0.427 and 0.415) and the *Setario-Buddleietum* (*Buddleia saligna*, *Setaria lindenbergiana*, *Solanum retroflexum*, *Boophane disticha*; g values between 0.851 and 0.556). However, the other associations have not been recognized by a distinct species group. Spots away from the diagonal in the table must identify them. There remain six small species groups consisting of a few *Rhoetea* species, but it is difficult to interpret their meaning in the light of Werger's (1973) classification.

The *Pentzietea incanae* are characterized by one large species group, which can be subdivided into two partial groups that seem to contrast the *Enneapogono-Rhigozion* against the *Zygophyllion*, or perhaps better, the *Melhanio-Hermannietum* against the *Nestlero-Pteronietum* (Fig. 8). The fusions occur between $g = 0.785$ and $g = 0.313$ though mainly above $g = 0.5$. Another small species group combines a few less constant *Enneapogono-Rhigozion* species.

The *Lycium prunus-spinosa* community is characterized clearly by a species group consisting of *Drosanthemum* sp., *Zygophyllum incrustatum*, *Lycium prunus-spinosa*, *Salsola glabrescens*, *Psilocaulon absimile*, *Chloris virgata* and *Tragus berteronianus*, fusing in that sequence at g-values between 0.690 and 0.371. Another smaller group containing partly the same species singles out one relevé from this community.

It is thus apparent, that the major groupings are readily identified by the computer program and characterized by species groups. Most, but not all, of the vegetation units of association rank are identified as well. Particularly some of those units which are not characterized by a strong group of characteristic species have not been identified by the program. Still, they are distinguished by Werger (1973) as associations, i.e. of equal rank and importance as other vegetation units possessing strong characteristic species groups (e.g. *Mayteno-Oleetum*, *Nanantho-Rhoetum*, *Sta-*

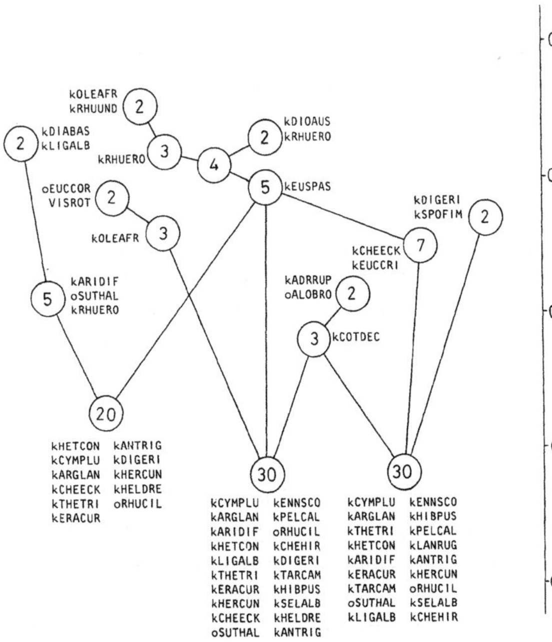

Fig. 7. *Rhoetea erosae (Rhoetalia ciliato-erosae)* plexus. For explanation see Fig. 4.

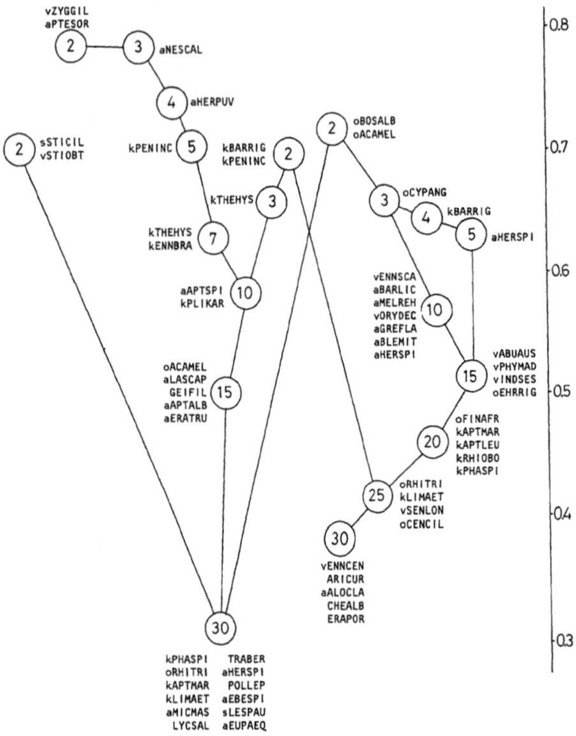

Fig. 8. *Pentzietea incanae* plexus. For explanation see Fig. 4.

chyo-Rhoetum, *Monechmatetum*). These vegetation units will be difficult to identify in the diagonal table. Other vegetation units cannot be identified by a species group, because they do not have such a group. They are typified by a characteristic combination of more widely spread species, i.e. by overlapping species groups (e.g. *Zizipho-Rhigozetum*, one of the grassland communities). The program allows a fairly easy recognition of these vegetation units. However, this occurs on the basis of 'overlapping spots' away from the diagonal in the diagonal table.

Patterns of hierarchization are difficult to identify in the output of the present program, though in some cases these are clearly indicated again by 'overlapping spots'.

Conclusion

In conclusion we may say that although not all details discussed by Werger (1973) are revealed in the classification suggested by the computer program, it is a powerful tool for a rapid tabular ordering of a large data set. The program approximates the structure and phytosociological groupings in the data set and it reveals phytosociological relations which were also revealed in a detailed analysis of the same data set by traditional means. However, in a traditional analysis very many more (undefined) criteria are used, consciously or not, in the classification process (cf. Kaplan 1964, Hull 1970, Holzner & Stockinger 1973, Werger 1974b).

Summary

The structure of phytosociological relationships between vegetation groupings as apparent from a classic phytosociological table is briefly explained and a method to describe this structure by a table method and by means of automatic data processing is outlined. It is based on a mathematical model using continuous clusters of species and arranging them in a diagonal table. This could replace the classical table work. The method is tested on a large South African data set, which has been previously classified and interpreted. It appears that the method seems to be an efficient tool in rapidly ordering large phytosociological data sets revealing a table structure which approximates the classification arrived at in the traditional manner.

References

Holzner, W. & Stockinger, F. 1973. Der Einsatz von Elektronenrechnern bei der pflanzensoziologischen Tabellenarbeit. Österr. Bot. Z.: 121: 303–309.

Hull, D.L. 1964–65. The effect of essentialism on taxonomy. Two thousand years of stasis. I + II. Brit. J. Phil. Sci. 15: 314–326. ibid. 16: 1–18.

Hull, D.L. 1970. Contemporary systematic philosophies. Ann. Rev. Ecol. Syst. 1: 19–54.

Kaplan, A. 1964. The conduct of inquiry. Chandler, San Francisco.

Maarel, E. van der, J.G.M. Janssen & J.W.M. Louppen. 1978. TABORD, a program for structuring phytosociological tables. Vegetatio 38: 143–156.

McIntosh, R.P. 1973. Matrix and plexus techniques. In: R.H. Whittaker (ed.), Ordination and classification of communities. Handb. Veg. Sc. 5: 157–191. Junk, The Hague.

Stockinger, F.J. & Holzner, W. 1972. Rationelle Methode zur Auswertung pflanzensoziologischer Aufnahmen mittels Elektronenrechner. In: E. van der Maarel & R. Tüxen (eds.), Grundfragen und Methoden in der Pflanzensoziologie, pp. 239–248. Junk, The Hague.

Stockinger, F.J. & Holzner, W. 1973a. Prinzipien einer Implementierung des pflanzensoziologischen Zerlegungsproblems. Mitt. Bot. Arbeitsgem. Oberösterr. Landesmus. Linz 4: 87–106.

Stockinger, F.J. & Holzner, W. 1973b. Grundlegende Fragen bei der Verwendung von EDV in der Pflanzensoziologie.

Symposium International Working Group for Data Processing in Phytosociology, Prague (manuscript).
Werger, M.J.A. 1973. Phytosociology of the Upper Orange River Valley, South Africa. V & R, Pretoria (Thesis Nijmegen).
Werger, M.J.A. 1974a. On concepts and techniques applied in the Zürich-Montpellier method of vegetation survey. Bothalia 11: 309–323.
Werger, M.J.A. 1974b. The place of the Zürich-Montpellier method in vegetation science. Folia Geobot. Phytotax. 9: 99–109.
Westhoff, V. & E. van der Maarel. 1973. The Braun-Blanquet approach. In: R.H. Whittaker (ed.), Ordination and classification of communities. Handb. Veg. Sc. 5: 619–726. Junk, The Hague.
Whittaker, R.H. 1962. Classification of natural communities. Bot. Rev. 28: 1–239.

Appendix 1. Species codes and species names of fig. 4–8

ABUAUS	Abutilon austro-africanum
ACAKAR	Acacia karroo
ACAMEL	Acacia mellifera
ACHASP	Achyranthes aspera
ADRRUP	Adromischus rupicola
ALOBRO	Aloe broomii
ALOCLA	Aloe claviflora
ANTRIG	Anthospermum rigidum
APTALB	Aptosimum albomarginatum
APTLEU	Aptosimum leucorrhizum
APTMAR	Aptosimum marlothii
APTSPI	Aptosimum spinescens
ARGLAN	Argyrolobium lanceolatum
ARICUR	Aristida curvata
ARIDIF	Aristida diffusa
ASPDEN	Asparagus denudatus
ASPSET	Asparagus setaceus
ASTFIL	Aster filifolius
ATRSEM	Atriplex semibaccata
BARLIC	Barleria lichtensteiniana
BARRIG	Barleria rigida
BERDIS	Berkheya discolor
BLEMIT	Blepharis mitrata
BOSALB	Boscia albitrunca
BRASER	Brachiaria serrata
BROWIL	Bromus willdenowii
CARSPI	Carex spicato-paniculata
CELAFR	Celtis africana
CENCIL	Cenchrus ciliaris
CHEECK	Cheilanthes eckloniana
CHEHIR	Cheilanthes hirta
CHEALB	Chenopodium album
CHEMUR	Chenopodium murale
CINLOB	Cineraria lobata
CLEBRA	Clematis brachiata
CLILIN	Cliffortia linearifolia
CLUPUL	Clutia pulchella
COTDEC	Cotyledon decus.-orbic. complex
CYMEXC	Cymbopogon excavatus
CYMPLU	Cymbopogon plurinodis
CYMVAL	Cymbopogon validus

Appendix 1. (continued)

CYPANG	Cyphocarpha angustifolia
DIABAS	Dianthus basuticus
DICANO	Dicoma anomala
DIGERI	Digitaria eriantha
DIOAUS	Diospyros austro-africana
DIOLYC	Diospyros lycoides
EBESPI	Eberlanzia spinosa
EHRRIG	Ehretia rigida
EHRERE	Ehrharta erecta
ENNBRA	Enneapogon brachystachyus
ENNCEN	Enneapogon cenchroides
ENNSCA	Enneapogon scaber
ENNSCO	Enneapogon scoparius
ERACAP	Eragrostis capensis
ERACUR	Eragrostis curvula
ERAGUM	Eragrostis gummiflua
ERAPLA	Eragrostis plana
ERAPOR	Eragrostis porosa
ERARAC	Eragrostis racemosa
ERATRU	Eragrostis truncata
EUCCOR	Euclea coriacea
EUCCRI	Euclea crispa
EUPAEQ	Euphorbia aequoris
EUSPAS	Eustachys paspaloides
FINAFR	Fingerhuthia africana
GEIFIL	Geigeria filifolia
GERPIL	Gerbera piloselloides
GREFLA	Grewia flava
HALLUC	Halleria lucida
HARFAL	Harpechloa falx
HAPSCA	Haplocarpha scaposa
HELCAE	Helichrysum caespititium
HELDRE	Helichrysum dregeanum
HELNUD	Helichrysum nudifolium
HELODO	Helichrysum odoratissimum
HELROS	Helichrysum rosum
HELRUG	Helichrysum rugulosum
HELLON	Helictotrichon longifolium
HERCUN	Hermannia cuneifolia
HERDEP	Hermannia depressa
HERPUV	Hermannia pulverata
HERQUA	Hermannia quartiniana
HERSPI	Hermannia spinosa
HETCON	Heteropogon contortus
HIBPUS	Hibiscus pusillus
HYPAET	Hypericum aethiopicum
INDSES	Indigofera sessilifolia
KIGAFR	Kiggelaria africana
KOECRI	Koeleria cristata
LANRUG	Lantana rugosa
LASCAP	Lasiocorys capensis
LESPAU	Lessertia pauciflora
LIGALB	Lightfootia albens
LIMAET	Limeum aethiopicum
LYCARE	Lycium arenicolum
LYCHIR	Lycium hirsutum
LYCSAL	Lycium salinicolum
MELREH	Melhania rehmannii
MELDEC	Melica decumbens
MICMAS	Microloma massonii
MYRAFR	Myrsine africana
NESKAL	Nestlera humilis
OLEAFR	Olea africana
ORYDEC	Orygia decumbens

Appendix 1. (continued)

OSYLAN	Osyris lanceolata
PELCAL	Pellaea calomelanos
PELQUA	Pellaea quadripinnata
PENSET	Pentaschistus setifolia
PENING	Pentzia incana
PHASPI	Phaeoptilum spinosum
PHYMAD	Phyllanthus maderaspatensis
PLIKAR	Plinthus karooicus
POLLEP	Polygala leptophylla
PTESOR	Pteronia sordida
RHAPRI	Rhamnus prinoides
RHIOBO	Rhigozum obovatum
RHITRI	Rhigozum trichotomum
RHUCIL	Rhus ciliata
RHUDEN	Rhus dentata
RHUDIV	Rhus divaricata
RHUERO	Rhus erosa
RHUPYR	Rhus pyroides
RHUUND	Rhus undulata
RHYTOT	Rhynchosia totta
RUBCOR	Rubia cordifolia
SELALB	Selago albida
SENLON	Senecio longiflorus
SETFLA	Setaria flabellata
SISBUR	Sisymbrium burchelli
SPOFIM	Sporobolus fimbriatus
STAAET	Stachys aethiopica
STICIL	Stipagrostis ciliata
STIOBT	Stipagrostis obtusa
SUTAUR	Sutera aurantiaca
SUTHAL	Sutera halimifolia
SUTKRA	Sutera kraussiana
THETRI	Themeda triandra
THEHYS	Thesium hystrix
TRABER	Tragus berteronianus
TRIGRA	Trichoneura grandiglumis
VERCAP	Vernonia capensis
VISROT	Viscum rotundifolium
WALDEN	Walafrida densiflora
ZYGGIL	Zygophyllum gilfillani

TRANSFORMATION OF COVER-ABUNDANCE VALUES IN PHYTOSOCIOLOGY AND ITS EFFECTS ON COMMUNITY SIMILARITY*, **

E. van der MAAREL***

Division of Geobotany, Toernooiveld, 6525 ED Nijmegen, The Netherlands.

Keywords:
Arrhenatheretum, Classification, Combined estimation, Cover-abundance scale, Ordination, Phytosociology, Salt marsh, Similarity, Transformation.

Introduction

The numerical treatment of phytosociological data is often based on estimates of cover and/or abundance according to the Braun-Blanquet and Domin scales. Since Schwickerath (1931, 1938, 1940) and Tüxen & Ellenberg (1937) published their transformations there has been discussion on the way the scale values should be used in calculations. Qualitative approaches, i.e. based on presence and absence have also been favoured (e.g. Williams & Lambert 1959, van der Maarel 1966) Dagnelie (1960) proposed a pseudoqualitative basis for various calculations by means of a 'coupure'. A coupure includes the deletion of lower values, usually according to a fixed criterion, e.g. the number of occurrences in a phytosociological table to be remained should be as close as possible to 50%. Dagnelie's approach remained largely unknown and apparently it has never been tested.

Transformation of cover-abundance values may involve the differential weighting of species in numerical phytosociological classification, or the standardization of values to zero mean (and unit variance) in the rows or columns (or both) of phytosociological data matrices. Such standardizations are usual in multivariate analysis and their implications for vegetation ecology have been discussed in many Anglo-American studies (e.g. Austin & Greig-Smith 1968, Noy-Meir 1973, Noy-Meir, Walker & Williams 1975, Orlóci 1978).

Since the use of numerical methods in European phytosociology is rapidly growing it may be useful to review the various transformations published so far and to give some proposals for unification.

Some practical examples of the effects of transformations on classification and ordination are then given with data used in various projects of the Working Group for Data-processing in Phytosociology. Such effects will include changes in patterns of (dis-) similarity between the phytosociological units – relevés or community types – involved, hence the title of this contribution. (see also Campbell 1978). Since I consider weighting of species performance as a more urgent transformation problem in phytosociology then standardization, the latter type of transformation will not be treated here.

During the preparation of this paper I became involved in two similar studies which have been published in Vegetatio (Jensén 1978, Campbell 1978) and which serve as additional sources for my paper.

* Contribution from the Working Group for Data-Processing in Phytosociology, based on an earlier report for that group (Van der Maarel 1972).

** Nomenclature of salt marsh species follows Lausi, Kortekaas, & Beeftink (1979), for names of *Arrhenatheretum* species see E. Oberdorfer, 1970, Pflanzensoziologische Exkursionsflora für Süddeutschland, 3. Aufl. Ulmer, Stuttgart.

*** This paper was partly prepared during my stays at the Växtekologiska Institutionen at Lund under NFR grant B 252605. I thank Prof. Nils Malmer, Head of this institute for his interest in my numerical approaches in phytosociology and Dr. Sven Jensén of the same institute for valuable discussions on data transformation. Of my own department I thank Sophie Hochstenbach for preparation of the numerical results and Jo Louppen for technical help and discussions concerning the computer programs involved. Finally I thank Drs. László Orlóci and Mike Austin for useful comments on the manuscript, and Dr. Dicky Clymo for discussions on transformation formulae and his permission to mention an unpublished formula.

On transformation

Angular transformation of cover data

In most cover-abundance scales cover is considered of major importance and in calculations the lower values of the scale, though defined as abundance categories, are often interpreted as cover values. (Tüxen & Ellenberg 1937). Dagnelie (1960) stated that the classes obtained in such a way are not equidistant. Dagnelie, following Goodall (1953–1954) suggested the angular transformation. This transformation takes a cover % value c as a proportion $p = c/100$, and converts it to y according to

$$y = 2 \arcsin \sqrt{p} \text{ for } p \leqq 0.5 \quad (1a)$$

$$y = 3{,}1416 - 2 \arcsin \sqrt{1-p}, \text{ for } p > 0.5 \quad (1b)$$

Dagnelie showed that the transformed values of the Braun-Blanquet scale (as converted to cover values by Tüxen & Ellenberg 1937 in the form adopted by Braun-Blanquet 1964, 1965) form a series which runs roughly parallel to the original series (Table 1). This may be considered an additional argument in the controversy on the transformations suggested by Tüxen & Ellenberg (1937) and Schwickerath (1931) in favour of the latter one. Van der Maarel (1966) developed a transformation which can be approximated with integers from 1–9. This transformation has been used by Fresco (1969) and van der Maarel & Fresco (1975).

The transformed scale is presented in Table 2 (col. 6) together with rounded values from angular transformations of the scales of Braun-Blanquet, Daubenmire (1968, modified form according to Bailey & Poulton 1968, see also Mueller-Dombois & Ellenberg 1974), Hult-Sernander, Domin (cf Shimwell 1971 for scales and references and Bannister 1966 for an angular transformation of the Domin scale), Doing (1954), and Barkman, Doing & Segal (1964, see also Londo 1976).

As may be seen from Table 2 it is difficult to transform low and often roughly estimated cover values corresponding with the abundance classes '*r*' and '+'. It is possible to treat the associated class values either as unity or to ignore them. Angular transformation has been regularly applied to cover % estimations (e.g. Bannister 1966, Smartt, Meacock & Lambert 1976) but it has never become a standard procedure.

Table 1. Braun-Blanquet cover-abundance values, corresponding cover % values according to Braun-Blanquet (1964), angular transformation values (after Dagnelie 1960) and values multiplied by 2.07 to arrive at values within the original range of the Braun-Blanquet scale.

Br. Bl. scale	corresponding cover % value	arc sin transf.	ibid. adapted to Br. Bl. scale
r	( )	( )	( )
+	0.1	0.06	0.12
1	5.0	0.45	0.93
2	17.5	0.86	1.78
3	37.5	1.31	2.73
4	62.5	1.82	3.77
5	87.5	2.42	5.00

Transformations of the Braun-Blanquet scale

As has been already remarked the angular transformation of the Braun-Blanquet cover values resembles the approach of Schwickerath (1931 a.f.) developed for the measurement of what he called 'Artmächtigkeit' or Gruppenmächtigkeit. This is a species weight, or 'species magnitude' (Braun-Blanquet 1965), or less appropriately 'species significance' (Krajina 1960). As a sort of species importance value (cf Curtis & MacIntosh 1951) it can be used in fidelity determination and calculation of spectra of syntaxonomical species groups (Segal & Westhoff 1959) etc.

Schwickerath used the Braun-Blanquet values 1–5 directly and replaced + by 0.25 (The value *r* was neglected). Tüxen & Ellenberg (1937) suggested a transformation for exactly the same procedures, i.e. to arrive at what they called a 'Gruppenwert'. Their transformation was based on the average cover % values in the classes 2–5, while 1 and + were arbitrarily replaced by 2.5 % and 0.1 % respectively.

Braun-Blanquet (1946, 1964, 1965) while adopting a somewhat different transformation seems to prefer the Tüxen-Ellenberg approach. He used (in his 1946 paper) a surprising motive. He disregarded Schwickerath's approach and the warning by Tüxen & Ellenberg (1937) to use the group value only for the comparison of facies and not to overemphasize ecologically important species occurring with low abundance values. Braun-Blanquet (1946) stated (in translation): 'until now the sociological significance of species in a plant community was estimated according to their constancy and their average abundance . . . . it has now become clear that the cover value of species presents a much better expression of their sociological significance'. Braun-Blanquet then defined cover value of a species as 100 times the sum of the average cover value of that species in a phytosociological table, divided by the number of relevés. Note that this is in fact a new index, which gives a measure for the performance of a species in a plant

Table 2. Angular transformation values for some currently used cover-abundance scales. Combined transformation values (see text) are added.

cover %	arcsin transf.	Hult-Sernander			Braun-Blanquet			Daubenmire Baily-Pouton			Domin			Doing			Barkman et al.			Braun-Blanquet diff. symbol 2			cover %
		arcs tr.	orig scale	comb tr.	arcs tr.	orig scale	comb tr.	arcs tr.	orig. scale	comb tr.	arcs tr.	orig scale	comb tr.	arcs tr.	orig scale	comb tr.	arcs tr.	orig scale	comb tr.	arcs tr.	orig. scale	comb. tr.	
																	1	r	1				
					1	r	1	1	1	1	1	+	1	1	r	1	1	+r	1	1	r	1	
					(0)												1	+p	1				
1					1	+	2				1	1	2	1	p	2	1	1p	2	1	+	2	1
					(0)												1	+a	2				
2	1										1	2	3				1	1a	3				2
		1	1	3										1	a	3				1	1	3	
3					1	1	3	1	2	3	1	3	4							(2)			3
					(2)												1	1b	4				
4														1	m	4	(2)						4
																	1	2m	5	1	2m	4	
																	(2)			(2)			
5	2										2	4	5										5
		2	2	5										2	01	5	2	2a	5	2	2a	5	
10					3	2	5	3	3	5													10
20	3	3	3	6							3	5	6	3	02	6	3	2b	6	3	2b	6	20
30	4										4	6	7	4	03	7	4	3a	7				30
40	5	5	4	7	5	3	7	5	4	7										5	3	7	40
											5	7	7	5	04	7	5	3b	7				
50	6													6	05	7							50
60														6	06	8	6	4a	8				60
	7				7	4	8	7	5	8	7	8	8							7	4	8	
70														7	07	8	7	4b	8				70
		8	5	8																			
80	8							8	6	9	8	9	9	8	08	9	8	5a	9				80
90	9				9	5	9							9	09	9				9	5	9	90
								9	7	9	9	10	9				9	5b	9				
100														9	10	9							100
1	2	3	4	5	6	7	8	9	10	11	12	13	14	15	16	17	18	19	20	21	22	23	24

community, which is not identical to the group values of Tüxen & Ellenberg (1937).

The Braun-Blanquet transformation and a similar one by Etter (1949, also mentioned by Becking 1957) has been used frequently in classical phytosociology (see Ellenberg 1956, Mueller-Dombois & Ellenberg 1974). Schwickerath's scale has nevertheless been preferred by Westhoff (1947), Meyer Drees (1949), Sissingh (1950), Londo (1971) and others.

In later approaches (e.g. Dagnelie 1960, Barkman et al 1964, van der Maarel 1966, Londo 1971) transformed scales similar to Schwickerath's were suggested, while in recent numerical treatments (e.g. Moore 1966, Schmid & Kuhn 1970) extra weight was given to the low abundance values of the scale. Coetzee & Werger (1973) proposed an intermediate transformation. A logarithmic transformation of the Braun-Blanquet (1964, 1965) average cover values is suggested by Jensén (1978) in the form of (1 + log cover %) and in some other forms. Logarithmic transformation was also applied by Hogeweg (1976) for reasons similar to that of Jensén, i.e. to find a balance between preponderance of very abundant versus rare species.

Table 3 presents the various transformations mentioned so far, as well as a scale suggested by Barkman (col. 11, pers. comm.) resembling that of Schmid & Kuhn (col. 12). The angular transformation as suggested earlier is added in col. 16.

A new numerical cover-abundance scale

At our department a differentiated Braun-Blanquet scale has been introduced for use at courses (Westhoff & van der

Table 3. Transformations of the Braun-Blanquet cover-abundance scale.

Braun-Blanquet orig. scale	Tüxen-Ellenberg 1937	Braun-Blanquet transf. 1964	Etter 1949	Londo 1971	van der Maarel 1966	Dagnelie 1960	Schwickerath 1931	Coetzee & Werger 1973	Barkman et al 1964	Barkman ms	Schmid-Kuhn 1970	Moore 1966	Jensén 1978	Combined transf.	angular transf. van der Maarel 1966	Braun-Blanquet
r	()	()	0	0	0	()	()	1	0	0.5	2	10	1,00	1	0(1)	r
+	0.1	0.1	1.0	0	0	0.2	0.25	5	2	1	4	10	1,00	2	1	+
1	2.5	5.0	10.0	0	0.5	1	1	10	3	2	6	11	1,69	3	2	1
2	15	17.5	20.0	1	1	2	2	20	6	4	7	12	2,25	5	3	2
3	37.5	37.5	37.5	3	2	3	3	30	8	8	8	13	2,58	7	5	3
4	62.5	62.5	62.5	5	4	4	4	40	9	10	9	14	2,79	8	7	4
5	87.5	87.5	87.5	7	5	5	5	50	10	10	9	15	2,94	9	9	5
1	2	3	4	5	6	7	8	9	10	11	12	13	14	15	16	17

Maarel 1978). In this scale a differentiation of the original Braun-Blanquet category 2 is adopted from Barkman et al. (1964): *2m* = cover 5%, abundance high, *2a* = cover 5–12%, *2b* = cover 12–25% Table 2, col. 21–23. This scale is an effective compromise between the original Braun-Blanquet scale, which is relatively insensitive in the range covered by category 2, and elaborate but laborious scales like the one of Barkman et al (1964).

This differentiated alpha-numerical Braun-Blanquet scale (a) may be replaced by the fully numerical 1–9 scale (b) as follows:

(a)	*r*	+	1	*2m*	*2a*	*2b*	3	4	5
(b)	1	2	3	4	5	6	7	8	9

This transformation is added to Table 2, col. 23. In comparison with the angular transformation, col. 21, this new scale can be considered a combination of a cover scale in angular transformation, with a weighting based on abundance. This weighting is maximal in the Braun-Blanquet interval 2 where high abundance values are supposed to be of relatively high importance. We may call this new scale a combined transformation, since the original idea in Braun-Blanquet's combined estimation is maintained.

The points on this scale are by no means 'equidistant'. Still, it expresses differences in the performance of species in an efficient way while keeping close to the original and ingenious idea of Braun-Blanquet. In many classification and ordination treatments at our department (cf. van der Maarel 1969, Westhoff & van der Maarel 1978, Kortekaas, van der Maarel & Beeftink 1976, van der Maarel, Janssen & Louppen 1978) this 'ordinal transform' scale, proved to be satisfactory.

Obviously the scale is not only applicable in transformations of existing data collected along other scales, it can also be used directly in the field. For many students without experience with the original Braun-Blanquet symbols the new scale appears easy to learn and moreover easy to use in the field. An additional advantage of the scale is that the raw data can be easily entered into the data-processing system in use. Therefore this scale may be recommended for further use.

Comparison of transformations.

The most important aspect in which the transformations listed in table 3 differ, is no doubt the ratio between the transformed values the Braun-Blanquet symbols + and 5 (symbol *r* is difficultly to include in the comparison because this category is often converted to zero cover, or it is not distinguished from category +). This ratio varies as follows: 0 (Londo 1971), 0,01 (Braun-Blanquet 1964 average cover values), 0,05 (Schwickerath 1931), 0,10 (Coetzee & Werger 1973, see also Campbell 1978), 0,22 (combined transformation), 0,52 (log transform), 0,67 (Moore 1966) and 1,00 in the presence-absence approach. Clearly the possible influence of species with low abundance or cover values against species with high cover values in any numerical comparison of relevés or species groups increases from very little to equal influence.

The shortening or lengthening of the range of the numerical values obtained in the various transformations may now be simulated by applying an exponential transformation of the general form

$$y = x^{w} \qquad (2)$$

where x = the original value and y its transform, for different values of w. It appeared that when taking the ordinal transform values from 1–9 as original values many of the existing transformations could be approached rather well with values for w varying from 0 to 4. See Table 4. (Of course, for each other series of starting values e.g. cover % values, a similar series of w-values can be found).

The deviations occurring within the range *2m–2a–2b* are acceptable. The Jensén and Coetzee & Werger scales are somewhat overestimated in the lower range, but this is consistent with the implied emphasis of these scales. With the exponential transformation an easy comparison can be made of the effects of various transformations on com-

Table 4. Approximation of the transformation scales from Schwickerath, Tüxen & Ellenberg, based on Braun-Blanquet (1964), Moore, Coetzee & Werger and Jensén, by exponential transformation of the 1–9 combined transformation scale. For easy comparison the exponential transforms are multiplied by a factor as to make the highest values corresponding to the Braun-Blanquet value 5 equal. The transformation function used here is of the form $y = ax^{w}$, which is explained in the Discussion. Average cover % values for *2a*, and *2b* are taken 8 % and 18 % and values for 1 and *2m* are arbitrarily chosen with regard to Tüxen & Ellenberg (1937) and Braun-Blanquet (1964). Schwickerath -and Moore- values for *2m*, *2a* and *2b*, and Coetzee & Werger values for *2m* are newly suggested.

Braun-Blanquet scale diff. symbol 2	Presence-absence transform.	$y = x^{0}$	Moore transform.	approx. $y = 8.67\,x^{0.25}$	Jensén transform. 1 + log. cover %	approx. $y = 0.98\,x^{0.50}$	combined transformation	$y = x^{1.00}$	Coetzee & Werger transform.	approx. $y = 1,852\,x^{1.50}$	Schwickerath transform.	approx. $y = 0.62\,x^{2.00}$	Tüxen & Ellenberg average cover %-values	approx. $y = 0.01333\,x^{4.00}$
r	1	1	10	8.7	1.00	0.98	1	1	1	2	( )	0.06	0.02	0.01
+	1	1	10	10.3	1.00	1.39	2	2	5	5	0.25	0.25	0.1	0.2
1	1	1	11	11.4	1.39	1.70	3	3	10	10	1	0.56	2.5	1.1
2m	1	1	11.5	12.2	1.69	1.96	4	4	13	15	1.5	1.0	5.0	3.4
2a	1	1	12	13.0	1.90	2.24	5	5	17	21	2	1.6	8.75	8.3
2b	1	1	12.5	13.6	2.25	2.32	6	6	24	27	2.5	2.2	18.75	17.3
3	1	1	13	14.1	2.58	2.60	7	7	30	34	3	3.0	37.5	32.0
4	1	1	14	14.6	2.70	2.77	8	8	40	42	4	4.0	62.5	54.6
5	1	1	15	15.0	2.94	2.94	9	9	50	50	5	5.0	87.5	87.5

munity similarity: either patterns of relevé similarity or of homotoneity levels of clusters (cf Westhoff & van der Maarel 1978). The quantitative data on species need only to be stored in ordinal transform values and on applying the appropriate value of the exponent w the required transformation is obtained. Incidentally, this may be considered an additional motive for choosing the 1–9 ordinal transform directly as a field scale. If the original Braun-Blanquet values are available, it is advised (table 2) to take ordinal transform 5 for Br Bl symbol 2. Other field scales, notably those of Hult – Sernander and Domin can be transformed to the 1–9 scale without loss of much accuracy. (Table 2).

It is then possible to apply a series of transformations in the sequence $w = 0$ (presence – absence), $w = 0{,}25$ or 0,50 (emphasis on presence), $w = 1$ (intermediate, presence and dominance balanced) $w = 1{,}50$ or 2,00 (emphasis on dominance) and $w = 4{,}00$ (strong emphasis on dominance). Indeed, we have programmed in our department such a serious of transformations which can be performed before applying one of the standard numerical analyses.

Dr. R.S. Clymo (Westfield College, London) devised a similar transformation function. It was primarily meant for cover % values, but it can used for all other values. Dr. Clymo agreed in presenting his formula in this paper for further comparisons. It reads:

$$y = \frac{1-e^{-ax}}{1-e^{-a}} \qquad (3)$$

where e is the base of natural logarithms, a is the transformation exponent and x the species score, which is supposed to range from 0 to 1. With this formula values for y are always between 0 and 1. When using cover values as proportions of unity a value of a near to 0 will result in a nearly linear relation between y and x, i.e. the x values remain largely untransformed; for large values of a a presence – absence transformation is approached, for $a = 4$ the transformed values are approximately Hult-Sernander values.*

* While completing another manuscript in which transformation of cover-abundance values is discussed (van der Maarel 1979) I came across exactly the same exponential transformation as I devised, viz. in an attempt by Baum (1977) to reduce the dimensionality of a dissimilarity matrix of taxonomically compared individuals. (This effect is also discussed in the present paper). Moreover Baum suggested an alternative reminding Clymo's function: $y = e^{-ax} - 1$!

Effects of transformation of cover – abundance values on the numerical relations between relevés

Multivariate methods used

As in the studies by Jensén (1978) and Campbell (1978) the effects of data transformations were investigated by applying some multivariate methods to concrete phytosociological datasets, viz. the *Arrhenatheretum* material provided by Mueller-Dombois & Ellenberg (1974), and the *Glauco-Puccinellietalia* and *Spartinetea* selections of the Working Group for Data – Processing, which were also used for the elucidation of the table rearrangement program TABORD (van der Maarel, Janssen & Louppen 1978). In all cases both a classification and ordination technique were used. The classification technique was Ward's clustering technique, and agglomerative clustering provided in the CLUSTAN package (Wishart 1975) which was used with Euclidean distance as a dissimilarity coefficient. The ordination technique was ORDINA (cf Roskam 1971, van der Maarel 1972a, van der Maarel, Janssen & Louppen 1978) a principal components analysis on a Euclidean distance matrix. The transformations used are based on the ordinal transform running from 1–9 and include values of the exponent $w = 0; 0{,}25; 1{,}00; 1{,}50; 2{,}00$ and 4,00. These transformations will be indicated as a, b, c, d, e and f respectively.

Any interpretation of the results obtained with various transformations should lead to the ranking of the transformations as to the effectiveness of the classification and ordination results they produce. Effectiveness can be evaluated in terms of the structure of the cluster dendrogram, discrimination between clusters at one particular level, separation of clusters of relevé points in an ordination space, the phytosociological and environmental interpretability of the results, and the agreement of the results with already established knowledge about the plant communities involved. The present study will concentrate on the latter.

Example Arrhenatheretum

This example involves 25 relevés of *Arrhenatheretum* grasslands in Southern Germany which has been used by Ellenberg (1956) to demonstrate the manual technique of constructing a phytosociological table and by Mueller-Dombois & Ellenberg (1974) to demonstrate various numerical treatments. We have also used this material to demonstrate our table rearrangement program TABORD (van der Maarel, Janssen & Louppen 1978). The table and ordination diagram presented in the latter paper is used as a reference for the following discussion.

The *Arrhenatheretum* material contains relevés of three types (subassociations), one with *Bromus erectus* and other xerophilous species (rel. 1, 3, 4, 9, 10, 15, 24), one with *Cirsium oleraceum*, *Deschampsia caespitosa* and other wetland species (rel. 6, 7, 8, 11, 14, 17, 19, 21), and one intermediate type in which *Helictotrichon pubescens* with *Trisetum flavescens* and other mesophilous species find an optimum (rel. 2, 5, 12, 13, 16, 18, 20, 22, 23). There is one deviating relevé joining the *Cirsium oleraceum* group but having five species of its own (ref. 19). These three types will be named *Bromus* –, *Cirsium* – and *Geum* type according to the indication by Mueller – Dombois & Ellenberg (1974). Fig. 1 shows the dendrogram resulting from Ward's clustering method with data transformation c, $w = 1$, i.e. the 1–9 ordinal transform. Fig. 2 presents the results of and ORDINA – ordination, for dimension 1 and 2, applied to each of the six transformations of the data matrix. For the interpretation of the results we make use of Table 5 which summarizes some relevant information on the relevés, and Table 6 which presents some characteristics of the clusters arrived at.

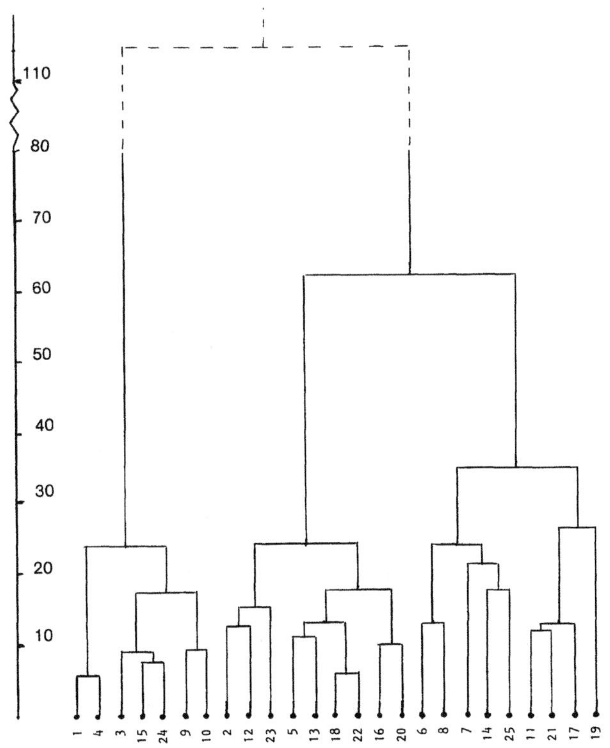

Fig. 1. Dendrogram of 25 *Arrhenatheretum* relevés clustered with Ward's method, with transformation c ($w = 1.00$) applied to the species scores.

Classification

The dendrograms for the six transformations are similar in overall structure. Fusions occur at regular intervals showing no obvious discontinuity where clusters could be isolated, except for the final step: In all cases two main groups result, one with the *Bromus erectus* relevés and one including all the others. This confirms the conclusion on the main structure of the data set reached by van der Maarel, Janssen & Louppen (1978). It is still interesting to compare the cluster sets obtained on the basis of different transformations. Two levels were arbitrarily chosen – one with 8 clusters and one with 3.

On the 8-cluster level (and the 9-cluster level with transformation a, because comparatively rapid fusions occur here, from 9 down to 3 clusters) the ordinal transformation c shows the best result in terms of the phytosociological interpretations of the resulting clusters: 7 out of 8 clusters have 'exclusive constants', i.e. species with a constancy $>$ 75 % in that cluster and a constancy of $<$ 75 % in all other clusters. The total number of such species is 20, only one less than in the b situation. Transformations b and d produce slightly less and the remaining three much less satisfactory results than c. Furthermore the c-transformation is the only one at which the deviating relevé 19 and no other relevés occur as a single – relevé cluster.

On the 3-cluster level the resulting grouping may be compared with the main division into three types by Ellenberg. Again the c- result is in complete accordance: all 25 relevés placed in the group they were originally assigned to. Now a reasonable second position is taken in by transformations d and e.

The two extreme transformations, i.e. 'presence-absence' and 'Tüxen – Ellenberg dominance weighting' produce unsatisfactory classifications.

Ordination

Fig. 2a–f shows how the configuration of points changes from widely spaced at transformations a and b to more concentrated at f. This concentration in the central part of the diagram is accentuated by the extreme position of relevé 2. The wide scattering of points in the first two diagrams should not be interpreted as an expression of the effectiveness of the corresponding ordinations. On the

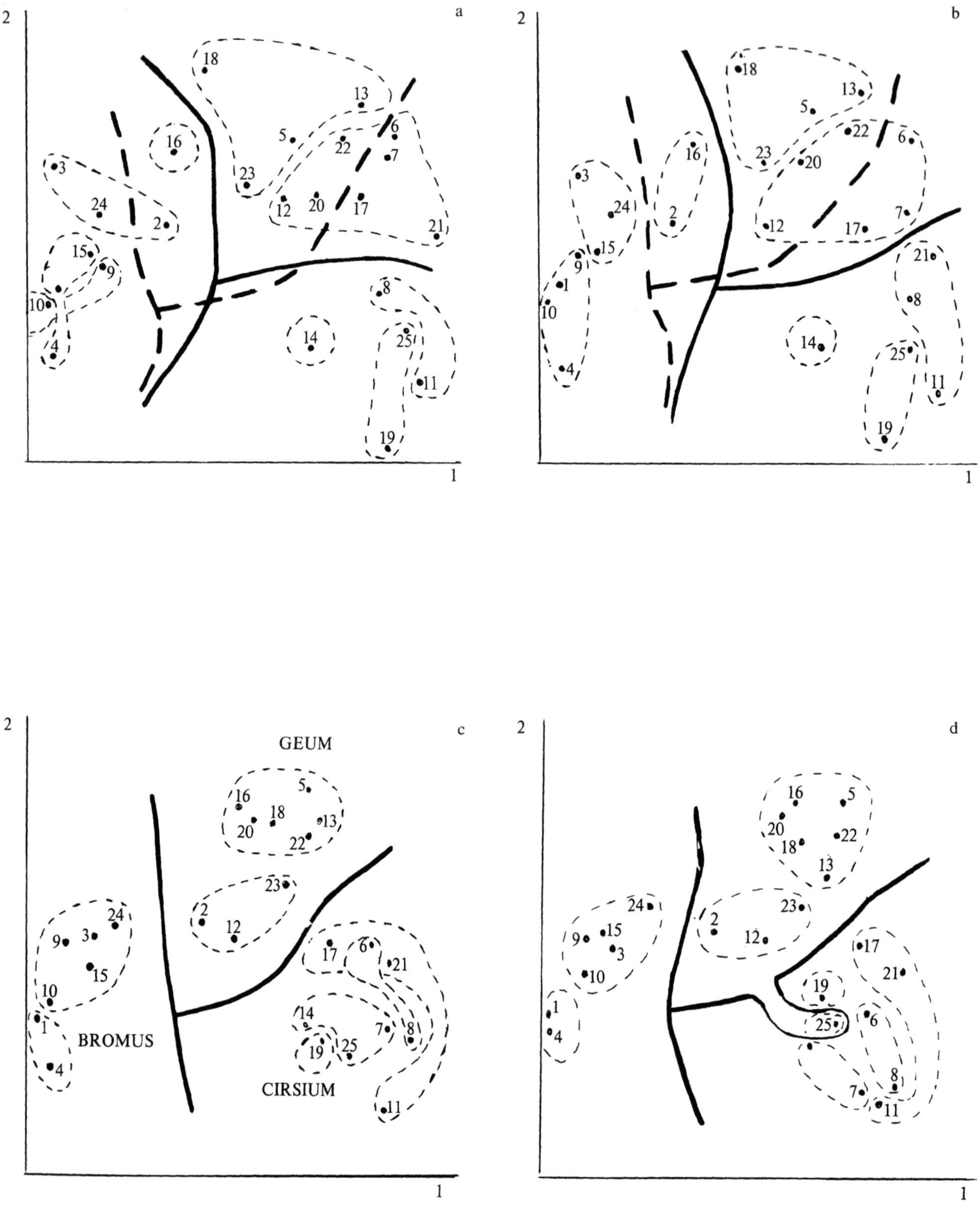
a
b
c
d
GEUM
BROMUS
CIRSIUM

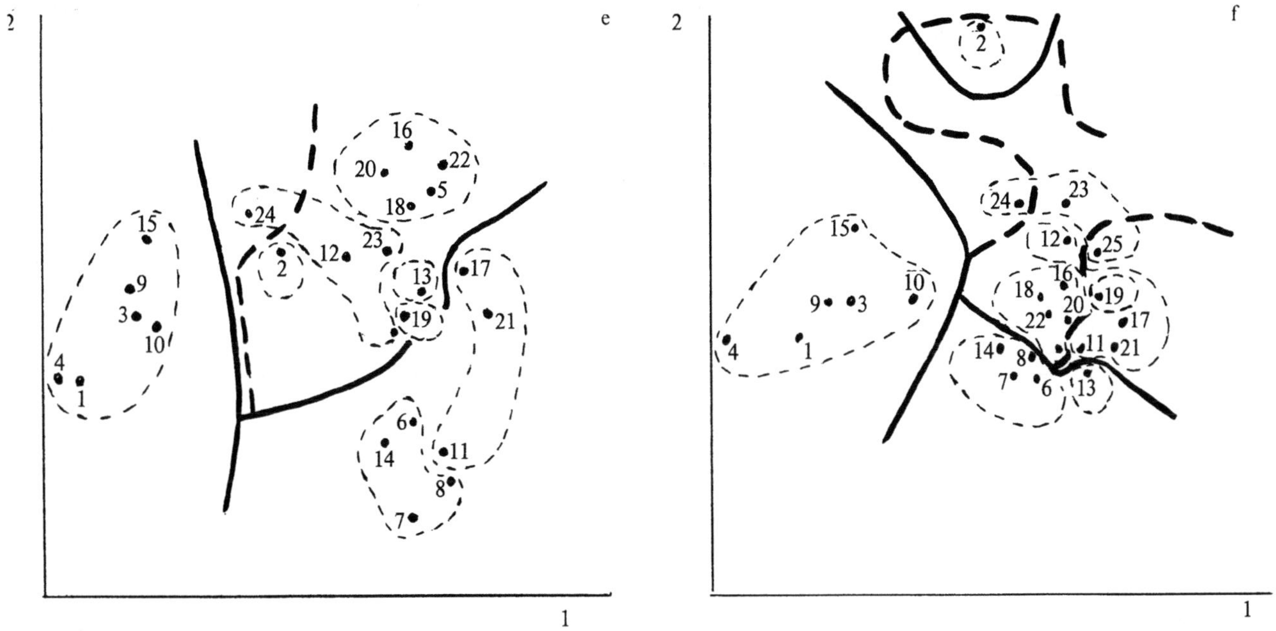

Fig. 2. a/f. Position along dimensions 1 and 2 of an ORDINA-ordination of 25 *Arrhenatheretum* relevés (positive and negative loadings reversed if necessary to obtain comparable structures.a: after exponential data transformation with $w = 0$, b: ibid. $w = 0.25$; c: ibid. $w = 1.00$, d: ibid. $w = 1.50$; e: ibid. $w = 2.00$; f: ibid. $w = 4.00$. Clusters are obtained with Ward's method and presented at the 8 cluster level (case *a* at the 9-cluster lvel) with boundaries in thin interrupted lines, at the 3-cluster level with thick lines. Correct cluster structure as to main division into three types by Ellenberg is indicated with thick interrupted lines (if deviating from 3-cluster system); the transformation *c* gave the correct situation.

contrary: the extracted variance per dimension is very low at lower values of *w* and increases with increasing weight on dominance. In the first case, when only presence and absence are concerned we can imagine that with a total species number of 94 many more dimensions are needed to explain most of the variation involved. The same trend towards lower dimensionality at high dominance levels has been reported by Jensén 1978 for species-poor lake vegetation, be it on a much higher level of extracted variance.* On the other hand by the heavy weighting of dominant species the total variation in the transformed data matrix is mainly contributed by, the 12 species occurring at least once with ordinal transform values 6, 7 or 8. 46 % extracted variance is reached with the first two dimensions. However, the variation due to the species with lower cover – abundance values is obscured. (Austin & Greig-Smith 1968 observed a similar trend with species occurring at high density levels).

The presence – absence ordination diagram shows a clear horse shoe configuration, along which the environmental transition occurs from the relatively very dry sites with the *Bromus* type (e.g. relevés 4 and 10) and the mesic sites with the *Geum* type (e.g. rel. 5, 13 and 18) to the moist sites with the *Cirsium* type (e.g. rel. 8 and 25) and the wet site of rel. 19 with *Glyceria fluitans* and *Phalaris arundinacea*. The most dissimilar pair of relevés in this case is 10 and 19.

If the results of the clustering are combined with that of the ordination, the transformations *b* and *c* can be considered the most effective ones, in the sense that in the middle range of the transformations the two different multivariate methods using the same dissimilarity matrix produce coherent structures.

Example Spartinetea

This example is concerned with 22 relevés taken from the selection *f* of 50 relevés used by the the Working Group for Data – Processing. The full selection is treated by van der Maarel, Janssen & Louppen (1978, tables 2 and 3). This selection was treated in the same way as the previous *Arrhenatheretum* material. Table 7 and Fig. 4 present

* (See Baum 1977!)

Table 5. Information on species composition of 25 Arrhenatheretum relevés which is relevant for the interpretation of Fig. 2 and Table 6. Transformations compared are as in Fig. 2.

Relevé nr.	1	2	3	4	5	6	7	8	9	10	11	12	13	14	15	16	17	18	19	20	21	22	23	24	25
Number of species	31	25	32	28	32	32	26	34	32	31	28	34	39	32	33	37	35	29	29	30	35	27	29	30	28
Nr. of species with value 7x8 or 9	1x7	1x8	1x7	1x8					1x7		1x7	1x7	1x7		1x7		1x7		1x7	1x7			1x7		1x7
Nr. of species with value exceeding other values of the species with 2 or more scale values	1	2					2	1			1		2	1				1	3					1	2
Nr. of species occurring 1x or 2x		1		1			2		1		2	3	2	2		3	2	2	7		2		1	3	2
Occurrence as single relevé cluster in 8-cluster-structure at transf. a-f		e f									f		e f	a b		a			c f						
Extreme position along ordination axis 1, 2 or 3		f:2		f:1										a:3					a:2 d:3 e:3						
Strong move in successive ordin.		x		x		x	x			x									x						
Nr. of misclassif. in 3-cluster syst.		3				2	2				1		1			1	2		1		2			2	3

Table 6. Characteristics of the clusters arrived at on the 8-cluster and 3-cluster levels with Ward's clustering method applied to 25 *Arrhenatheretum* relevés (Ellenberg 1956, Mueller-Dombois & Ellenberg 1974) and figures for extracted variance per dimension in ORDINA-ordinations, for six data transformations.

Transformation	a	b	c	d	e	f
Value of w	0	0.25	1.00	1.50	2.00	4.00
Nr. of clusters compared at 8 cluster level	9	8	8	8	8	8
Nr. of clusters with at least one constant and exclusive species	5	6	7	7	5	5
Total nr. of constant-exclusive species	9	21	20	17	12	12
Nr. of clusters compared at 3 cluster level	3	3	3	3	3	4
Nr. of relevés classified correctly according to Ellenberg's classification	19	20	25	24	24	19
Extracted variance dimension 1%	19,5	21,5	25,3	26,0	26,8	31,0
2	9,0	10,0	13,1	13,5	13,0	14,9
3	6,6	6,7	9,2	10,2	10,4	10,7
1+2+3	35,1	38,2	47,6	49,7	50,2	56,6
Most dissimilar pair of relevés	10-19	10-19	1-19	1-19	1-19	2-4

information on the relevés. The comparison of ordination and classification results is based on the results obtained from the total selection by applying the program TABORD. In this treatment 11 clusters were derived. Two of these could be subdivided in view of the quantitative differences in the occurrence of some species. The result is a system of 13 elementary clusters.

On a higher level 6 main clusters may be distinguished, four of which being characterized by the dominance of one *Spartina* species. The clusters dominated by *Spartina alterniflora* (rel. 5, 6) and *S. patens* (rel. 7, 8) are clear cut and so is the cluster with the three *S. townsendii* dominated relevés (rel. 9, 10 and 11), but two of the *S. maritima* relevés (rel. 3 and 4) are only slightly less similar to the remaining group of *S. townsendii* relevés than to the 'pure' *S. maritima* relevés (rel. 1 and 2) The structures with 13 clusters and 6 clusters are described in table 7. (On the 13-cluster level the number of clusters involved in the comparison is 12 in four cases, and at the 6-cluster level it is seven in three cases, see Fig. 3).

Table 8 presents information on the clusters in a similar way as Table 6. Since the clusters on the 13–cluster level are small no characteristics of the clusters are given. At the 6-cluster level a relevé is considered correctly 'placed' if

Table 7. Information on 22 *Spartinetea* relevés. M = *Spartina maritima* cluster, Mp = *Spartina maritima-Puccinellia maritima* cluster, A = *Spartina alterniflora* cluster, P = *Spartina patens* cluster T = *Spartina townsendii* cluster, Ts = *Spartina townsendii- Suaeda maritima* cluster; *Spm* = *Spartina maritima; Pum* = *Puccinellia maritima; Sap* = *Salicornia perennis: Spa*= *Spartina alterniflora; Spp* = *Spartina patens; Spt* = *Spartina townsendii; Sae* = *Salicornia europaea; Sum* = *Suaeda maritima; Ath* = *Atriplex hastata.*

Relevé nr.	1	2	3	4	5	6	7	8	9	10	11	12	13	14	15	16	17	18	19	20	21	22
Nr. in selection f	2	4	6	9	12	15	16	17	21	23	24	27	29	32	33	34	39	40	44	45	47	50
13-cluster level TABORD	1	1	8	2	3	3	4	4	5	5	5	9	10	7b	7b	10	11	11	6a	6b	7a	7a
6-cluster level	M	M	M_p	M_p	A	A	P	P	T	T	T	T_s	T_s	T_s	T_s	T_s	T_s	T_s	T_s	T_s	T_s	T_s
Prevailing species	Spm	Spm	Pum Sap Spm	Sap Pum Spm	Spa	Spa	Spp	Spp	Spt	Spt	Spt	Pum Sae Spt	Pum Sae Spt	Sae Spt	Sae Spt	Sae Pum Spt	Sum Spt	Sum Spt	Ath Pum Spt	Ath Spt	Spt Sum	Spt Sum
Nr. of species	1	4	6	6	1	4	3	6	3	2	1	6	5	4	5	7	6	4	8	3	5	4
Species with value 7, 8 or 9	9	8	8,7	8	8	9	9	9	7	9	9	8	7	8	7	7	8	9	7	9		
Nr. of species occurring 1x						11		2											2	1		
Occurrence as single-relevé cluster at transf. a-f in 13 cluster structure		a				a	a,b	a,b											a f	a f		
Extreme position along ordination axis 1, 2 or 3						a:3 b:3	f:1 2	f:1 2		f:1	f:1											
Strong move in successive ordinations			x	x	x	x	x	x		x	x									x		
Nr. of misclassifications in 6 cluster system			3	3					3	1	1									1		

the cluster to which it is assigned is either identical with one of the six clusters as determined by the dominant *Spartina* species, or included in such a cluster.

If we consider the similarity between the classification at the two levels with the respective classifications considered best on the basis of earlier phytosociological interpretations, the optimum transformations for the 13-cluster level are *d* and *e*, while for the 6 cluster level *c*, *d* and *e* are acceptable, but *c* is better.

The best agreement between the classification and ordination structures is found at transformation *d*, although not as good as in the best case of the *Arrhenatheretum* example. As could be expected in this example transformations with an emphasis on dominance, and also the ordinal transformation produce acceptable results while, as in the *Arrhenetheretum* example, those with an emphasis on presence, fail. However, the transformation with the strongest emphasis on dominance, i.e. transformation *f*, is also less effective.

This pattern can be explained as follows: In some clusters the dominant *Spartina* species occurs with others species, in some clusters it does not (Table 7). In a presence – absence transformation the companion species change the position of the corresponding relevé, as is the case with for example relevé 4 as compared with rel. 2 and rel. 6 as compared with rel. 5. On the other hand clusters with the same prevailing species which varies in cover – abundance value, will become more heterogeneous with higher

Table 8. Characteristics of the clusters arrived at with Ward's method applied to 22 *Spartinetea* relevés and figures for ordination results. (See Table 6).

Transformation	a	b	c	d	e	f
Value of w	0	0.25	1.00	1.50	2.00	4.00
Nr. of clusters compared at 13 cluster level	13	12	13	12	12	12
Nr. of identical clusters	2	3	8	10	10	8
Nr. of relevés placed correctly according to 13 cluster structure	3	4	13	17	17	12
Nr. of clusters compared at 6 cluster level	7	6	7	6	7	6
Nr. of identical clusters.	1	4	4	3	4	3
Nr. of relevés placed correctly according to 4 cluster structure	8	8	17	15	15	11
Nr. of clusters with at least one constant-exclusive species	3	5	6	6	7	5
Extracted variance dimens ion 1%	33,7	33,2	24,3	20,4	18,9	20,2
2	15,8	17,6	20,0	19,5	18,3	18,0
3	11,7	11,5	11,9	12,8	13,6	13,8
1+2+3	61,2	62,3	56,2	52,7	50,8	52,0
Most dissimilar pair of relevés	8-19	8-19	3-8	3-8	8-20	8-20

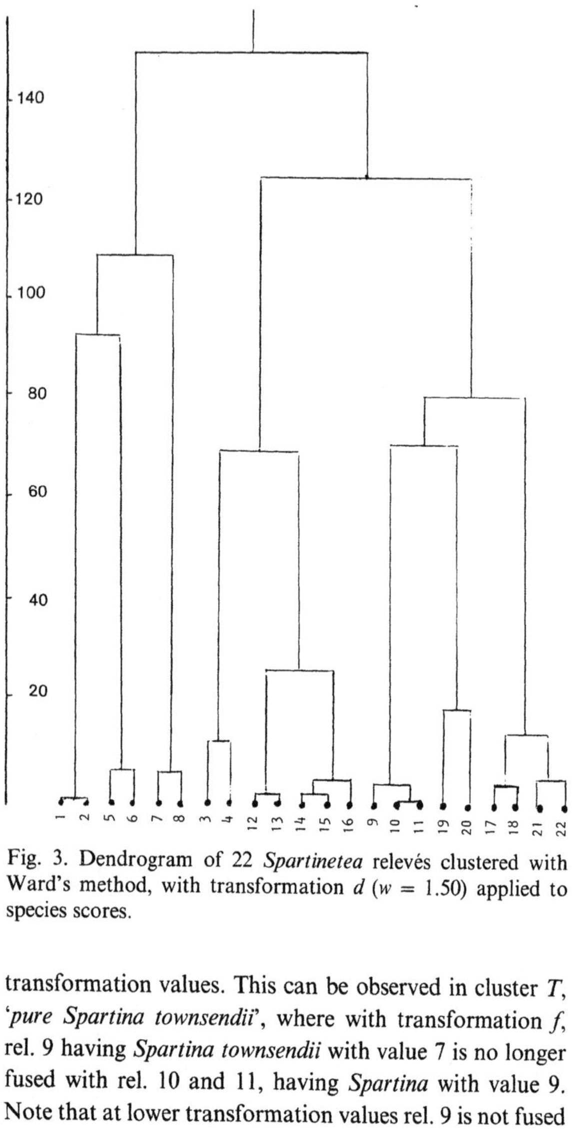

Fig. 3. Dendrogram of 22 *Spartinetea* relevés clustered with Ward's method, with transformation *d* ($w = 1.50$) applied to species scores.

transformation values. This can be observed in cluster *T*, '*pure Spartina townsendii*', where with transformation *f*, rel. 9 having *Spartina townsendii* with value 7 is no longer fused with rel. 10 and 11, having *Spartina* with value 9. Note that at lower transformation values rel. 9 is not fused with rel. 10 and 11 either, because it has two extra species.

The effect of the transformations on the general structure of the ordination diagram is different from that obtained with the *Arrhenatheretum* selection. The percentage of extracted variance increases only from transformation *a* to *b* and then decreases again. This could be explained by the interesting fact that the most dissimilar pair of relevés at lower transformation values is 18 and 19 and at higher values 8 and 20, while relevés 19 and 20 are both characterized by *Atriplex hastata*, which species does not occur in any other relevé involved. Relevés 8 and 20 each have one species (*Juncus maritimus* and *Scirpus maritimus* respectively) with cover – abundance value 5, lacking in all other relevés. Weighting these species more heavily as the total variation in the data-set increases, provides an explanation for why more axes are needed.

Example Glauco – Puccinellietalia

This example includes 23 relevés taken from selection *e* of 58 relevés used by the Working Group for Data – Processing (see van der Maarel et al 1978, tables 7 and 8). The same treatments were applied as in the other examples. Information on the relevés and the treatments is presented in Tables 9 and 10. Only one classification was used as a reference. This is an 8-cluster classification which is based on (1) the results of a TABORD version used in the TABORD treatment of selection *e* (van der Maarel, Janssen & Louppen 1978), (2) Feoli's (1977) results and (3) the original phytosociological classification applied by the authors of the tables from which selection *e* was taken (see Feoli 1977 for details)). Six of these clusters can be named according to the original syntaxonomical units, two deviating relevés, 22 and 23, have been named clusters X and Y (see Table 9). The overall similarity between the 8-cluster classification and the series of classifications with Ward's analysis is less than in the two previous examples, mainly because of the deviating position of relevés 21, 22 and 23 which are obviously misclassified in the 8-cluster

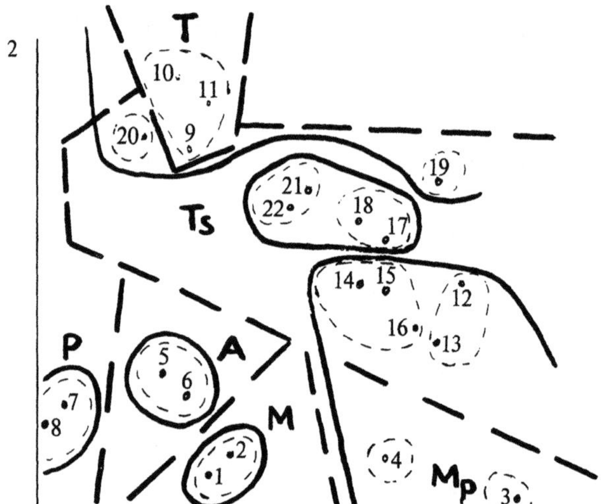

Fig. 4. Position along dimensions 1 and 2 of an ORDINA-ordination of 22 *Spartinetea* relevés with transformation *d*. See for the selection and abbreviations Table 7. The lines refer to the 12-cluster-and 6-cluster configuration with transformation *d* (thin interrupted and thick lines) and to the 6-cluster structure as obtained with TABORD (thick interrupted lines).

Table 9. Information on 23 *Glauco-Puccinellietalia* relevés.

Relevé nr.	1	2	3	4	5	6	7	8	9	10	11	12	13	14	15	16	17	18	19	20	21	22	23
Nr. in selection c	2	3	11	17	18	19	20	20	27	28	30	31	35	38	41	42	45	47	49	50	53	54	55
Prevailing species	Pum	Pum Liv	Jug	Scr	Scr	Pum Ast	Pum	Jug Glm	Fer	Fer	Fer Atm	Fer Atm	Pum	Jug	Jum	Jum	Pum	Jug	Cae Gem	Scm	Jug Gem	Jug	Pas Arm
8 cluster level	P	P	J	S	S	P	P	J	F	F	F	F	P	J	M	M	P	C	C	C	J	X	Y
Nr. of species	8	9	6	14	14	3	5	9	10	9	12	15	3	19	16	9	4	15	11	10	9	12	13
Species with value 7, 8 en 9	9	8,7	9	7	-	9,7	9	9,7	8	8	2x7	2x7	9,7	9	8	8	9	8	2x7	8	3x8 2x7	8,7	8,7
Nr. of species occurring 1x			1	4	2		1		1		4	4			2	1		1		1			2
Occurrence as single relevé cluster at transf. a-f in 8 cluster structure											a									f	c d e		e f
Extreme position along ordination axis, 1, 2 or 3											a:3 b:2												e:3
Strong move in successive ordinations																						x	x
Nr. of misclassifications in 8 cluster system			2	2	2	1	1	2	1	1			1	2	2	2	1	2	1	1	6	6	4

system. Transformation *e* is slightly superior to the others, but transformation *d* is not far behind. The agreement between the classification and ordination results is very good at transformation *d*, less at *e* and *f*, and much less at *a*, *b* and *c*. Apparently the structure of this data-set, with the total number of high cover-abundance scores high, is most optimally approached with a moderate emphasis on dominance, similarly as in the *Spartinetea* example.

Table 10. Characteristics of the clusters arrived at with Ward's method applied to 23 *Glauco-Puccinellietalia* relevés and figures for ordination results. (See Table 6).

Transformation	a	b	c	d	e	f
Value of w	0	0.25	1.00	1.50	2.00	4.00
Nr. of clusters compared at 8 cluster level.		8	8	8	8	8
Nr. of identical clusters.	0	1	3	5	5	4
Nr. of relevés placed correctly according to 8 cluster structure						
Nr. of clusters with at least one constant-exclusive species	4	4	7	7	7	7
Nr. of clusters compared at 4 cluster level.	4	5	4	4	5	5
Nr. of relevés placed correctly	6	17	22	23	19	19
Extracted variance dimension 1,%	20,4	21,1	23,3	26,1	29,4	40,7
" 2	14,1	14,3	15,2	16,8	18,1	22,6
" 3	10,4	11,1	12,4	11,1	10,6	8,6
" 1+2+3	44,9	46,5	50,9	54,0	58,1	71,9
Most dissimilar pair of relevés	11-15	11-23	11-23	11-21	11-21	8-17

Since no elaborated phytosociological classification of the *Glauco – Puccinellietalia* selection on a higher level of abstraction is available, only a superficial comparison can be made at the 4-cluster level. On this level cluster *P*, *Puccinellion maritimae* remains on its own. Of the other clusters, which could all be assigned to the *Armerion maritimae*, cluster *F*, *Armerio – Festucetum* of dry places, is also a cluster of its own, clusters *S*, *J*, *C* and *M*, representing related communities of wetter places, fuse, while the deviating clusters *X* and *Y* form a new cluster in which all three former groups are represented (See also Van der Maarel et al. 1978: table 8 and fig. 3). Because of the more complicated structure of this material the criterion for correct placing of a relevé was extended: (1) respective clusters are identical; (2) cluster forms part of a standard cluster to which it is compared (1 and 2 as in previous examples); (3) cluster overlap with standard cluster for more than 50 % of the relevés. The results (Table 10) show that the optimal transformation is *d* and transformation *c* is second. Thus as compared with the 8-cluster level we see a shift towards the lower transformation values.

Discussion

Weighting and standardization

Transformation of species performance data through weighting of cover – abundance figures is an important issue in numerical phytosociological treatments. This paper confirms in a general way earlier findings and statements about the dependence of both classification and ordination results on the transformation to which the data are subjected.

In view of the similarity between combined cover – abundance scales and the angular transforms of the cover % scales, and the possibility to approach various transformations with the exponential function, one might consider any cover – abundance scale as a transformation of some theoretical scale for species performance in phytocoenoses. This applies also to the still commonly used traditional field scales of Braun-Blanquet and Domin.

Standardization is a particular, but in our view a secondary form of data transformation. It is equally considered as a central problem in multivariate analyses (Dagnelie 1960, Austin & Greig-Smith 1968, Noy-Meir 1973, Noy-Meir & Whittaker 1978). Some effects of standardization resemble those of weigthing, as has become especially clear from the general studies on the ecological meaning of transformation by Noy-Meir, Walker & Williams (1975). From their description of the features of standardization the following may be summarized (see also Van der Maarel 1979):

– Standardization by species, e.g. by the sum of the performance values for each species in all relevés, implies the underweighting of differences in frequency of a species in a data – set and hence the overweighting of the rarer species.

– Standardization by site (we would say by relevé), e.g. by the sum of the performance values of all species for each relevé, implies the underweighting of differences in total cover (or cover and abundance) in the relevés. It 'weights' the relevés with a low cover (cover-abundance) total and consequently the species with high cover – abundance values are overweighted relative to other values in the relevé.

– No standardization implies a weighting (by nature of the multivariate methods used) of common species, especially when occuring together in group of relevés.

To these general features we may add that through standardization by species the influence of the dominant species is diminished. Thus the similarity between the effects of standardization and weighting is clear: standardization by species has effects similar to that of weighting with a low value of the transformation exponent w, and standardization by relevé resembles weighting with a high value of w. In this connection the use of an intermediate transformation has a similar effect as non – standardization.

Interestingly in all our examples and in those discussed by Jensén (1978) and Campbell (1978) the intermediate transformations lead to ecologically satisfactory results. Noy – Meir et al (1975) conclude that non – standardization, having a similar effect,' may seem appropriate for many phytosociological studies.' Indeed, when the general structure of a phytosociological data set is our main concern and we wish to include both relationships between sites and between species in our approach, as is essential in phytosociology, non-standardization or, similarly, intermediate weighting seems effective.

Obviously standardization and weighting are two different types of transformation, although some similarities in effect exist. The main difference is that with weighting the amplitude of the scale is changed and hence, in terms of standardization, the variance. On the other hand, standardization makes the data- set more uniform. Standardization by species may wipe out the influence of differences in species frequency, irrespective of the amplitude in cover – abundance value. As Noy-Meir et al (op. cit.) demonstrate this may be effective in a phytosociological approach where groups of faithful and differential species are sought (cf Goodall 1953, 1969). Standardization by relevé diminishes the influence of differences in total cover and abundance in a site and this may be ineffective where (Noy-Meir et al., op. cit.) 'the study is concerned primarily neither with species nor with sites but rather with the bulk of vegetation, the total plant cover or biomass, and the patterns of its distribution between species and stands'. Clearly, in such cases no standardization would be appropriate. It may be worthwhile studying the effects of standardization, in a phytosociological context first by sites and then by species, after an optimal weighting has been found, or in combination with different weightings. (cf Gauch, Whittaker & Wentworth 1977).

Finally weighting and standardization should be considered in relation to the type of resemblance function used in further multivariate analyses (cf Orlóci 1978). To mention one example: as van der Maarel (1979) pointed out the Euclidean distance is a very attractive resemblance measure, but it has the disadvantage that strong differences in the performance value of one or a few species may outweigh the similarity between relevés as to the joint occurrence of many species with equal but low performance values. In this case standardization by relevé, i.e. by performance value total, would cope with this problem. The disadvantage of this standardization as such (see above) could be overcome by a prior intermediate weighting. An additional advantage of this standardization is that the Euclidean distance becomes fixed in range, i.e. between 0 and 1.

In order to manipulate with weighting and standardization in relation to each other we may extend the exponential function by making use of a transformation function suggested by Orlóci (1978). Orlóci's function is of the form

$$y_{ij} = a_i x_{ij} \tag{4}$$

where y_{ij} is the transformed score x_{ij} of species i in relevé j and a_i is a standardization factor specific to species i. (Orlóci used other symbols and spoke of weighting here, but the examples he gave for the substitution of a, such as $a_i = 1/x_i$ refer to standardization in the sense used in the present paper).

The combined transformation function then becomes:

$$y_{ij} = a_i x_{ij}^w \tag{5}$$

(As was shown in Table 4 the factor a can also be used for adapting ranges of scales!)

Effects of transformation on different levels of hierarchic classification

It has been suggested (Smartt, Meacock & Lambert 1976, M.B. Dale and E. van der Maarel in pers. comm.) that in an hierarchical classification the weighting may be changed along with the level of abstraction. Broad groupings of relevés, the higher syntaxonomical units such as orders and classes, are characterized by combinations of species on a presence-absence basis. In the lower groupings, the lower syntaxonomical units such as associations and alliances, the quantitative differences between species become important. The finest subdivisions, the variants and facies (see Westhoff & van der Maarel 1978), are characterized mainly by dominance or prominence of one species: they may come out through strong weighting of cover-abundance values.

The results of the present study provide some evidence for differences in efficiency of the transformation used and the level of abstraction in the hierarchical classification, but there is no general tendency. In the *Arrhenatheretum* case the higher level in the hierarchy is optimally approached by a range of transformation values which are slightly higher than in the lower level situation. In the *Spartinetea* example there is hardly any difference. At both levels of classification a comparatively strong weighting gives optimal results. In the *Glauco-Puccinellietalia* example higher transformation values are slightly more efficient on the lower level of classification.

These results may still fit into the general picture if we take into account that the three examples represent different levels of syntaxonomical complexity of their own. The *Glauco-Puccinellietalia* example, as the name indicates, shows variation on the syntaxonomical order level and here we may expect what was actually found. The *Arrhenatheretum* example represents a data-set which is syntaxonomically already homogeneous enough to be approached with a moderate dominance weighting. Since the subassociations put together in this material are mainly characterized by groups of differential species and hardly by quantitative differences between species, the shift towards lower transformations is understandable. The *Spartinetea*- example represents an extreme situation: grosso modo each facies of a *Spartina* species or one other halophyte represents almost a separate syntaxonomical class.

Since the chosen examples are small in number and small in size we cannot derive more than just indicative results, but they justify the general suggestion that in hierarchical classifications, either divisive or agglomerative ones, various transformations should be tried on various levels of abstraction.

Transformation and intrinsic features of species performance

In his discussion of coupures Dagnelie (1960, see Introduction) used three arguments in favour of deleting certain performance values in a phytosociological data matrix. One of the arguments is biological. According to it low cover-abundance values are biologically less meaningful, at least in relation to higher values occurring elsewhere in the matrix. In view of the newer results obtained with transformations it is by no means certain that low quantities of a species occurring with high values elsewhere in a table do not have an indicator value. In many cases, e.g. in the *Spartinetea* treatment by Kortekaas, van der Maarel & Beeftink 1976, separate community types can be based on species occurrences with low values.

This brings us back to the pros and cons of the various scales mentioned in this contribution. Unless standardizations would be applied, full cover scales or scales transforming low values to zero cover, are ecologically ineffective because we would loose information on species which never attain high cover-abundance values by the nature of their intrinsic features, but are often good character-species, such as many tiny *Orchidaceae*, *Cyperaceae*, *Brassicaceae* and *Scrophulariaceae* (see van der Maarel 1966). Therefore I would suggest to retain all information carried in the combined scales such as those of Braun-Blanquet and Domin. Transformations with a strong weight on the mere presence, such as Moore's scale and in extremis, the presence-absence transformation appear to perform badly in most phytosociological cases known so far. Transformation with a heavy weight on high cover values, notably the original and still used group value transformation of Tüxen & Ellenberg, perform equally bad.

Finally we may interpret the relative success of intermediate transformations as an indication for a general, ecologically significant species performance. If we seek a parallel with other biological processes, a logarithmic transformation seems obvious, although the exact type of transformation is not yet clear (see Hogeweg 1976, Smartt et al. 1976, Jensén 1978). However, abundance and cover may not be ruled by the same biological processes.

From the exponential formula presented here, we could also deduct a parallel with exponentially developing processes and conclude that species performance may be expressed as an exponential phenomenon within a limited range for the value of the exponent. Maybe, this range is dependent on both an abundance-type and a dominance – type of species performance.

Summary

Various cover and cover-abundance scales are compared, together with some current transformations, including the angular transformation and logarithmic transformations. A new cover-abundance scale, being a fully numerical, extended Braun-Blanquet scale is introduced under the name ordinal scale. The transformation of cover-abundance data is discussed in terms of a general transformation function $y = x^w$ and it is shown that by applying different values of w to the ordinal scale all other scales can be approached, ranging from the presence-absence transformation for $w = 0$ to to the Tüxen & Ellenberg cover scale for $w = 4$.

The effect of the transformation of cover-abundance values on community similarity is shown in three examples: (1) 25 relevés of *Arrhenatheretum* grassland, being the exampletable used by Mueller-Dombois & Ellenberg; (2) 22 relevés of *Spartinetea* communities; and (3) 23 relevés of *Glauco-Puccinellietalia* communities, the latter two sets being taken from selections devised by the Working Group for Data-Processing. In all examples classification and ordination results obtained with intermediate transformations were superior to those obtained with either a strong weighting on mere presence, or an emphasis on dominance. This comparison was based on previous phytosociological experience.

Weighting and standardization are compared with each other. It is concluded (1) that standardization by species has effects similar to those of weighting of rare species with low cover-abundance values; (2) that standardization by relevé is similar to weighting the dominant species; and (3) that no standardization and intermediate weighting are hence similar in effect.

It is stated that the relations between weighting and standardization are insufficiently studied and need more attention. A general formula for applying both forms of transformation is presented: $y = a\,x^w$.

It is supposed that the effect of the amount of weighting will depend on the heterogeneity level of the data-set under study. Indeed some indications were found that weighting the dominant species may help distinguishing lower syntaxonomical units, especially facies, and also higher units, even up to the class level if the communities are characterized by one or a few dominant species.

Some remarks are made on the possible ecological background of the effects of weighting. It is argued that biological (reproduction) processes responsible for the performance of a species are of an exponential nature. Two performance types may be distinguished, viz. the abundance type and the dominance type. Both types may be approached with logarithmic scales.

References

Austin, M.P. & P. Greig-Smith, 1968. The application of quantitative methods to vegetation survey. II. Some methodological problems of data from rain forest. J. Ecol. 56: 827–844.

Bailey, A.W. & C.E. Poulton, 1968. Plant communities and environmental relationship in a portion of the Tillamook burn, Northwestern Oregon. Ecology: 49: 1–13.

Bannister, P. 1966. The use of subjective estimates of cover – abundance as the basis for ordination. J. Ecol. 54: 665–674.

Baum, B.R. 1977. Reduction of dimensionality for heuristic purposes. Taxon 26: 791–795.

Becking, R.W. (1957). The Zürich-Montpellier school of phytosociology. Bot. Rev. 23: 411–488.

Braun-Blanquet, J. 1946. Über den Deckungswert der Arten in den Pflanzengesellschaften der Ordnung Vaccinio-Piceetalia. Jahresber. Naturforsch. Ges. Graubündens 130: 115–119.

Braun-Blanquet, J. 1964. Pflanzensoziologie, Grundzüge der Vegetationskunde. 3. Aufl. Springer, Wien, New York. XIV + 865 pp.

Braun-Blanquet, J. 1965. Plant sociology: the study of plant communities. Transl. rev. and ed. by C.D. Fuller & H.S. Conard. Hafner, London, 439 pp.

Campbell, B.M.: 1978. Similarity coefficients for classifying relevés. Vegetatio 37: 101–109.

Coetzee, B.J. & M.J.A. Werger, 1973. On hierarchical syndrome analysis and the Zürich-Montpellier table method. Bothalia 11: 159–164.

Curtis, J.T. & R.P. McIntosh. 1951. An upland forest continuum in the prairie – forest border region of Wisconsin. Ecology 32: 476–496.

Dagnelie, P. 1960. Contribution à l'étude des communautés végétales par l'analyse factorielle. Bull. Service Carte Phytogéogr. CNRS B-5: 7–71, 93–195.

Daubenmire, R. 1968. Plant communities. A textbook of plant synecology. Harper & Row, New York. XI + 300 pp.

Doing Kraft, H. 1954. l'Analyse des carrés permanents. Acta Bot. Neerl. 3: 421–425.

Ellenberg, H. 1956. Aufgaben und Methoden der Vegetationskunde. Ulmer, Stuttgart. 136 pp.

Etter, H. 1949. De l'analyse statistique des tableaux de végétation. Vegetatio 1: 147–154.

Feoli, E. 1977. A criterion for monothetic classification of phytosociological entities on the basis of species ordination. Vegetatio 33: 147–152.

Fresco, L.F.M. 1969. Q-type factor analysis as a method in synecological research. Acta Bot. Neerl. 18: 477–482.

Goodall, D.W. 1953–1954. Objective methods for the classification of vegetation. I, II and III. Austr. J. Bot. 1: 39–63, 1: 434–456, 2: 304–324.

Goodall, D.W. 1969. A procedure for recognition of uncommon species combinations in sets of vegetation samples. Vegetatio 18: 19–35.

Hogeweg, P. 1976. Topics in biological pattern analysis. Thesis Utrecht, 208 pp.

Jensén, S. 1978. Influences of transformation of cover values on classification and ordination of lake vegetation. Vegetatio 37: 19–31.

Kortekaas, W.M., E. van der Maarel & W.G. Beeftink. 1976. A numerical classification of European Spartina communities. Vegetatio 33: 51–60.

Krajina, V.J. 1960. Can we find a common platform for the different schoolt of forest – type classification? Silva Fenn. 105: 50–59.

Lausi, D., W.G. Beeftink & W.M. Kortekaas, 1979. Survey of salt marsh relevés, selected for the Working – Group for Data – Processing in Phytosociology. In prep.

Londo, G. 1971. Patroon en proces in duinvalleivegetaties langs een gegraven meer in de Kennemerduinen. (With summary). Thesis Nijmegen. 279 pp.

Londo, G. 1976. The decimal scale for relevés of permanent quadrats. Vegetatio 33: 61–64.

Maarel, E. van der 1966. Over vegetatiestructuren, – relaties en – systemen, in het bijzonder in de duingraslanden van Voorne (With summary). Thesis Utrecht, 170 pp.

Maarel, E. van der 1969. On the use of ordination models in phytosociology. Vegetatio 19: 21–46.

Maarel, E. van der 1972. On the transformation of cover – abundance values in phytosociology. Report Botanical Laboratory, Nijmegen, 12 pp.

Maarel, E. van der 1972a. Ordination of plant communities on the basis of their plant genus, family and order relationships. In E. van der Maarel & R. Tüxen (ed) Grundfragen und Methoden der Pflanzensoziologie. Ber. Int. Symp. Rinteln 1970, p. 183–192. Junk, Den Haag.

Maarel, E. van der 1979. Multivariate methods in phytosociology with reference to the Netherlands. In M.J.A. Werger (ed.) The science of vegetation, p. 161–225. Junk, The Hague.

Maarel, E. van der & L.F.M. Fresco. 1975. Zusammenhänge zwischen Vegetation und Substrat erläutert an Ordinationsmodellen. In: Vegetation und Substrat. Ber. Int. Symp. Rinteln 1969, p. 233–253. Cramer, Vaduz.

Maarel, E. van der, J.G.M. Janssen & J.M.W. Louppen. 1978. TABORD, a program for structuring phytosociological tables. Vegetatio 38: 143–156.

Maarel, E. van der, L. Orlóci & S. Pignatti. 1976. Data – processing in phytosociology, retrospeet and anticipation. Vegetatio 32: 65–72.

Meyer Drees, E. 1949. Combined taxation and presence in analysing and comparing association tables. Vegetatio 2: 43–46.

Moore, J.J. 1966. PHYTO. Re-arranging a phytosociological array according to the principles of Braun-Blanquet. Programs in Fortran II. Mimeogr. Paper Dublin. 4 pp.

Mueller-Dombois, D. & H. Ellenberg. 1974. Aims and methods of vegetation ecology. Wiley, New York, XXX + 547 pp.

Noy-Meir, I. 1973. Data transformation in ecological ordination. I. Some advantages of non – centering. J. Ecol. 61: 329–341.

Noy-Meir, I.D. Walker & W.T. Williams. 1975. Data transformations in ecological ordination. II. On the meaning of data standardization. J. Ecol. 63: 779–800.

Orlóci, L. 1978. Multivariate analysis in vegetation research. 2nd ed. Junk, The Hague. IX + 451 pp.

Noy-Meir, I. & R.H. Whittaker. 1978. Recent developments in continuous multivariate techniques. In R.H. Whittaker (ed), Ordination of plant communities, p. 337–378, Junk, The Hague.

Roskam, E. 1971. Programme ORDINA: Multidimensional

ordination of observation vectors. Programme Bull. 16, Psychology Lab. Nijmegen, 8 pp.
Schmid, P. & N. Kuhn. 1970. Automatische Ordination von Vegetationsaufnahmen in pflanzensoziologischen Tabellen. Naturwissenschaften 57: 462.
Schwickerath, M. 1931. Die Gruppenabundanz (Gruppenmächtigkeit); ein Beitrag zur Begriffsbildung in der Pflanzensoziologie. Englers Bot. Jahrb. 64: 1–16.
Schwickerath, M. 1938. Neue Beiträge zur Kenntnis der Gruppenmächtigkeit der Assoziation. Englers Bot. Jahrb. 68: 497–514.
Schwickerath, M. 1940. Die Artmächtigkeit. Fedde Rep. Beih. 121: 48–52.
Segal, S. & V. Westhoff. 1959. Die vegetationskundliche Stellung von Carex buxbaumii in Europa, besonders in den Niederlanden. Acta Bot. Neerl. 8: 304–329.
Shimwell, D.W. 1971. The description and classification of vegetation. Sidgwick & Jackson, London, XIV + 322 pp.
Sissingh, G. 1950. Onkruid-associaties in Nederland (With French summary). Thesis Wageningen, 224 pp. + bijl.
Smart, P.F.M., S.E. Meacock & J. M. Lambert. 1976. Investigations into the properties of quantitative vegetational data. II. Further data type comparisons. J. Ecol. 64: 47–78.
Tüxen, R. & H. Ellenberg. 1937. Der systematische und ökologische Gruppenwert. Ein Beitrag zur Begriffsbildung und Methodik der Pflanzensoziologie. Mitt. Flor.-Soz. Arbeitsgem. 3: 171–184.
Westhoff, V. 1947. The vegetation of dunes and salt marshes on the Dutch islands of Terschelling, Vlieland and Texel Thesis Utrecht, 131 pp.
Westhoff, V. & E. van der Maarel. 1978. The Braun-Blanquet approach. 2nd ed. In: R.H. Whittaker (ed). Classification of plant communities, p. 287–399. Junk, The Hague.
Williams, W.T. & J.M. Lambert. 1959. Multivariate methods in plant ecology. I. Association analysis in plant communities. J. Ecol. 47: 83–101.
Wishart, D. 1975. CLUSTAN 1c user manual. Computer Centre University College London, 124 pp.

RELEVÉ RANKING BASED ON A SUM OF SQUARES CRITERION*

Enrico FEOLI & Laura FEOLI CHIAPELLA**

Istituto ed Orto Botanico, Università degli Studi, Trieste, Italy I 34100

Keywords:
Ranking, Relevé, Selection

Introduction

Recently, there is a growing interest in the treatment of very large sets of phytosociological data with the aim to accomplish a meaningful synthesis in the classification of plant communities of large geographical areas (van der Maarel, Orlóci & Pignatti 1976). The numerical classification of European *Spartinetea*, by Kortekaas, van der Maarel & Beeftink (1976), is an example. Notwithstanding the fact that there are automated procedures available to classify very large sets of relevés (e.g. Janssen 1975, Briane, Lazare & Salanon 1977, van der Maarel, Janssen & Louppen 1978 among others (cf. Orlóci 1978), an efficient reduction of the number of relevés may be desirable for easier computations and interpretation of the results. The reduction of the number of relevés may be achieved in different ways. One is through random selection, and another by subjective choice of those relevés containing the species which are considered important. In both cases the problem arises how to evaluate to what extent the selected set of relevés may be considered representative of a given phytosociological table. In this paper, we suggest a method which allows to measure the efficiency of the selection on the basis of a sum of squares criterion. An example is given based on the use of data from Majella's summits (Feoli Chiapella & Feoli 1977).

* Contribution from the Working Group for Data-Processing in Phytosociology, International Society for Vegetation Science.
** The work was completed at the Department of Plant Sciences of the University of Western Ontario, London, Canada. We wish to thank Prof. L. Orlóci for the hospitality and the helpful discussions. The work was supported by Italian C.N.R., within the project 'Promozione qualità dell'ambiente' subproject 'Metodologie matematiche e basi di dati'.

Method

The method suggested here for relevés is an adaptation of the method proposed by Orlóci (1973, 1978, p. 27) for species. In this case the dispersion criteria are computed for the relevés which are ordered according to their independent share of the total sum of squares. If the relevés are standardized with the standard deviation of the species scores, the method estimates the resemblance between the relevés in a way similar to the correlation coefficient. If the relevés are left unstandardized the measure of resemblance is analogous to the covariance. The relevé with the highest contribution to the total sum of squares is considered the most representative one for its phytosociological table. The independent share of the total sum of squares of a relevé is used as a measure of the contribution of the relevé to the total resemblance structure of the phytosociological table.

The cumulative percentage of the total sum of squares may be a useful parameter on which to base decisions. Experience with many phytosociological tables has shown that the relevés accounting for 60 % of the total sum of squares retain 80–90 % of the species of the tables. On the basis of this fact we suggest that the relevés whose cumulative percentage of sum of squares reach about 60 % may be sufficient to represent in a satisfactory way a phytosociological table in its most important aspects.

Example

The performance of the method is illustrated using the four phytosociological tables presented by Feoli Chiapella & Feoli (1977). Program RANK (Orlóci 1978) was applied to each table. The independent share of the cumulative sum

Table 1. Results of relevé selection from the four phytosociological tables of Majella's summits (Feoli Chiapella & Feoli 1977) based on cumulative percentages of sum of squares. See the main text for explanations.

Associations	Gnaphalio-Plantaginetum atratae	Leontopodio-Elynetum	Saxifrago-Papaveretum julici	Crepidi-Leontodontetum montani
Relevé number	4 14 6 1	24 20 28	52 31 35 58 32	41 39 38 56
Cumulative % of sum of squares	27 43 54 63	24 50 60	23 36 44 52 60	27 44 54 63

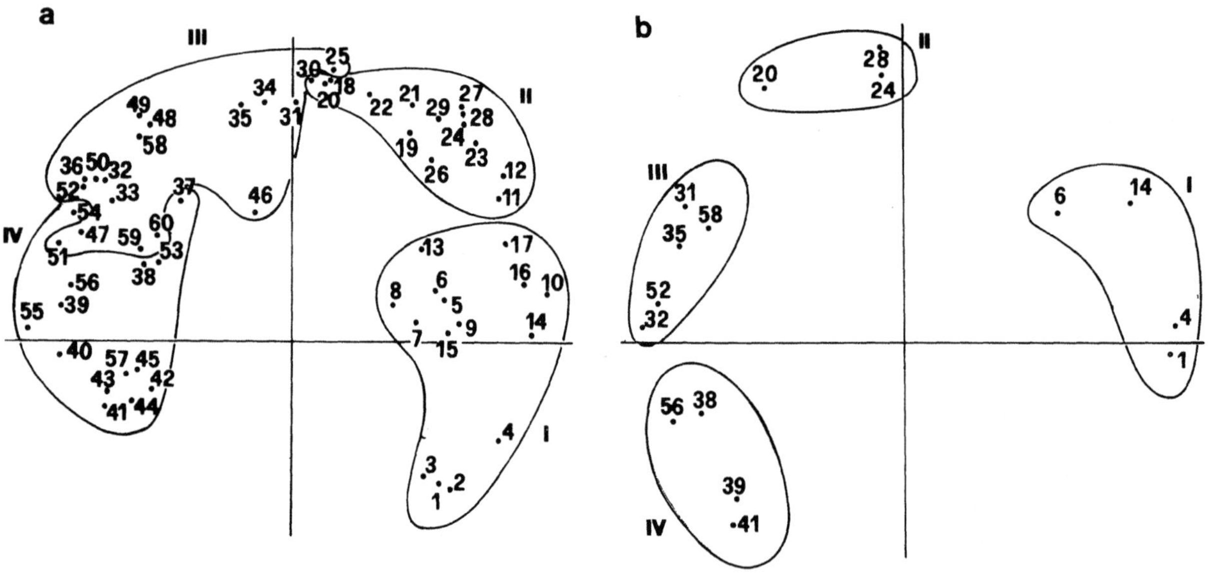

Figure 1. (a) Ordination of 60 relevés of Majella's summits after Feoli-Chiapella & Feoli (1977); (b) Ordination of a reduced set of relevés extracted from the set of 60 according to the cumulative percentages of sum of squares given in Tab. 1. Numbers to points identify relevés. The roman numbers identify: I *Gnaphalio-Plantaginetum atratae*, II *Leontopodio-Elynetum*, III *Saxifrago-Papaveretum julici*, IV *Crepidi-Leontodontetum montani*.

of squares of the relevé (the first, the first two, first three etc.) are presented in Table 1. The selected relevés (up to 5 per table) are drawn from all the phytosociological variants. This proves that the method can identify relevés which are representative for the main phytosociological aspects in the tables. In order to test if the selected relevés reproduce the same ordination pattern of the total set of data, the ordination method suggested by Feoli Chiapella & Feoli (1977) has been applied. The results, presented in Fig. 1, show a very close accordance between the ordination of the selected relevés and the ordination of all relevés. This fact then proves that the relevé selection based on the sum of squares criterion gives a representative subset of data reproducing efficiently the original data structure.

Summary

When a very large number of phytosociological types have to be compared, a reduction of the number of relevés is desirable. In this paper a method of relevé selection from given phytosociological tables is suggested. The method is based on a sum of squares criterion. The advantage, in comparison with other selection procedures, is that this method provides a means on the basis of which the efficiency of a relevé selection can be objectively measured.

References

Briane, J.P., J.J. Lazare & R. Salanon. 1977. Le traitement des très grands ensemπles de données en analyse factorielle des correspondances. Proposition d'une méthodologie appliquée à la Phytosociologie. Laboratoire de Taxonomie végétale expérimentale et numérique associé au C.N.R.S. 121: 38 pp.

Feoli Chiapella, L. & E. Feoli. 1977. A numerical phytosociological study of the summits of the Majella massive (Italy). Vegetatio 34: 21–39.

Janssen, J.G.M. 1975. A simple clustering procedure for preliminary classification of very large sets of phytosociological relevés. Vegetatio 30: 67–71.

Kortekaas, W.M., E. van der Maarel & W.G. Beeftink. 1976. A numerical classification of European Spartina communities. Vegetatio 33: 51–60.

Maarel, E. van der, J.G.M. Janssen & J.M.W. Louppen. 1978. TABORD, a program for structuring phytosociological tables. Vegetatio 38: 143–156.

Maarel, E. van der, L. Orlóci & S. Pignatti. 1976. Data-processing in phytosociology, retrospect and anticipation. Vegetatio 32: 65–72.

Orlóci, L. 1973. Ranking characters by a dispersion criterion. Nature 244: 371–373.

Orlóci, L. 1978. Multivariate Analysis in Vegetation Research. 2nd ed. Junk, The Hague, IX+451 pp.

HIERARCHICAL CLASSIFICATION OF EUROPEAN SALT MARSH VEGETATION BASED ON NUMERICAL METHODS*, **

Duilio LAUSI & Enrico FEOLI***

Istituto ed Orto Botanico, Università di Trieste, Via Valerio 30, Trieste, I 34100, Italy

Keywords:
Classification, Europe, Hierarchy, Numerical method, Ordination, Ranking, Salt marsh, Syntaxonomy.

Introduction

While the importance of the numerical methods in phytosociology has been widely recognized, specifically the application of these methods has led phytosociologists to turn interest to theoretical vegetation science (cf. Lausi 1972, Lausi 1973, Westhoff & van der Maarel 1978, van der Maarel 1975). The aim of the present paper is to contribute further to this topic by discussing a hierarchical classification of European salt marsh communities based on numerical methods. The main problem are the quantitative establishment of hierarchical levels and the comparison of such levels with the higher levels in the classical syntaxonomy of salt marshes.

The developments of the present study are all based on the following ideas:

(1) Each type of vegetation, i.e. a homogeneous set of relevés, should be characterized by frequently occurring species. Obviously, such species have potential value for the characterization of the type.

(2) If a species is frequently occurring in one type only, it has a high classificatory power at low hierarchical levels.

(3) If a species occurs frequently in many types, it is expected to have classificatory power at higher hierarchical levels.

(4) If a species is frequent in all types in which it occurs it has potential value for the differentiation between types.

This reasoning is logically consistent with the notions of set theory. A species can characterize a vegetation type only if it reaches a certain level of frequency in that type. A species which occurs with a low frequency in any type will have a low characterizing and differentiating potential. It should be observed that a faithful species in the sense of Braun-Blanquet does not need to be frequent in the type to which it is said to be faithful!

When many synoptic tables (cf. Westhoff & van der Maarel 1978) are involved in the study it is possible to obtain a frequency distribution for each species in the usual way, using Raunkiaer's frequency classes. We consider a species as frequent if it reaches Raunkiaer's class V, i.e. if it occurs in more than 80 % of the relevés of the table under consideration. Of course, species with a frequency distribution which is confined to class V are extremely rare. Generally, species have frequency distributions over all the five classes. Using a similar data set Cristofolini, Lausi & Pignatti (1970) found that the frequency distribution of these species considered as 'companion species' in the literature has a maximum in class I, while the frequency distribution of species considered as 'characteristic' has two well-defined maxima, one in class I and the other in class V (fig. 1). The previous reasoning which suggests a method for the selection of characterizing species is supported by these earlier results.

* Contribution from the Working Group for Data-Processing in Phytosociology, International Society for Vegetation Science, which is based on two conference papers presented at the meetings of the group at Prague, September 1973, and Montpellier, September 1975. The themes of these papers were Ordination of species, and Classification of associations, both based on frequency data from synoptic tables of salt marsh vegetation.

** For nomenclature of species see Table 2.

*** We would like to thank Dr. W.G. Beeftink, Dr. E. van der Maarel, Prof. L. Orlóci and Prof. S. Pignatti for the stimulating discussions and Messrs S. Sedmak and A. Zampar for their technical assistance. Special thanks are due to Dr. E. van der Maarel for the critical revision of the manuscript. The work was supported by the Italian CNR.

Material

The material used in the present study consists of 162 phytosociological tables of various salt marsh associations contained in the *selection c* of the data archive of the Working Group for Data Processing in Phytosociology (Beeftink 1972; van der Maarel, Orlóci & Pignatti 1976). This selection includes 2605 relevés with 650 species. This material is considered geographically representative for the large area involved in this study, i.e. Europe and North Africa. Although regionally incomplete, this material may be considered sufficient for the aims of the present study which concentrates on similarity relations at higher hierarchical levels. The numerical analyses have been applied to the synoptic forms of the original tables.

Concerning coding of the species, no taxa below the rank of species have been distinguished in this work, since in the archive of the Working Group only the species level has been codified (Pignatti 1976; Kortekaas, Lausi & Beeftink, in prep.). This has consequences in two specific cases. *Agrostis stolonifera* is included in *A. maritima* since it is not always clear from the various publications involved in the selection whether this taxon was considered a separate species or a member of an aggregate. For *Suaeda fruticosa*, however, it was possible to distinguish the subspecific level with the taxa: *Suaeda fruticosa* (L.) Forsk in tab. 135, and *S. fruticosa* (L.) Forsk subsp. *braevifolia* Moq. in the other tables.

Methods

Selection of characterizing species

The classificatory potential of a given species (SW) is estimated as follows:

$$SW = (F(\mathrm{V})_{obs} - F(\mathrm{V})_{exp})/\sigma(\mathrm{V}) \qquad (1)$$

where $F(\mathrm{V})obs$ and $F(\mathrm{V})exp$ are the observed and expected frequencies in class V and $\sigma(\mathrm{V})$ is the standard deviation of the mean difference between observed and expected frequencies in class V for all 650 species. The observed frequency is given by the number of phytosociological tables in which the species occurs. The expected frequency in the 5th class is given by the following formula:

$$F(\mathrm{V})_{exp} = F(\mathrm{I})_{obs}^{0,82\times 4} \qquad (2)$$

This estimation is derived from the general function

$$F(j)_{exp} = F(\mathrm{I})_{obs}^{0,82(j-1)} \qquad j = 1,\ldots,5 \qquad (3)$$

This function has been empirically fitted to the average frequency distribution of a companion species. Such a distribution is unimodal with a maximum in class I and exponentially decreasing values for higher frequency classes.

The species with $SW \geqq 1$ have been provisionally considered sufficiently reliable characterizing species since the observed class V value deviates at least σ from the

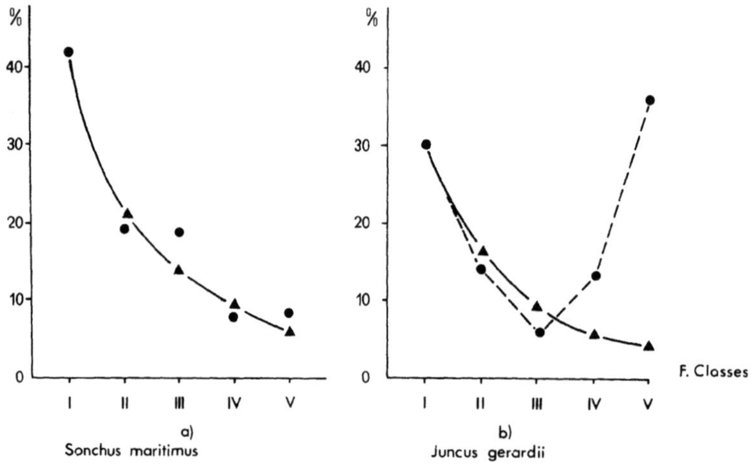

Fig. 1a. Unimodal frequency distribution in the usual five frequency classes of a comparison species (*Sonchus maritimus sw* = 0.09).
b. Bimodal frequency distribution in the usual five frequency classes of a character-species (*Jucus gerardii: SW* = 1.27).
● Observed frequency values.
▲ Expected frequency values according to formula (2).

expected one. In such cases the corresponding frequency distribution is always bimodal with a second maximum in class V. As a rough but useful parameter to characterize the shape of the frequency distribution we can use the entropy value H as calculated for the five frequency classes over all tables in which the species occurs. This parameter is a measure of evenness in our case. Species with an equal distribution over the classes have a high H value, with a maximum $H_{max} = {}^2\log 5$, whereas H decreases the more the frequency distribution is confined to one class only. Examples of the frequency distributions of a species with $SW < 1$ and one with $SW \geqq 1$ are presented in fig. 1. The species with $SW \geqq 1$ have been selected for automatic classification of the synoptic tables.

Classification of synoptic tables

The classification is based on average linkage clustering (*WPVAC*, following Sneath & Sokal 1973) and on the monothetic classificatory method (*MC*) of Feoli (1977). The first was performed by the computer program *CLUSTAN* I *B* of Wishart (1969), the second by the program *MLTAX*1 of Feoli & Lagonegro (1978). Average linkage clustering has been applied to frequency (*ALCAF*) and binary data (*ALCAB*) with the *Similarity ratio* as similarity function (cf. Westhoff & van der Maarel 1978, van der Maarel, Janssen & Louppen 1978). This method has been chosen because, as compared to other polythetic agglomerative methods, it is less affected by the continual variation in the data and free from the influence of group size (cf. Feoli Chiapella & Feoli 1977).

The *MC* method has been chosen since it can be easily interpreted in terms of set theory and since it allows to obtain groups of co-occurring species during the classification procedure. *MC* splits the entire set of synoptic tables according to the presence of certain species or species groups each of which is considered a single character. Naturally *MC* has been applied only to binary data. The results of the classifications performed have been evaluated by Concentration analysis (Feoli & Orlóci 1979), a method measuring the linkage between species and groups of relevés or groups of synoptic tables. The program used for Concentration analysis (*CA*) is RQT of Orlóci (1978).

Other methods of multivariate analysis have been applied to the data viz. among the classification procedures the method of Parks (1970), Sum of Squares Agglomeration (Orlóci 1978), Single linkage clustering (Sneath & Sokal 1973); among the ordination methods: Principal components analysis, Factor analysis, Discriminant analysis and Canonical correlation analysis. The results of all these methods are not discussed here since in general they are very similar in their results.

Recognition of significant hierarchical levels

As was stated before we had to concentrate on the higher levels of resemblance between the synoptic tables studied. The first level has been obtained as follows: to make effective use of the selected characterizing species each cluster on a certain level should have at least one of those species in frequency class V. This appeared to be at the 19 cluster level.

Then the structure of the remaining part of the dendrogram was analyzed with information statistics. It is supposed that a good structure; i.e. a clear division into subsets will be reflected by a high value of the mutual information content, as defined by the difference of total and joint information content (Kullback 1959, Orlóci 1978). As a more sensitive parameter we used the ratio between mutual information and equivocation information, the latter being defined as the difference between joint and mutual information.

The value of this ratio was plotted against the number of clusters at each dichotomic level up from the 19 cluster level and peaks in the resulting curve have been used to define significant hierarchical levels. In this way the dichotomic structure of the dendrogram is considered more important than the internal similarity levels of clusters. This will be further discussed in another paper.

Results

Selection of characterizing species

The species with $SW \geqq 1$ are listed in table 2a, in order of their SW value, and together with the entropy values H of the frequency distribution of the species in the five frequency classes for the tables in which they occur. It appears that species with $SW \geqq 1$ and with high entropy are the ones that are often considered as character species of the higher syntaxa (alliance, order, class). This means that the method of species selection used here has a good potential to fit the judgement of the phytosociologists having previous experience with this vegetation. Notwithstanding the strong reduction of the total species number with 93 %!), in all of the tables at least one of the selected species is retained. The mean number of the selected

Table 1. Data sources: numerical codes for the synoptic phytosociological tables and Authors.

Table No.	Author
1 - 4	Adriani,M.J. (1945)
5 - 23	Pignatti,S. (1966)
24 - 37	Gillner,V. (1960)
38 - 43	Wendelberger,G. (1950)
44 - 52	Corillion,M.R. (1953)
53	Tüxen,R.& V.Westhoff (1963)
58	Soó,R. (1947)
62 - 63	Braun-Blanquet, J. &Cl.De Ramm (1957)
64	Rivas-Goday, S. (1955)
65	Ri -Goday,S. (1945)
66 - 69	Fonces,F.C. (1945)
70	Molinier,René & G. Tallon (1965)
71 - 75	Bolós, O. de (1962)
76 - 81	Braun-Blanquet,J. & O. de Bolós (1957)
82 - 88	Wilkón-Michalska, J. (1963)
89 - 96	Hocquette,M.,J.-M.Géhu & M. Fauquet (1965)
97 - 99	Bodrogközy, G. (1959)
100 - 103	Bodrogközy, G. (1965b)
104 - 105	Bodrogközy, G. (1960)
106 - 111	Bodrogközy, G. (1962)
112 - 119	Bodrogközy, G. (1965a)
120 - 124	Bodrogközy, G. (1966)
125 - 127	Bodrogközy, G. (1965c)
128 - 132	Moore, J.J. (unpubl. tables)
133	Ivimey-Cook,R.B. &M.C.F.Proctor (1966,unpubl.rel.)
134 - 138 ( )	Westhoff, V. (unpubl. rel.)
158 - 162	Frei, M. (1937)
163 - 165	Braun-Blanquet, J. & R. Tüxen (1952)
166 - 172	Oberdorfer, E. (1952)
173 - 176	Nordhagen,R. (1954)
177 - 181	Nègre,R. (1956)
182 - 189	Wolff,W.J. (1968)

Table 2a

Selected character-species with bimodal frequency distribution. *SW*: diagnostic weight of the species according to formula (1). *H*: entropy of the species according to formula (1). *H*: entropy of the species according to Shannon's formula.

Table 2b

Species with unimodal frequency distribution with a low *SW* value and with a high entropy *(H)* value.

2a

SW	H	Species
3.98	0.00	Molinia caerulea (L.)Moench
3.98	0.00	Potentilla arenaria Borkh.
3.18	0.72	Mesembryanthemum nodiflorum L.
2.65	0.00	Frankenia corymbosa Desf.
2.65	0.92	Cyperus pannonicus Jacq.
2.46	0.92	Blysmus rufus (Huds.) Lk.
2.34	1.51	Puccinellia limosa (Schur)Holmb.
1.99	1.00	Diplotaxis ollivieri Maire
1.99	1.00	Carex divisa Huds.
1.84	1.96	Suaeda fruticosa (L.)Forsk
1.82	1.50	Puccinellia phryganodes(Trin)S&M
1.76	1.00	Spartina alterniflora Loisel.
1.76	1.00	Sanguisorba officinalis L.
1.76	1.00	Frankenia reuteri Boiss.
1.76	1.00	Limonium ovalifolium Poir.
1.76	1.00	Matricaria pubescens Sch.
1.61	1.68	Juncus articulatus L.
1.55	1.97	Arthrocnemum fruticosum (L.)Moq.
1.54	1.87	Schoenoplectus tabernemontani (C.C.Gmel) Palla
1.52	2.15	Plantago maritima L.
1.49	1.45	Chenopodium glaucum L.
1.48	2.10	Festuca rubraL. ssp.litoralis
1.46	2.15	Armeria maritima (Mill.)Willd.
1.43	2.08	Agrostis maritima(Lamk.)Meyer
1.33	1.58	Carex glareosa Schk.
1.32	2.18	Glaux maritima L.
1.27	1.92	Carex extensa Good.
1.27	2.08	Juncus gerardii Loisel.
1.18	1.92	Beckmannia eruciformis (L.)Host
1.18	1.92	Achillea collina J. Becker
1.16	2.20	Leontodon autumnalis L.
1.13	2.12	Juncus maritimus Lamk.
1.13	1.58	Carex paleacea Wg.
1.13	1.58	Alopecurus bulbosus Gouan
1.13	1.58	Convolvulus arvensis L.
1.09	1.88	Festuca pseudovina Hackel
1.09	2.15	Salicornia europaea L.
1.08	2.01	Arthrocnemum perenne (mill.)Moss
1.07	2.18	Puccinellia maritima(Huds.)Parl.
1.04	1.97	Pholiurus pannonicus (Host.)Trin.
1.04	1.81	Camphorosma annua Pall.
1.04	2.11	Eleocharis uniglumis (Lk.)Schult
1.03	2.23	Triglochin maritimum L.

2b

SW	H	Species
0.99	1.50	Beta macrocarpa Guss.
0.99	1.50	Stellaria humifusa Rottb.
0.99	1.50	Carex subspathacea Wormsk.
0.96	2.23	Potentilla egedii Wormsk.
0.96	1.95	Plantago tenuiflora W. et K.
0.96	1.95	Bromus mollis L.
0.94	2.11	Arthrocnemum glaucum (Delisle) Ung.-Sternb.
0.94	2.07	Sphenopus divaricatus (Gouan)Rohb.
0.92	1.84	Sagina procumbens L.
0.92	1.84	Schoenus nigricans L.
0.88	1.94	Frankenia pulverulenta L.
0.88	2.12	Sagina maritima G.Don
0.84	1.62	Spartina stricta Roth
0.82	1.50	Festuca fenas Lag.
0.82	2.00	Eleocharis pauciflora (Light.) Lk
0.81	2.04	Carex distans L.
0.80	2.18	Plantago coronopus L.
0.78	2.05	Alopecurus pratensis L.
0.77	2.21	Halimione portulacoides (L.)Aellen
0.76	1.75	Centaurium tenuiflorum (Hoffgg. etLk.) Fritsch
0.76	1.50	Suaeda splendens (Pourr)Gren.etGodr.
0.76	1.50	Petrosimonia crassifolia Rge.
0.76	1.50	Melilotus officinalis (L.)Lam.
0.76	1.50	Bupleurum semicompositum L.
0.74	2.26	Limonium vulgare Mill.
0.74	2.15	Limonium humile Mill.
0.72	1.55	Samolus valerandi L.
0.70	2.29	Aster tripolium L.
0.69	1.89	Plantago crassifolia Forsk.
0.67	2.17	Lepturus incurvatus (L.)Trin.
0.66	2.18	Phragmites communis Trin.
0.66	2.06	Juncus subulatus Forsk.
0.66	1.92	Scorzonera parviflora Jacq.
0.66	1.84	Camphorosma monspeliaca L.
0.64	2.32	Medicago lupulina L.
0.64	1.94	Trifolium fragiferum L.
0.64	1.66	Atriplex tatarica L.
0.60	2.12	Scirpus maritimus L.
0.58	2.26	Puccinellia distans (L.)Parl.
0.57	2.01	Polypogon monspeliensis L.
0.56	1.76	Festuca arundinacea Schreb.
0.55	2.23	Lepidium cartilagineum Mayer
0.53	2.19	Puccinellia palustris Hayek

species for each table is 4, that is about 26 % of the mean number of species in the tables. The a priori fixed arbitrary limit for the species weight ($SW \geqq 1$) seems appropriate to select a number of species maintaining a good level of information on the general variation of the data structure.

The following observations can be made on table 2:
(1) The species with the highest values of *SW* are partly species with their main distribution outside the proper salt marshes e.g. *Molinia caerulea*. Their value is entirely a differentiating one. Species as *Mesembryanthemum nodiflorum*, however, are highly characteristic species on lower levels of the syntaxonomical hierarchy.
(2) Many well-known saltmarsh species such as *Armeria maritima* and *Triglochin maritimum* have moderate *SW* values but high *H* values. Apparently they have a wide distribution with all frequency classes covered but still with a maximum at class V.
(3) Among the diagnostic species used in the literature for defining mediterranean syntaxa only a relatively low number has been retained on the basis of the selection criterion.

Apparently many of those species have not been selected because they do not occur with frequency class V in any table. Some of them which show a high entropy value *H* are listed in Table 2 b. The sociological meaning of many species listed here becomes clear when we consider their use as local or regional character-species in the Mediterranean area to which they are confined. This phenomenon may be explained by the relatively high beta diversity (Whittaker 1972) in mediterranean salt marshes, i.e. a great differentiation of phytocoena. As compared to northern and central Europe in mediterranean regions the floristic and vegetational diversity continuously increased during the Ice-Age. In the coolest period, central-asiatic taxa such as *Artemisia* and *Chenopodiaceae* migrated to the mediterranean regions which probably coincided with an expansion of salt steppes. The syntaxonomical importance of these evolutionary events has been generally recognized in the literature e.g. Braun-Blanquet & Bolós (1957) with high weight given to historical and chorological considerations when dealing with mediterranean syntaxa. This is reflected in nomenclature by the use of epithets such as mediterranean, galloprovincialis, orientale, occidentale etc..

Species groups

The classification according to MC has resulted in seven main groups, some of which can be subdivided:

1.1 *Suaeda fruticosa, Frankenia reuteri, Limonium ovalifolium*
1.2 *Diplotaxis ollivieri, Matricaria pubescens, Frankenia corymbosa, Mesembryanthemum nodiflorum*
2.1 *Puccinellia limosa, Camphorosma annua*
2.2 *Pholiurus pannonicus, Festuca pseudovina, Beckmannia eruciformis, Eleocharis uniglumis, Achillea collina, Convolvulus arvensis*
3.1 *Blysmus rufus, Juncus articulatus*
3.2 *Triglochin maritimum, Glaux maritima, Juncus gerardii, Festuca rubra litoralis, Armeria maritima, Plantago maritima, Juncus maritimus, Leontodon autumnalis*
3.3 *Puccinellia maritima, Salicornia europaea*
3.4 *Arthrocnemum perenne, Arthrocnemum fruticosum*
3.5 *Carex extensa*
4 *Carex paleacea, Schoenoplectus tabernaemontani*
5 *Molinia coerulea, Sanguisorba officinalis*
6 *Puccinellia phryganodes, Carex glareosa*
7 *Alopecururs bulbosus, Carex divisa, Spartina alterniflora*

The remaining 4 species (*Chenopodium glaucum, Cyperus pannonicus, Potentilla arenaria, Camphorosma monspeliaca*) don't take part in the groups but remain isolated.

The species groups 1.1, 1.2, 2.1, 2.2, 3.2, 3.3, 3.4 have been used in Concentration analysis since they are the most frequent ones in the clusters of high hierarchical level.

Classification of synoptic tables

The results of *ALCAF* and *ALCAB* are presented respectively in Figure 2 and 3. These figures contain the complete dendrograms, the species distribution in the clusters, the names of the tables, the regions in which the relevés were made. Author names and table numbers are presented in Table 1. The detailed dendrogram obtained by *MC* is not presented since it is very similar to that obtained by *ALCAB*.

Analysis of the occurrence of the species in the main clusters of *ALCAF* allows to define 8 main groups of synoptic tables with 18 subgroups. *ALCAB* allows to define 11 main groups with 19 subgroups, and *MC* 7 main groups with 27 subgroups. The distribution of the types in the subgroups of the three classifications is very similar, in fact the mutual information (Kullback 1959, Orlóci 1978) is very high: I (*ALCAB; ALCAF*) = 290.47 nit (information units based on natural logarithms); I(*MC; ALCAF*) = 308.48 nit and I(*MC; ALCAB*) = 317.40 nit. (Note that the value 2 I approximates a χ^2 value). This allows the conclusion that at low hierarchical levels (subgroups) the results of the three classifications are in good accordance with each other.

In order to facilitate the comparison between the three classifications their results are synthetically presented in

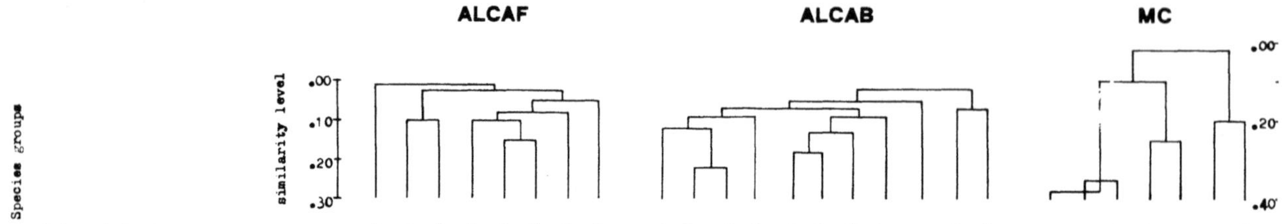

Species groups		ALCAF								ALCAB											MC						
	Codes of clusters	1	2	3	4	5	6	7	8	1	5a 6	4	5b	3	2a	2b	2c		7	8	1	5 6	4	3	2	7	8
	Codes of tables from	128	106	124	90	11	68	78	177	128	18	90	6	16	106	110	97	64	78	177							
	Codes of tables to	16	119	104	135	170	169	81	181	94	73	92	65	104	126	58	119	170	81	181							
	N° of tables	38	22	18	29	17	21	7	5	55	38	13	3	17	15	3	5	2	7	4	62	35	8	8	31	7	5
2.2	Eleocharis uniglumis	13	20	50						14				24	33						14			100			
2.2	Pholiurus pannonicus		32	11		6								12	47									75	13		
2.2	Festuca pseudovina		73	28	7	6				5				35	73		100							88	52		
2.2	Beckmannia eruciformis		4	28										24	13									63			
2.2	Achillea collina		18	11											13		80							13	16		
2.2	Convolvulus arvensis		18														80								13		
	Potentilla arenaria		10														40								3		
2.1	Puccinellia limosa		68	28	7					2				35	80	100								62	61		
2.1	Camphorosma annua		64			6								12	87									38	42		
5	Sanguisorba officinalis			11										12							2				3		
5	Molinia coerulea			6										6													
	Chenopodium glaucum	3	4	17						2				6		33					2	6			10		
	Cyperus pannonicus			17										6		66									10		
4	Schoenoplectus tabernaemontani	1	10	44	3					11				29	7						11	6			13		
4	Carex paleacea			17						5																	
6	Puccinellia phryganodes	10								7											7						
6	Carex glareosa	8								5											5						
3.1	Juncus articulatus	10		11		6				9				6							10				3		
3.1	Blysmus rufus	13		6						11											10						
3.5	Carex extensa	29			3	12	19			18	13	8		18							19	20					
3.2	Agrostis maritima	90	27	83	10	23	19			75	16			100	20			50			70	31		75	29		
3.2	Plantago maritima	71	36	39	31	18		14		71		8		59	27		40		14		61	11	100		48	14	
3.2	Juncus gerardii	81	10	22	17	18	24			64	21		33	29		67					61	20			16		
3.2	Triglochin maritimum	90		17	41	24	14			93	16		33								89				6		
3.2	Juncus maritimus	42			10	47	43	14		29	42		100	12					14		40	37				14	
3.2	Festuca rubra litoralis	68		17	10					56		8									52						
3.2	Glaux maritima	76		17	31					75											63				3		
3.2	Armeria maritima	55		6	24					56											47				3		
3.2	Leontodon autumnalis	37		6						25				6							22						
3.3	Salicornia europaea	40	4		100	30	90	28		47	76	100			7	33			28		47	74	100		10	28	
3.3	Puccinellia maritima	31		6	59		14			44	8	54									39	3			3		
3.4	Arthrocnemum perenne	5			10		67			11	37										8	34	25				
3.4	Arthrocnemum fruticosum	3			21	35	100	28			95								14		11	66	25			28	
7	Alopecurus bulbosus	5			3					7											5						
7	Carex divisa	3								4											3						
7	Spartina alterniflora				3					4											3						
1.1	Suaeda fruticosa				3			86	40	2									100	25						86	40
1.1	Limonium ovalifolium							29			3								14							33	
1.1	Frankenia reuteri							29											43							33	
1.2	Frankenia corymbosa								60									14	50								60
1.2	Mesembryanthemum nodiflorum								80											100							80
1.2	Matricaria pubescens								40											50							40
1.2	Diplotaxis ollivieri								40											50							40

Fig. 4. Comparison between *ALCAF*, *ALCAB* and *MC* based on the frequency distribution of the species in the most evident clusters of tables. In *ALCAF* and *ALCAB* the codes of tables follow the same sequence as in Fig. 2 and Fig. 3. See text.

Figure 4. In this figure the clusters are coded in such a way that they maximally overlap with the clusters of *ALCAF*. In *ALCAB* some clusters of *ALCAF* (see the codes) appear to be split up into two or three clusters. Cluster 2 is subdivided in 2a, characterized by *Puccinellia limosa* and *Camphorosma annua*, 2b, characterizd by *Puccinellia limosa*, *Juncus geradii* and *Salicornia europea*, and 2c, characterized by *Festuca pseudovina*, *Achillea collina* and *Convolvulus arvensis*. Cluster 5 is subdivided in 5a, characterized by *Juncus maritumus* and *Arthrocnemum fruticosum* (see Fig. 3) and 5b, characterized by *Juncus maritumus* (see Fig. 3).

Fig. 4 shows the good accordance between the three classifications regarding the species distribution in the main clusters, despite the different hierarchical structure of the dendrograms. Cluster 1 is characterized by the high presence of group 3.2, cluster 2 by species groups 2.1 and 2.2, cluster 3 by some species of group 3.2, as well as species of groups

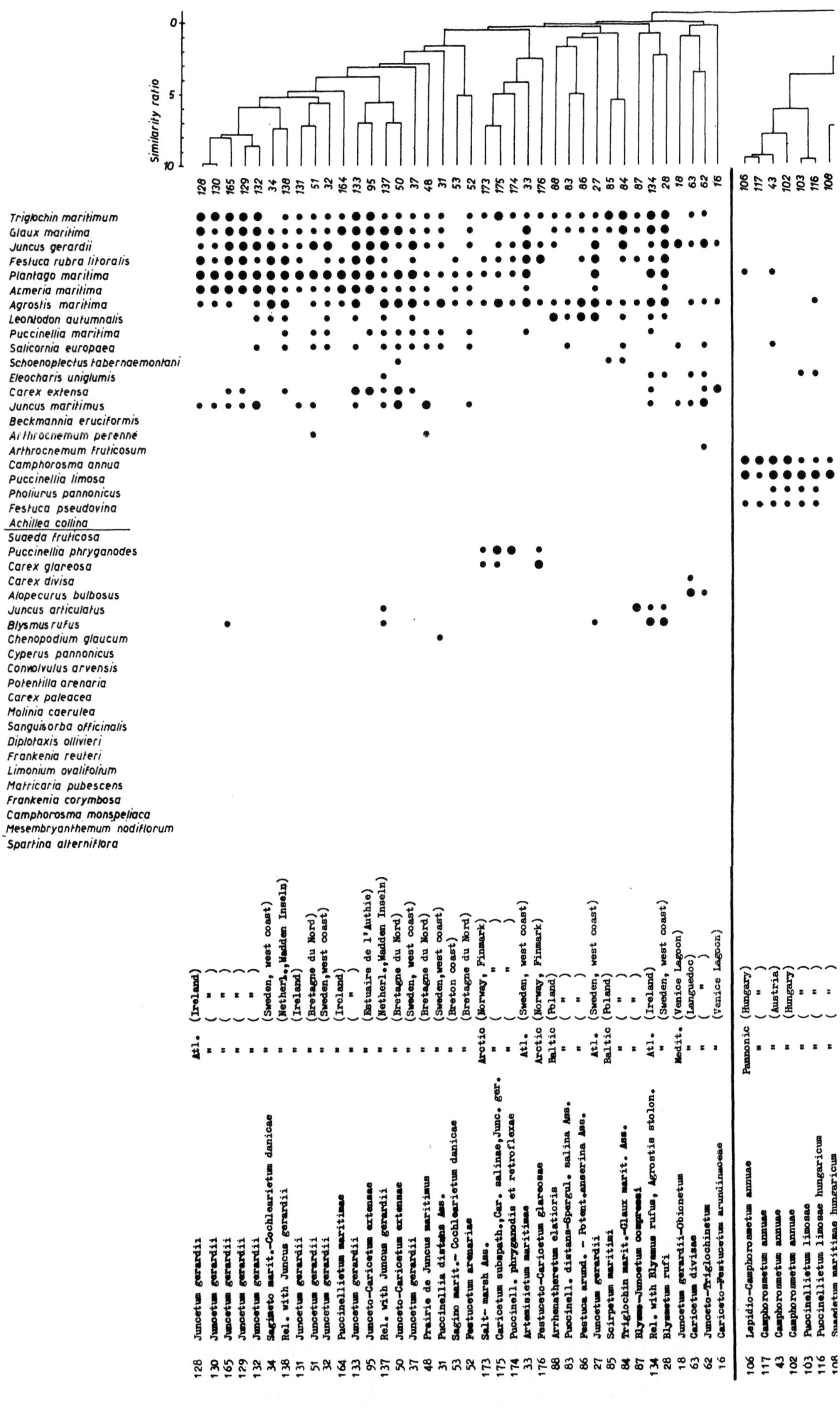
Similarity ratio
0
5
10
128 130 165 129 132 34 138 131 51 32 164 133 95 137 50 37 48 31 53 52 173 175 174 33 176 88 83 86 27 85 84 87 134 28 18 63 62 16
106 117 43 102 103 116 108
Triglochin maritimum
Glaux maritima
Juncus gerardii
Festuca rubra litoralis
Plantago maritima
Armeria maritima
Agrostis maritima
Leontodon autumnalis
Puccinellia maritima
Salicornia europaea
Schoenoplectus tabernaemontani
Eleocharis uniglumis
Carex extensa
Juncus maritimus
Beckmannia eruciformis
Arthrocnemum perenne
Arthrocnemum fruticosum
Camphorosma annua
Puccinellia limosa
Pholiurus pannonicus
Festuca pseudovina
Achillea collina
Suaeda fruticosa
Puccinellia phryganodes
Carex glareosa
Carex divisa
Alopecurus bulbosus
Juncus articulatus
Blysmus rufus
Chenopodium glaucum
Cyperus pannonicus
Convolvulus arvensis
Potentilla arenaria
Carex paleacea
Molinia caerulea
Sanguisorba officinalis
Diplotaxis ollivieri
Frankenia reuteri
Limonium ovalifolium
Matricaria pubescens
Frankenia corymbosa
Camphorosma monspeliaca
Mesembryanthemum nodiflorum
Spartina alterniflora
128 Juncetum gerardii Atl. (Ireland)
130 Juncetum gerardii " (")
165 Juncetum gerardii " (")
129 Juncetum gerardii " (")
132 Juncetum gerardii " (")
34 Sagineto marit.-Cochlearietum danicae " (Sweden, west coast)
138 Rel. with Juncus gerardii " (Netherl.,Wadden Inseln)
131 Juncetum gerardii " (Ireland)
51 Juncetum gerardii " (Bretagne du Nord)
32 Juncetum gerardii " (Sweden,west coast)
164 Puccinellietum maritimae " (Ireland)
133 Juncetum gerardii " (")
95 Junceto-Caricetum extensae " (Estuaire de l'Authie)
137 Rel. with Juncus gerardii " (Netherl.,Wadden Inseln)
50 Junceto-Caricetum extensae " (Bretagne du Nord)
37 Juncetum gerardii " (Sweden, west coast)
48 Prairie de Juncus maritimus " (Bretagne du Nord)
31 Puccinellia distans Ass. " (Sweden,west coast)
53 Sagino marit.- Cochlearietum danicae " (Breton coast)
52 Festucetum arenariae " (Bretagne du Nord)
173 Salt- marsh Ass. Arctic (Norway, Finmark)
175 Caricetum subspatha.,Car. salinae,Junc. ger. " (")
174 Puccinell. phryganodis et retroflexae " (")
33 Artemisietum maritimae Atl. (Sweden, west coast)
176 Festuceto-Caricetum glareosae Arctic (Norway, Finmark)
88 Arrhenatheretum elatioris Baltic (Poland)
83 Puccinell. distans-Spergul. salina Ass. " (")
86 Festuca arund. - Potent.anserina Ass. " (")
27 Juncetum gerardii Atl. (Sweden, west coast)
85 Scirpetum maritimi Baltic (Poland)
84 Triglochin marit.-Glaux marit. Ass. " (")
87 Blysmo-Juncetum compressi " (")
134 Rel. with Blysmus rufus, Agrostis stolon. Atl. (Ireland)
28 Blysmetum rufi " (Sweden, west coast)
18 Junceto gerardii-Obionetum Medit. (Venice Lagoon)
63 Caricetum divisae " (Languedoc)
62 Junceto-Triglochinetum " (")
16 Caricato-Festucetum arundinaceae " (Venice Lagoon)
106 Lepidio-Camphorosmetum annuae Pannonic (Hungary)
117 Camphorosmetum annuae " (")
43 Camphorosmetum annuae " (Austria)
102 Camphorosmetum annuae " (Hungary)
103 Puccinellietum limosae " (")
116 Puccinellietum limosae hungaricum " (")
108 Suaedetum maritimae hungaricum " (")

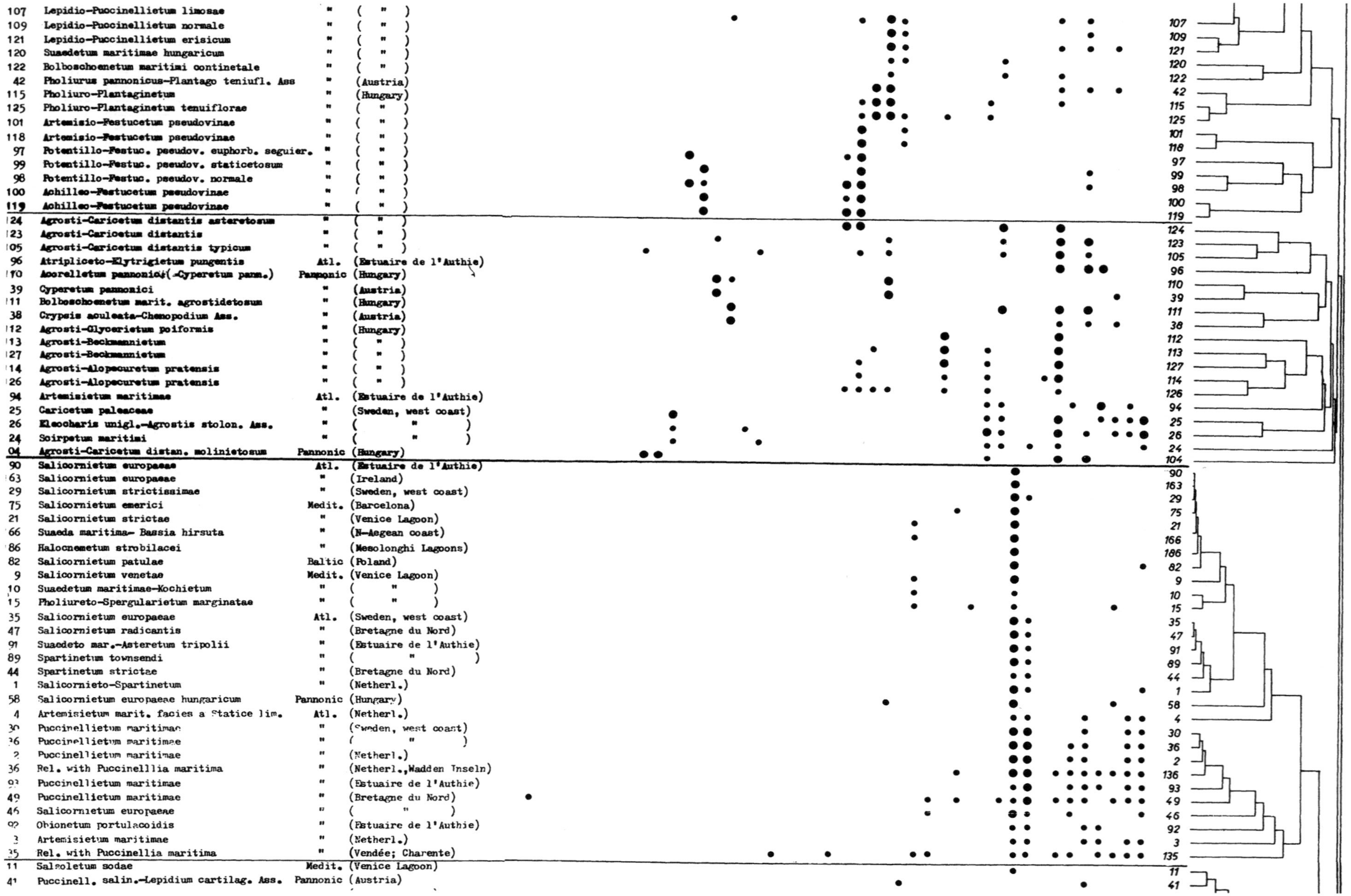
107 Lepidio-Puccinellietum limosae " (")
109 Lepidio-Puccinellietum normale " (")
121 Lepidio-Puccinellietum erisicum " (")
120 Suaedetum maritimae hungaricum " (")
122 Bolboschoenetum maritimi continetale " (")
42 Pholiurus pannonicus-Plantago teniufl. Ass " (Austria)
115 Pholiuro-Plantaginetum " (Hungary)
125 Pholiuro-Plantaginetum tenuiflorae " (")
101 Artemisio-Festucetum pseudovinae " (")
118 Artemisio-Festucetum pseudovinae " (")
97 Potentillo-Festuc. pseudov. euphorb. seguier. " (")
99 Potentillo-Festuc. pseudov. staticetosum " (")
98 Potentillo-Festuc. pseudov. normale " (")
100 Achilleo-Festucetum pseudovinae " (")
119 Achilleo-Festucetum pseudovinae " (")
124 Agrosti-Caricetum distantis asteretosum " (")
123 Agrosti-Caricetum distantis " (")
105 Agrosti-Caricetum distantis typicum " (")
96 Atripliceto-Elytrigietum pungentis Atl. (Estuaire de l'Authie)
110 Acorelletum pannonici(-Cyperetum pann.) Pannonic (Hungary)
39 Cyperetum pannonici " (Austria)
111 Bolboschoenetum marit. agrostidetosum " (Hungary)
38 Crypsis aculeata-Chenopodium Ass. " (Austria)
112 Agrosti-Glycerietum poiformis " (Hungary)
113 Agrosti-Beckmannietum " (")
127 Agrosti-Beckmannietum " (")
114 Agrosti-Alopecuretum pratensis " (")
126 Agrosti-Alopecuretum pratensis " (")
94 Artemisietum maritimae Atl. (Estuaire de l'Authie)
25 Caricetum paleaceae " (Sweden, west coast)
26 Eleocharis unigl.-Agrostis stolon. Ass. " (")
24 Scirpetum maritimi " (")
104 Agrosti-Caricetum distan. molinietosum Pannonic (Hungary)
90 Salicornietum europaeae Atl. (Estuaire de l'Authie)
163 Salicornietum europaeae " (Ireland)
29 Salicornietum strictissimae " (Sweden, west coast)
75 Salicornietum emerici Medit. (Barcelona)
21 Salicornietum strictae " (Venice Lagoon)
166 Suaeda maritima- Bassia hirsuta " (N-Aegean coast)
186 Halocnemetum strobilacei " (Mesolonghi Lagoons)
82 Salicornietum patulae Baltic (Poland)
9 Salicornietum venetae Medit. (Venice Lagoon)
10 Suaedetum maritimae-Kochietum " (")
15 Pholiureto-Spergularietum marginatae " (")
35 Salicornietum europaeae Atl. (Sweden, west coast)
47 Salicornietum radicantis " (Bretagne du Nord)
91 Suaedeto mar.-Asteretum tripolii " (Estuaire de l'Authie)
89 Spartinetum townsendi " (")
44 Spartinetum strictae " (Bretagne du Nord)
1 Salicornieto-Spartinetum " (Netherl.)
58 Salicornietum europaeae hungaricum Pannonic (Hungary)
4 Artemisietum marit. facies a Statice lim. Atl. (Netherl.)
30 Puccinellietum maritimae " (Sweden, west coast)
36 Puccinellietum maritimae " (")
2 Puccinellietum maritimae " (Netherl.)
136 Rel. with Puccinelllia maritima " (Netherl., Wadden Inseln)
93 Puccinellietum maritimae " (Estuaire de l'Authie)
49 Puccinellietum maritimae " (Bretagne du Nord)
46 Salicornietum europaeae " (")
92 Obionetum portulacoidis " (Estuaire de l'Authie)
3 Artemisietum maritimae " (Netherl.)
135 Rel. with Puccinellia maritima " (Vendée; Charente)
11 Salsoletum sodae Medit. (Venice Lagoon)
41 Puccinell. salin.-Lepidium cartilag. Ass. Pannonic (Austria)
107
109
121
120
122
42
115
125
101
118
97
99
98
100
119
124
123
105
96
110
39
111
38
112
113
127
114
126
94
25
26
24
104
90
163
29
75
21
166
186
82
9
10
15
35
47
91
89
44
1
58
4
30
36
2
136
93
49
46
92
3
135
11
41

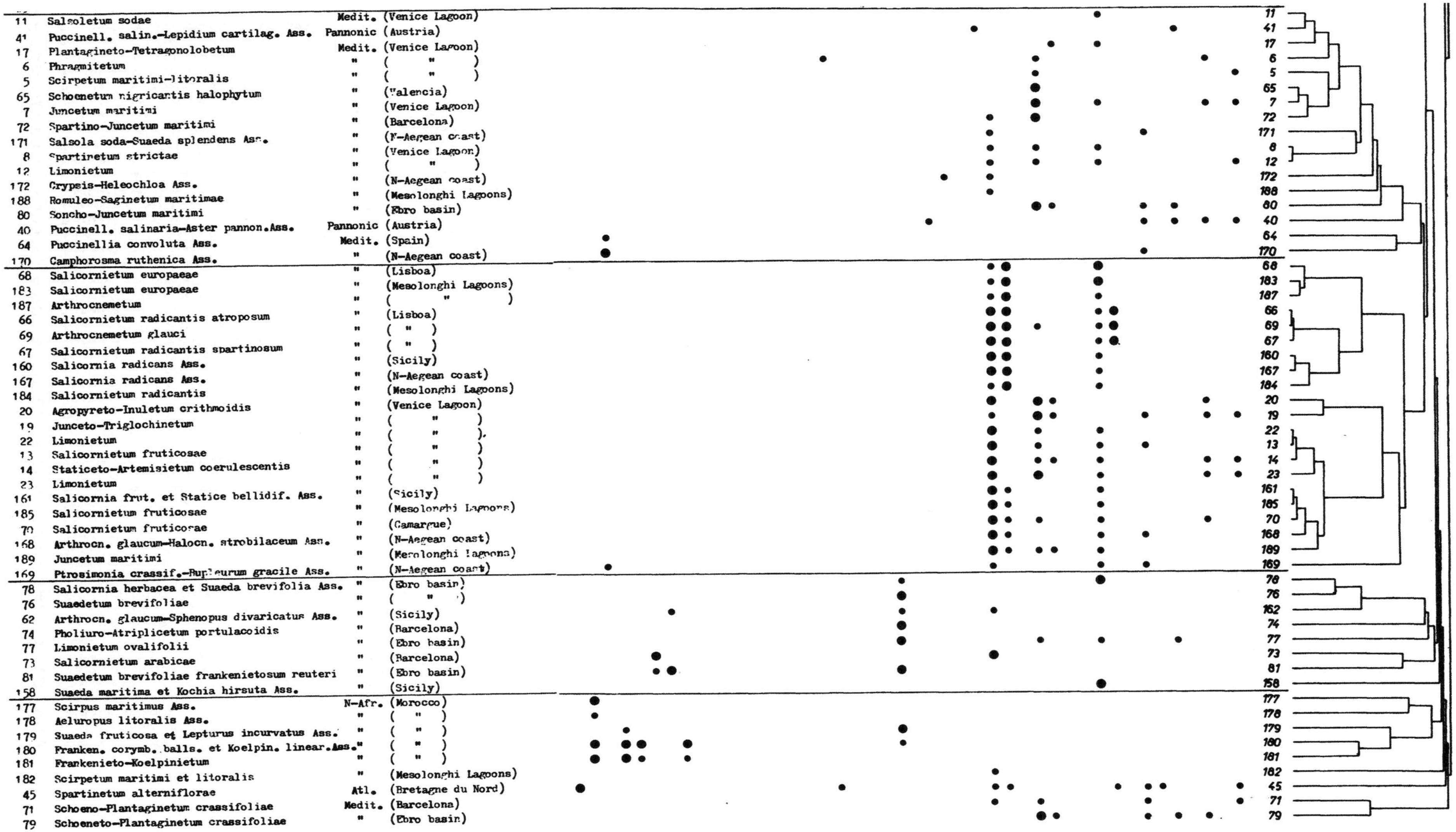

Fig. 2 (see text)

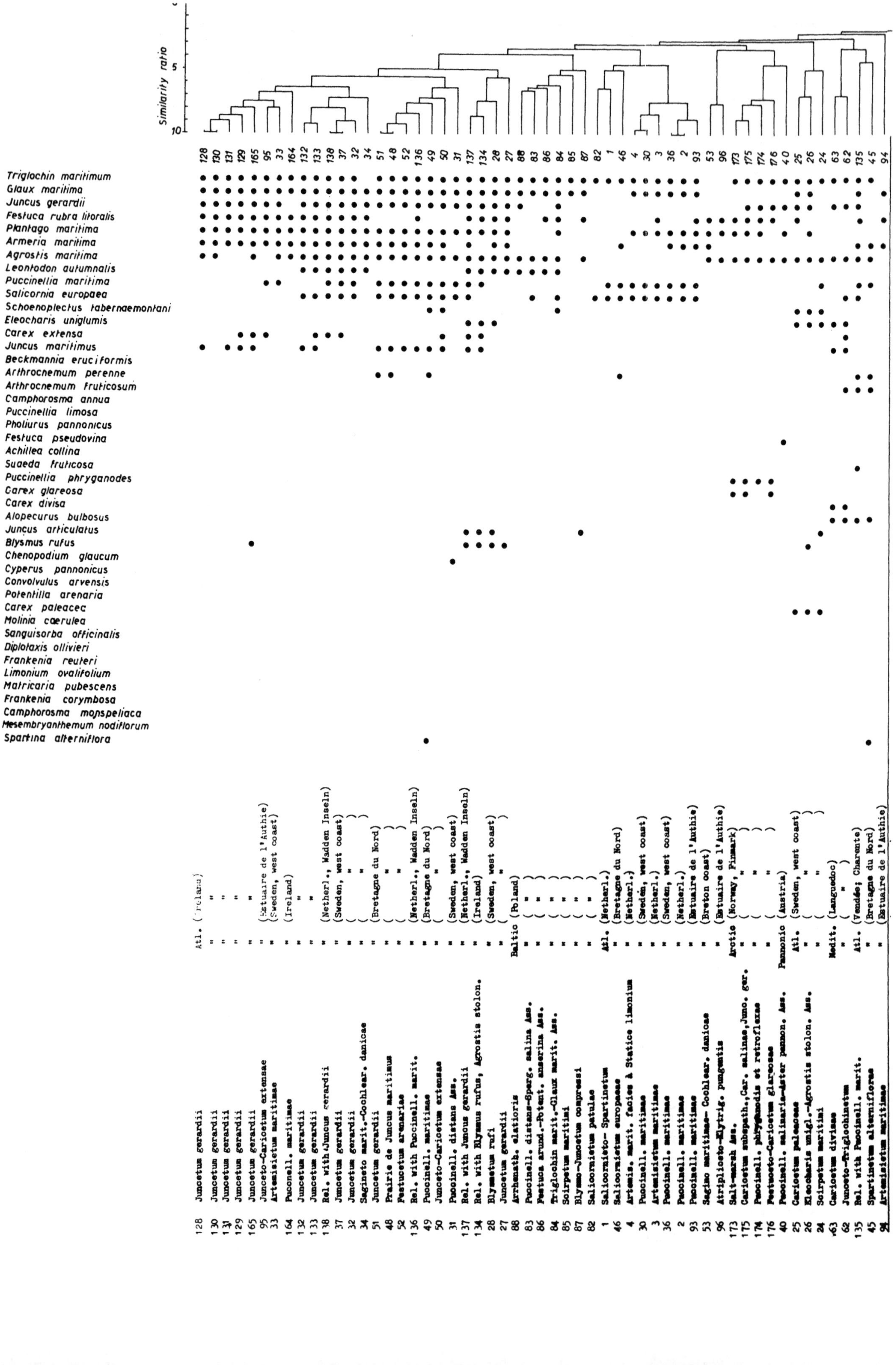

Similarity ratio
5
10
128 130 131 129 165 95 33 164 132 133 138 37 32 34 51 48 52 136 49 50 31 137 134 28 27 88 83 86 84 85 87 82 1 46 4 30 3 36 2 93 53 96 173 175 174 176 40 25 26 24 63 62 135 45 94
Triglochin maritimum
Glaux maritima
Juncus gerardii
Festuca rubra litoralis
Plantago maritima
Armeria maritima
Agrostis maritima
Leontodon autumnalis
Puccinellia maritima
Salicornia europaea
Schoenoplectus tabernaemontani
Eleocharis uniglumis
Carex extensa
Juncus maritimus
Beckmannia eruciformis
Arthrocnemum perenne
Arthrocnemum fruticosum
Camphorosma annua
Puccinellia limosa
Pholiurus pannonicus
Festuca pseudovina
Achillea collina
Suaeda fruticosa
Puccinellia phryganodes
Carex glareosa
Carex divisa
Alopecurus bulbosus
Juncus articulatus
Blysmus rufus
Chenopodium glaucum
Cyperus pannonicus
Convolvulus arvensis
Potentilla arenaria
Carex paleacea
Molinia caerulea
Sanguisorba officinalis
Diplotaxis ollivieri
Frankenia reuteri
Limonium ovalifolium
Matricaria pubescens
Frankenia corymbosa
Camphorosma monspeliaca
Mesembryanthemum nodiflorum
Spartina alterniflora
128 Juncetum gerardii Atl. (Ireland)
130 Juncetum gerardii " "
131 Juncetum gerardii " "
129 Juncetum gerardii " "
165 Juncetum gerardii " "
95 Junceto-Caricetum extensae " (Estuaire de l'Authie)
33 Artemisietum maritimae " (Sweden, west coast)
164 Puccinell. maritimae " (Ireland)
132 Juncetum gerardii " "
133 Juncetum gerardii " "
138 Rel. with Juncus gerardii " (Netherl., Wadden Inseln)
37 Juncetum gerardii " (Sweden, west coast)
32 Juncetum gerardii " (")
34 Sagineto marit.-Cochlear. danicae " (")
51 Juncetum gerardii " (Bretagne du Nord)
48 Prairie de Juncus maritimus " (")
52 Festucetum arenariae " (")
136 Rel. with Puccinell. marit. " (Netherl., Wadden Inseln)
49 Puccinell. maritimae " (Bretagne du Nord)
50 Junceto-Caricetum extensae " (")
31 Puccinell. distans Ass. " (Sweden, west coast)
137 Rel. with Juncus gerardii " (Netherl., Wadden Inseln)
134 Rel. with Blysmus rufus, Agrostis stolon. " (Ireland)
28 Blysmetum rufi " (Sweden, west coast)
27 Juncetum gerardii " (")
88 Arrhenath. elatioris Baltic (Poland)
83 Puccinell. distans-Sperg. salina Ass. " (")
86 Festuca arund.-Potent. anserina Ass. " (")
84 Triglochin marit.-Glaux marit. Ass. " (")
85 Scirpetum maritimi " (")
87 Blysmo-Juncetum compressi " (")
82 Salicornietum patulae " (")
1 Salicornieto- Spartinetum Atl. (Netherl.)
46 Salicornietum europaeae " (Bretagne du Nord)
4 Artemis. marit. facies à Statice limonium " (Netherl.)
30 Puccinell. maritimae " (Sweden, west coast)
3 Artemisietum maritimae " (Netherl.)
36 Puccinell. maritimae " (Sweden, west coast)
2 Puccinell. maritimae " (Netherl.)
93 Puccinell. maritimae " (Estuaire de l'Authie)
53 Sagino maritimae- Cochlear. danicae " (Breton coast)
96 Atripliceto-Elytrig. pungentis " (Estuaire de l'Authie)
173 Salt-marsh Ass. Arctic (Norway, Finmark)
175 Caricetum subspath., Car. salinae, Junc. ger. " (")
174 Puccinell. phryganodis et retroflexae " (")
176 Festuceto-Caricetum glareosae " (")
40 Puccinell. salinaria-Aster pannon. Ass. Pannonic (Austria)
25 Caricetum paleaceae Atl. (Sweden, west coast)
26 Eleocharis unigl.-Agrostis stolon. Ass. " (")
24 Scirpetum maritimi " (")
63 Caricetum divisae Medit. (Languedoc)
62 Junceto-Triglochinetum " (")
135 Rel. with Puccinell. marit. Atl. (Vendée; Charente)
45 Spartinetum alterniflorae " (Bretagne du Nord)
94 Artemisietum maritimae " (Estuaire de l'Authie)

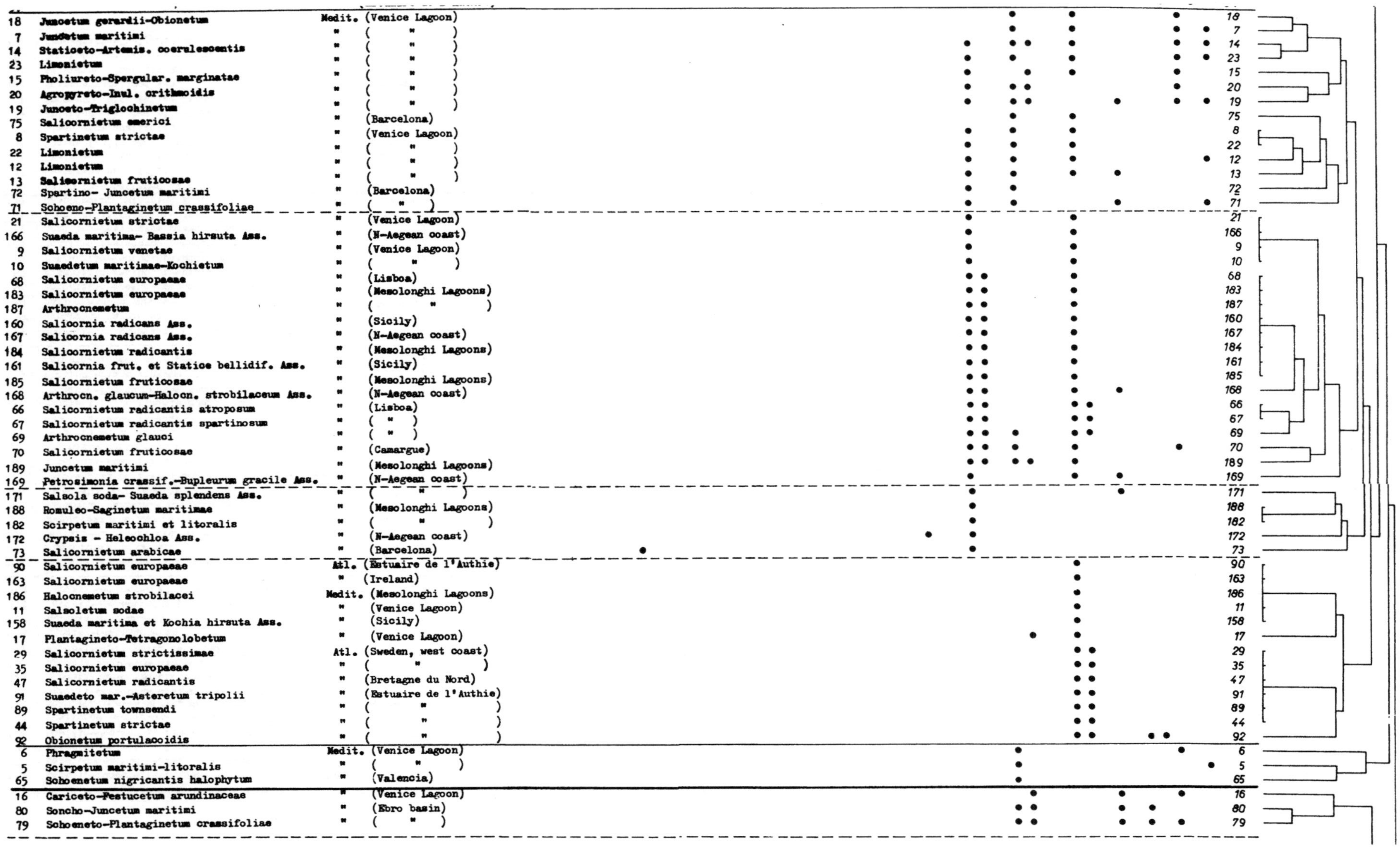

18 Juncetum gerardii-Obionetum Medit. (Venice Lagoon) 18
7 Juncetum maritimi " (") 7
14 Statioeto-Artemis. coerulescentis " (") 14
23 Limonietum " (") 23
15 Pholiureto-Spergular. marginatae " (") 15
20 Agropyreto-Inul. crithmoidis " (") 20
19 Junceto-Triglochinetum " (") 19
75 Salicornietum emerici " (Barcelona) 75
8 Spartinetum strictae " (Venice Lagoon) 8
22 Limonietum " (") 22
12 Limonietum " (") 12
13 Salicornietum fruticosae " (") 13
72 Spartino- Juncetum maritimi " (Barcelona) 72
71 Schoeno-Plantaginetum crassifoliae " (") 71
21 Salicornietum strictae " (Venice Lagoon) 21
166 Suaeda maritima- Bassia hirsuta Ass. " (N-Aegean coast) 166
9 Salicornietum venetae " (Venice Lagoon) 9
10 Suaedetum maritimae-Kochietum " (") 10
68 Salicornietum europaeae " (Lisboa) 68
183 Salicornietum europaeae " (Mesolonghi Lagoons) 183
187 Arthrocnemetum " (") 187
160 Salicornia radicans Ass. " (Sicily) 160
167 Salicornia radicans Ass. " (N-Aegean coast) 167
184 Salicornietum radicantis " (Mesolonghi Lagoons) 184
161 Salicornia frut. et Statice bellidif. Ass. " (Sicily) 161
185 Salicornietum fruticosae " (Mesolonghi Lagoons) 185
168 Arthrocn. glaucum-Halocn. strobilaceum Ass. " (N-Aegean coast) 168
66 Salicornietum radicantis atroposum " (Lisboa) 66
67 Salicornietum radicantis spartinosum " (") 67
69 Arthrocnemetum glauci " (") 69
70 Salicornietum fruticosae " (Camargue) 70
189 Juncetum maritimi " (Mesolonghi Lagoons) 189
169 Petrosimonia crassif.-Bupleurum gracile Ass. " (N-Aegean coast) 169
171 Salsola soda- Suaeda splendens Ass. " (") 171
188 Romuleo-Saginetum maritimae " (Mesolonghi Lagoons) 188
182 Scirpetum maritimi et litoralis " (") 182
172 Crypsis - Heleochloa Ass. " (N-Aegean coast) 172
73 Salicornietum arabicae " (Barcelona) 73
90 Salicornietum europaeae Atl. (Estuaire de l'Authie) 90
163 Salicornietum europaeae " (Ireland) 163
186 Halocnemetum strobilacei Medit. (Mesolonghi Lagoons) 186
11 Salsoletum sodae " (Venice Lagoon) 11
158 Suaeda maritima et Kochia hirsuta Ass. " (Sicily) 158
17 Plantagineto-Tetragonolobetum " (Venice Lagoon) 17
29 Salicornietum strictissimae Atl. (Sweden, west coast) 29
35 Salicornietum europaeae " (") 35
47 Salicornietum radicantis " (Bretagne du Nord) 47
91 Suaedeto mar.-Asteretum tripolii " (Estuaire de l'Authie) 91
89 Spartinetum townsendi " (") 89
44 Spartinetum strictae " (") 44
92 Obionetum portulacoidis " (") 92
6 Phragmitetum Medit. (Venice Lagoon) 6
5 Scirpetum maritimi-litoralis " (") 5
65 Schoenetum nigricantis halophytum " (Valencia) 65
16 Cariceto-Festucetum arundinaceae " (Venice Lagoon) 16
80 Soncho-Juncetum maritimi " (Ebro basin) 80
79 Schoeneto-Plantaginetum crassifoliae " (") 79

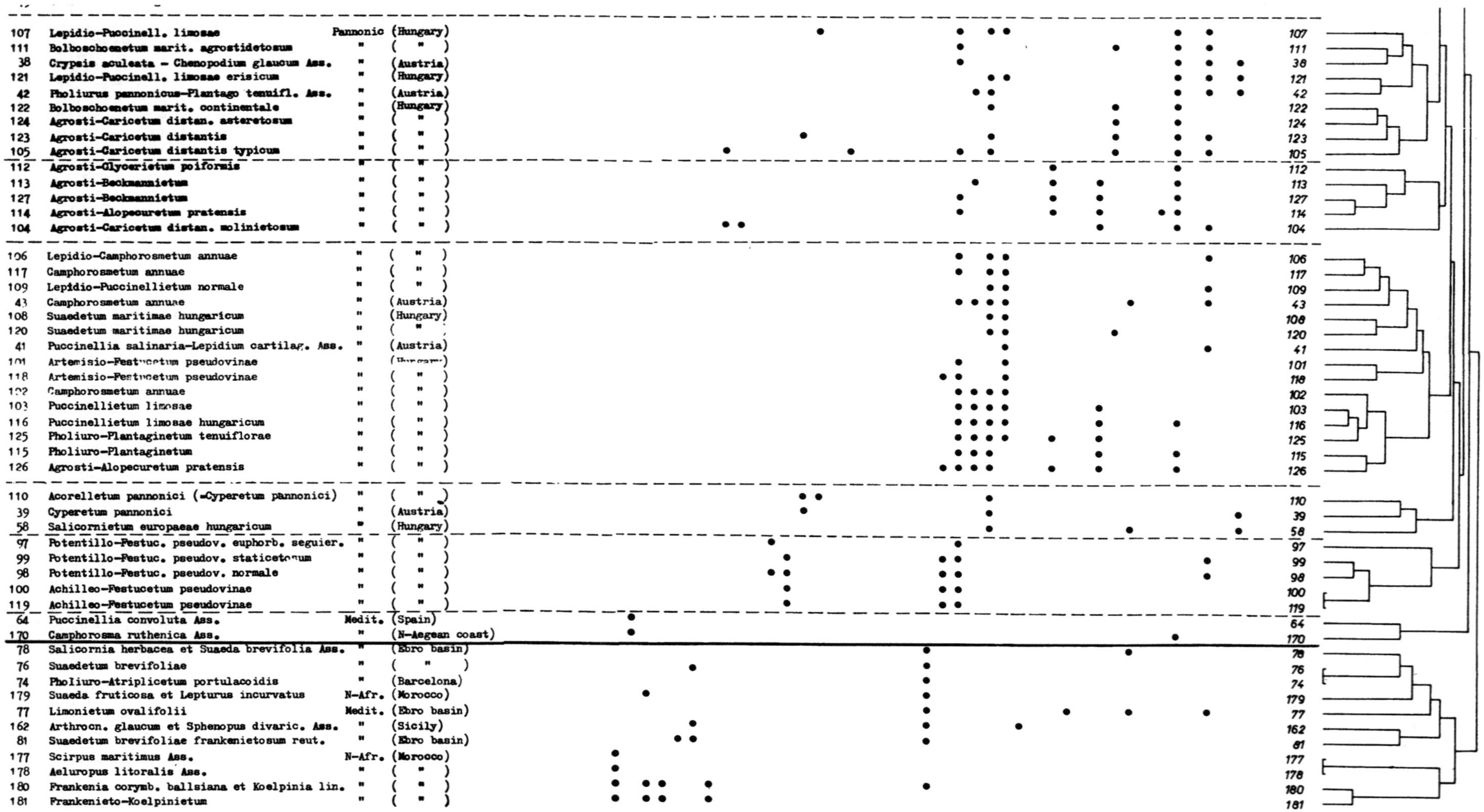

Fig. 3 (see text)

2.2 and 2.1, cluster 4 by species group 3.3, clusters 5 and 6 are characterized in different proportions by species groups 3.4 and 3.3 as well as by some species of group 3.2 (especially by *Juncus maritimus*), cluster 7 is characterized by species group 1.1 and cluster 8 by species group 1.2 Concerning the hierarchical structure of the dendrograms the smallest differences are found between *ALCAB* and *MC* which are both based on binary data. On the other hand there are some remarkable differences between these and *ALCAF* (frequency data).

The greatest difference between the three dendrograms concerns the position of cluster 1. In *ALCAF* this cluster appears strongly separated from all others, while in *ALCAB* and *MC* it is linked with clusters 4, 5 and 6. Another difference concerns the cluster 7 that is linked to clusters 4, 5, 6 in *ALCAF*, while in *ALCAB* and *MC* it is linked with cluster 8. The different fusion patterns may be due to the great influence of the frequencies in determining the similarities between the tables.

In *ALCAF*, cluster 1 is separated from all the others because it has many species of group 3.2 (the largest group of species) with high frequencies (see Figure 2). Moreover in *ALCAF*, cluster 1 contains the lowest number of tables in comparison with *ALCAB* and *MC*, because several tables with many species of group 3.2 but with a high frequency of *Salicornia europaea* and *Puccinellia maritima* have been assigned to cluster 4 (see Fig. 2, tables 30 to 135). Cluster 7 is linked with clusters 4, 5 and 6 since some tables have *Salicornia europaea* and *Arthrocnemum fruticosum* with high frequency.

The results of Concentration analysis allow an evaluation of the structure of the three dendrograms in terms of correspondence between species groups and clusters of tables. In Figure 5 the *CA* ordination patterns of species groups and clusters of *ALCAF*, *ALCAB* and *MC* are presented. The *Chi*-square values are significant at a very high probability level in all cases. The contingency coefficients (*C*) and the canonical correlation values (*R*) between species groups (X_1) and clusters of tables (Y_1) are a little higher for *ALCAB* and *MC* than for *ALCAF*. From Fig. 4 and 5 a good correspondence between the three ordination patterns and the classifications of *ALCAB* and *MC* becomes evident. The hierarchical dendrogram structure of *ALCAF* has no correspondence with the ordination patterns of *CA*, since cluster 1 doesn't result separated from clusters 4, 5 and 6.

The hierarchical dendrogram structure of *ALCAF* could not be confirmed from a floristic point of view, because cluster 1 is characterized by species which are largely in common with other clusters. On the basis of the results of *CA* it seems logical to chose the results of *ALCAB* or *MC* as a hierarchical classification of the European salt marshes involved. Since the two classifications are very similar and since the detailed results of *ALCAB* have already been presented in Fig. 3, the syntaxonomical interpretation will be based only on the results of *ALCAB*.

Syntaxonomical interpretation of clusters

In order to discuss the syntaxonomy derived from *ALCAB*,

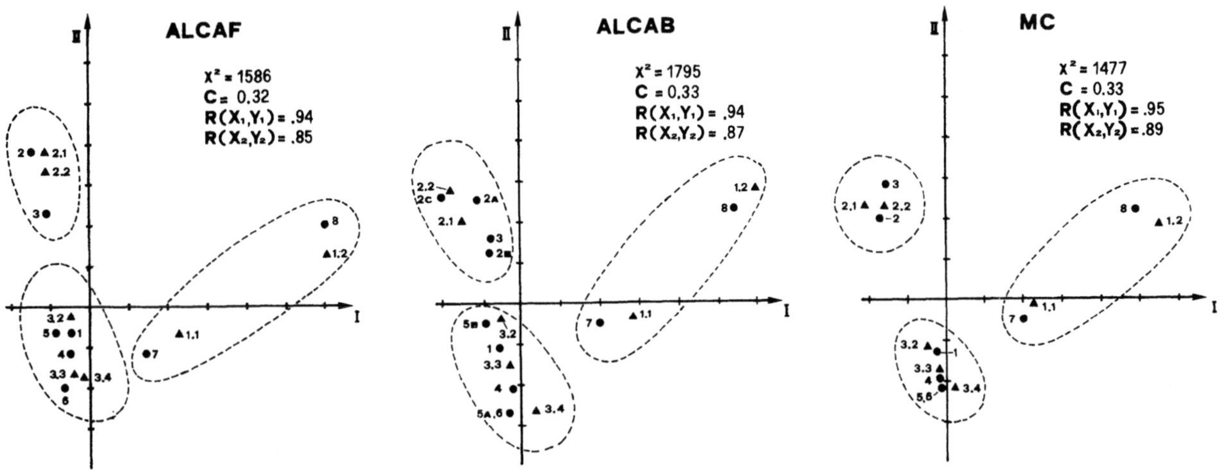

Fig. 5. Ordination of the clusters of tables and groups of species of Fig. 4 according to Concentration analysis. *C* is the contingency coefficient and *R* the canonical correlation between the axis of species and the axis of clusters. See text.
▲ species groups.
● clusters of tables.

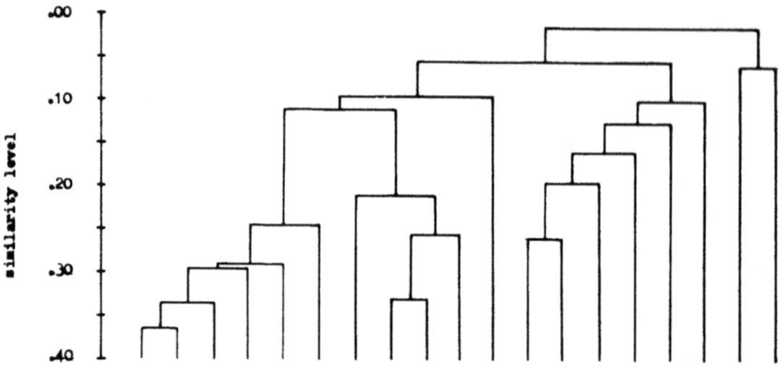

Species groups	Cluster code	1	2	3	4	5	6	7	8	9	10	11	12	13	14	15	16	17	18	19
	Code of tables from	128	88	82	173	25	63	90	18	21	171	6	16	107	112	106	110	97	78	177
	to	27	85	93	176	24	45	92	71	169	73	65	79	105	104	126	58	119	81	181
	N° of tables	25	5	9	4	3	4	13	14	19	5	3	3	9	5	15	3	5	7	4
	Mean number of selected species	10	5	5	6	8	9	2	4	3	2	2	4	5	4	4	3	3	2	3
	Mean number of species in the tables	25	44	13	24	12	40	8	17	14	20	16	29	18	32	17	8	55	19	15
	Mean % of retained species	45	14	39	29	64	30	26	28	29	12	14	14	20	13	31	38	7	15	20
	σ of the % of retained species	20	9	10	11	6	21	12	12	13	10	10	3	15	4	14	0	4	5	13
2.2	Eleocharis uniglumis					100	50								80	33				
	Pholiurus pannonicus										20			11	20	50				
	Festuca pseudovina													44	40	73		100		
	Beckmannia eruciformis														80	13				
	Achillea collina															13		80		
	Convolvulus arvensis																	80		
	Potentilla arenaria																	40		
2.1	Puccinellia limosa													67		80	100			
	Camphorosma annua													22		87				
5	Sanguisorba officinalis													11	20					
	Molinia coerulea														20					
	Chenopodium glaucum													11			33			
	Cyperus pannonicus													11			67			
4	Schoenoplectus tabernaemontani		20			100								56	6					
	Carex paleacea					100														
6	Puccinellia phryganodes				100															
	Carex glareosa				75															
3.1	Juncus articulatus	12				33								11						
	Blysmus rufus	20				33														
3.5	Carex extensa						25	8	29	5			100							
3.2	Agrostis maritima	88	80		100	100	100		21	10	20		100	100	100	15				
	Plantago maritima	100	20	67	75	33		8					67	77	20	26		40	14	
	Juncus gerardii	96	40		75	67	75		50	5		33	67	56			67			
	Triglochin maritimum	96	100	100	100	100	100		42			33								
	Juncus maritimus	56					50		93	16		100	67						14	
	Festuca rubra litoralis	76	40	10	100	33	25	8												
	Glaux maritima	100	100	67		67	25													
	Armeria maritima	92		44			50													
	Leontodon autumnalis	40	80																	
3.3	Salicornia europaea	52	40	100			50	100	71	100							33		29	
	Puccinellia maritima	56		89		33	25	54		16										
3.4	Arthrocnemum perenne	12		10			50			74										
	Arthrocnemum fruticosum						100		80	100	100								14	
7	Alopecurus bulbosus						100													
	Carex divisa						50													
	Spartina alterniflora	4					25													
1.1	Suaeda fruticosa						25												100	24
	Limonium ovalifolium										20								14	
	Frankenia reuteri																		43	
1.2	Frankenia corymbosa																		14	50
	Mesembryanthemum nodiflorum																			100
	Matricaria pubescens																			50
	Diplotaxis ollivieri																			50
	N° of species in the clusters	15	9	8	7	12	17	5	7	8	4	3	5	12	9	9	5	5	8	5
	Within cluster similarity x 100	50	50	45	70	50	33	47	40	45	40	30	50	33	48	31	40	30	25	25

Fig. 6. Synthetic structure of the dendrogram of Fig. 3 used for syntaxonomical interpretation. For each species it is reported the frequency in each cluster. See text.

Fig. 3 has been summarized in Figure 6. In this figure the main species groups are ordered according to their projection on a line interpolating the points in the ordinations models (Fig. 5) produced by *CA*. The dendrogram of Fig. 6 shows 19 main clusters of tables characterized by species groups. (See also Fig. 3 for information on the types).

– Cluster 1 is characterized by the high frequency of many species of group 3.2 but not by own species. This cluster lumps types that could be assigned to the *Armerion* alliance. It contains the highest mean number of selected species per table i.e. 10. All the types have an atlantic distribution (Figure 3).

Cluster 2 is characterized by *Leontodon autumnalis* and by the lack of some species of the group of *Armeria maritima* and *Juncus maritimus* (3.2). The tables of this cluster are all from baltic regions (Poland), the greater part of them could be identified as the alliance *Puccinellio maritimae-Spergularion salinae* Beeftink 1965.

– Cluster 3 is characterized by the high frequency of *Puccinellia maritima* and *Salicornia europaea* and by the lack of *Agrostis maritima, Juncus gerardii, Juncus maritimus* and *Leontodon autumnalis*. Almost all the types of this cluster could be assigned to the *Puccinellion maritimae* R. Tüxen 1937, which has mainly an atlantic distribution.

– Cluster 4 is characterized by *Puccinellia phryganodes* and *Carex glareosa* and by other species of group 3.2. This cluster could be identified as *Puccinellion phryganodes* Hadač 1946, which seems to include the *Caricion glareosae* Nordhagen 1954. The distribution of the types of this alliance is mainly arctic and subarctic.

– Cluster 5 contains only three types from the south west coast of Sweden. It is characterized by *Carex paleacea* which has been suggested by Dahl & Hadač (1941) as character species of a special alliance: *Magnocaricion paleaceae*. This cluster contains the highest percentage of selected species (64 %).

– Cluster 6 is characterized by *Alopecurus bulbosus* and *Carex divisa* (species group 7). Two types of this cluster have been assigned to *Juncion maritimi*. The cluster contains the highest total number of selected species (17); it includes tables from both the west mediterranean and the atlantic coasts.

– Cluster 7 is characterized by the high frequency of *Salicornia europaea* and *Puccinellia maritima* (species group 3.2). The tables of this cluster are from both the atlantic and mediterranean coasts. Notwithstanding this wide distribution, the similarity within the cluster is very high. The mean species number (8) is very low. The types have been assigned to *Thero-Salicornion* or to *Thero-Suaedion*, both being characterized by annual species. However, from the cluster structure the alliance *Thero-Suaedion* doesn't appear as a distinct syntaxonomical unit.

– Cluster 8 is characterized by *Juncus maritimus, Arthrocnemum fruticosum* and *Salicornia europaea*. All the tables of this cluster are from mediterranean areas, especially from the Venice Lagoon. Three of them are from Spain (Barcelona). The types have been assigned to different alliances: *Juncion maritimi, Puccinellio-Salicornion* Pignatti 1966, *Thero-Salicornion* and *Plantaginion crassifoliae*. Notwithstanding this heterogeneous syntaxonomical assignment cluster 8 is relatively homogeneous. It could be subdivided into two clusters only: one characterized by *Juncus gerardii* and by *Carex extensa* and the other by the lack of these two species.

– Cluster 9 is characterized by species group 3.4 (*Arthrocnemum perenne* and *A. fruticosum*) as well as by the high frequency of *Salicornia europaea*. It is a very homogeneous cluster with tables from all over the mediterranean coasts. In this cluster we find types assigned mainly to *Salicornion fruticosae* and some of them to *Thero-Salicornion, Staticion galloprovincialis* or *Staticion orientale*.

– Cluster 10 is characterized only by *Arthrocnemum fruticosum*. It has the lowest mean number of selected species (2) but a relatively high mean total number of species (20). The tables are mainly from North-Aegean coasts; only one is from Spain. This cluster includes types which have been assigned to *Thero-Salicornion* and to *Salicornion arabicae* or types which may be considered transitional to *Saginion maritimae* or *Nanocyperion*.

– Cluster 11 is characterized only by *Juncus maritimus*. It contains the least halophilous associations of the data set. Two of them have been assigned to *Phragmition*, the other could be assigned to *Plantaginion crassifoliae*. The tables of this cluster are from the mediterranean area.

– Cluster 12 is characterized by the high frequency of *Carex extensa, Agrostis maritima, Plantago maritima, Juncus gerardii* and *Juncus maritimus*. The tables, all from the mediterranean area, don't contain *Arthrocnemum fruticosum* and *Salicornia europaea*, for this reason they appear separated from the other mediterranean tables and become connected with the continental tables with *Agrostis maritima* and *Juncus gerardii*, even if at a low similarity level. The types of this cluster have been assigned to *Juncion maritimi* or *Plantaginion crassifoliae*.

– Cluster 13 is characterized by the high frequency of *Agrostis maritima, Plantago maritima* and with lower frequency by *Juncus gerardii* and *Puccinellia limosa*. Some types (38–42; Fig. 3) of this cluster could be assigned to *Juncion gerardii* Wendelberger 1943, while all of them have indicator species of *Agrostion, Cypero-Spergularion salinae, Puccinellion limosae, Bolboschoenion, Phragmition* and *Beckmannion eruciformis*. All the tables of this cluster are from the pannonic area.

– Cluster 14 is characterized by *Beckmannia eruciformis* and *Eleocharis uniglumis* and could be identified as *Beckmannion eruciformis*. As in cluster 13 the tables contain indicator species of *Agrostion, Ruppion, Phragmition, Puccinellion limosae, Juncion gerardii* and *Festucion pseudovinae*. The types of this cluster have a pannonic distribution.

– Cluster 15 is characterized by *Puccinellia limosa, Camphorosma annua, Festuca pseudovina* and *Pholiurus pannonicus*. It could represent the pannonic alliance *Puccinellion limosae*. In this cluster a subgroup of tables are characterized by indicator species of the *Beckmannion eruciformis*.

– Cluster 16 is characterized by *Puccinellia limosa, Cyperus pannonicus* and *Juncus gerardii*, it could represent the alliance *Cypero-Spergularion salinae* of pannonic distribution. As cluster 7 (*Thero-Salicornion*) it has the lowest number of species in the tables.

– Cluster 17 is characterized by *Festuca pseudovina, Achillea collina, Convolvulus arvensis* and *Potentilla arenaria*. The tables have all been assigned to *Festucion pseudovinae* Soó 1968 of pannonic distribution. The mean species number in the tables is the highest of the data set, which can be explained by the meadow character of the type.

– Cluster 18 is characterized by *Suaeda fruticosa* subsp. *braevifolia* and *Frankenia reuteri*. The tables of the cluster originate from salt steppe vegetation types of the Ebro basin (Spain) and Morocco (North Africa). The majority of them have been assigned to *Suaedion braevifoliae* Br.-Bl. et De Bolós 1957.
– Cluster 19 is characterized by *Mesembryanthemum nodiflorum*, *Frankenia corymbosa*, *Diplotaxis ollivieri* and *Matricaria pubescens*. This cluster includes salt desert types of North-Africa (Morocco) which have never been assigned to any syntaxa of a higher level.

By comparing the 19 clusters from a syntaxonomical point of view it appears that several of them correspond completely to already described alliances. Clusters 1, 2, 3, 4, 5 of atlantic, baltic and arctic distribution, and clusters 14, 15, 16, 17 of pannonic distribution have the best correspondence with known alliances. Details on the geographical distribution of the alliances corresponding to these clusters were given by Beeftink (1972) on the basis of a larger data set including the tables used in the present study.

From the mediterranean area only cluster 18 has a good correspondence with an alliance: *Suaedion braevifoliae*. For each of the others: 7, 8, 9, 10, 11, 12, there are types assigned to different alliances, even if each cluster shows a high within-similarity. This fact could be explained by the presence of species with a local importance in such types but without a high frequency in the tables. Such species have not been selected in this study since they do not have enough potential to characterize sets of tables, even if their fidelity would allow the characterization of alliances restricted to the Mediterranean area such as the *Plantaginion crassifoliae*, *Staticion galloprovincialis*, *Staticion orientale* and *Juncion maritimi*. On the basis of the selection of species those alliances do not fit into any cluster.

According to the species selection the salt marsh vegetation of the Mediterranean shows three main trends of variation. The first one is indicated by cluster 7, which is characterized by annual species. The second trend is indicated by cluster 8, containing species with mainly atlantic distributions. This trend is mainly found in the Venice Lagoon. It suggests a connection between atlantic and mediterranean salt marsh vegetation. This connection can be easily explained by ecological factors such as a high variation of the tide level, wet halic soils and low winter temperatures which are characteristic for the north Adriatic Sea. The third trend, indicated by cluster 9, is correlated with more strictly mediterranean characteristics such as arid halic soils and higher temperatures in all seasons.

These three mediterranean vegetational trends correspond to clusters which have a relatively high homogeneity as measured on the basis of the selected species. For this reason these clusters should have the syntaxonomical rank of alliance, as judged from the classificatory scheme produced by *ALCAB*. In fact this level may be adopted for all 19 clusters produced by *ALCAB*, because most of them correspond to already well-defined alliances. At this hierarchical level no real discrepancies with the syntaxonomical literature have been found, notwithstanding the strong reduction in species used in the classification process.

Clearly there is a considerable variation as to the homogeneity of the numerical units arrived at, because the collection of phytosociological tables as such is rather heterogeneous in itself. Some tables involved have only one real halophyte as a constant species and include transitions to non-halophytic vegetation. It is remarkable that tables of the latter kind have a marginal position in clusters, or belong to isolated clusters, such as clusters 10 and 11.

Numerical syntaxonomy

The main purpose of numerical syntaxonomy is the quantification of the various syntaxonomical levels by measures of similarity. As Kortekaas, van der Maarel & Beeftink (1976) and Westhoff & van der Maarel (1978) pointed out, units which are considered as belonging to one hierarchical level do not vary much in similarity level. On the basis of Jaccard's coefficient (or similarity ratio for presence-absence values) (which gives values about 10 % lower than the Sørensen and Kulczinski coefficients) the association level corresponds with a similarity level of 0.40–0.60, the alliance level with 0.30–0.40 and the order level with 0.10–0.30. No values have been reported for the class level.

The dendrogram of Fig. 3, as synthesized in Fig. 6, shows fusions within the 19 main clusters on levels between 0.25 and 0.70, the majority varying only between 0.30 and 0.50. These levels correspond to the alliance level according to the general findings reported above, which may be considerd further evidence for adopting the alliance level for our 19 clusters. The dendrogram of Fig. 6 can also be used to create groups of clusters at the order and the class levels respectively. It appears that the order level cannot be defined easily: the number of clusters drops from 16 at similarity level 0.30 to only 6 at 0.10. To obtain the most effective fusions the mutual information between species and clusters was calculated for the various dichotomic levels of the dendrogram, and compared with the equivocation information. (see above). Two relative maxima in the value of the ratio of these parameters were found at 0.42 and 0.49 respectively. The first maximum is found with the following fusion pattern: 1+2+3+4 at fusion level 0.29), 8+9 (0.33), 12+13 (0.26), all other clusters remaining isolated. The next best pattern is: clusters 1 to 10 (level 0.11), 12 to 16 (0.13), 11, 17, 18 and 19 isolated. These two patterns were then adopted as order and class level respectively.

By comparing the fusions with the syntaxonomical literature (Beeftink 1968; Soó 1968; Braun Blanquet, Roussine & Nègre 1952; Braun Blanquet 1974; Pignatti 1966) it is possible to find the following correspondences.

The fusion of clusters 1, 2, 3, 4 (respectively *Armerion maritimae*, *Puccinellio maritimae-Spergularion salinae*, *Puccinellion maritimae* and *Puccinellion phryganodis* incl. *Caricion glareosae*) lumps *Glauco-Puccinellietalia* and *Carici-Puccinellietalia*. The fusion of clusters 8, 9 links *Juncetalia maritimi* and *Salicornietalia fruticosae*. The fusion of clusters 12, 13 links *Juncetalia* types of mediterranean distribution with the continental order *Festuco-Puccinellietalia* Soó 1968.

Concerning the remaining clusters, the 5th, connected expecially with cluster 4, corresponds to the *Magnocaricion paleàceae* of boreal distribution; the 6th links mediterranean and atlantic types assigned to *Juncetalia*; the 7th corresponds to *Thero-Salicornietalia* including *Thero-Suaedetalia*; the 10th contains contact types of *Salicornietalia fruticosae*, *Juncetalia* and *Saginetalia*; the 11th contains contact types with *Phragmitetalia*. Clusters 13–15, containing types all assigned to *Festuco-Puccinellietalia* Soó 1968 (Syn. *Puccinellietalia* Soó 1940), lead to the suggestion to subdivide this order in three different ones. The 17th cluster could correspond to *Artemisio-Festucetalia pseudovinae* Sóo 1968 *Hydrostachietalia* Ţopa non Grossheim).

Cluster 16 (*Cypero-Spergularin salineae*) have been assigned by Soó (1968) to *Thero-Salicornietalia*, but to *Puccinellietalia* Soó 1940 (Syn. *Festuco-Puccinellietalia* Soó 1968) by Beeftink (1968).

Cluster 18 and 19 could represent two new orders or classes characterized respectively by *Suaeda braevifolia* and *Mesembryanthemum nodiflorum*, the first including the west mediterranean salt steppes of Europe, the second salt desert types of North-Africa.

At the class level the atlantic and mediterranean salt marsh vegetation (from cluster 1 to cluster 10) give origin to only one class including the proposed classes: *Astereoa tripolii*, *Juncetea maritimi*, *Salicornietea fruticosae* and *Thero-Salicornietea*. The pannonic salt marshes give origin to one class too, corresponding to *Festuco-Puccinellietea* Soó 1968. This class includes also the pannonic types assigned to *Thero-Salicornietea*.

This result is quite in accordance with Concentration Analysis applied to the results of *ALCAF*, *ALCAB* and *MC*. In the ordination patterns of *CA* (Fig. 5) the class of mediterranean and atlantic vegetation (clusters 1, 4, 5, 6) appears well separated from the class of pannonic vegetation (clusters 2, 3), both represented by clusters with a high density.

Discussion

The numerical procedures applied in this study lead to a hierarchical system of European salt marsh communities in which the levels are quantified by similarity measures. The hierarchical system is based on a logical, reproducible structure and it offers new data on which judgements about the classification of salt marshes can be derived.

The lowest level of the system obtained accords very well with the alliance level, as used in the literature. At the order level many differences exist between the various systems. The present numerical system tends to combine some orders as distinguished in the syntaxonomical literature whereas other orders, which have been traditionally considered as such tend to be subdivided. At the class level our system suggests the unification of some classes as they are used in literature.

The system seems to satisfy both edaphic-ecological and historic-geographical criteria which are the main ecological criteria for a syntaxonomical classification (Westhoff & van der Maarel 1978). Although the edaphic-ecological criterion appears to prevail the synchorological differentiation is also reflected in the system. In fact it is possible to find a good accordance between the clusters and the geographical origin of the corresponding tables (Fig. 3).

The numerical system shows the main trends of variation of the halophytic vegetation through three groups of species: the euryhalic group (3.2), the stenohalic group of mainly coastal-mediterranean distribution (3.4) and the more or less euryhalic continental pannonic groups (2.1, 2.2). Fig. 6 shows the pattern of variation in the importance of these groups in the various clusters. Species group 3.2, fully represented in the cluster corresponding to the alliance *Armerion*, may be considered the core of the system. It is connected with group 3.4 by species group 3.3 (*Salicornia europaea* and *Puccinellia maritima*). This connection indicates the unification of atlantic and mediterranean salt marshes at the class level. The central position of species groups 3.2 and 3.3 is evident also in all ordinations produced by Concentration analysis (Fig. 5). Through closer investigation of the figures presented in this study much more information on the ecology and chorology of the elaborated types of vegetation could be derived. However, this cannot be the aim of the present contribution.

The application of numerical methods does not only lead to graphical displays of data structures from which trends of variation can be deducted, it also creates logically reproducible reference systems for the quantification of ecological concepts and thus opens new perspectives for phytosociology. It is true that still no communis opinio exists as to the choice of coefficients and techniques, but in any particular set of options one logical structure arises which can lead to objective comparisons.

When considering the wide variety of opinions amongst phytosociologists as to the syntaxonomy of vegetation types one concludes towards the urgent need to provide an objective reference system. Obviously the objectives of any syntaxonomical research should be expressed in clear and logical terms, and this is not so much a matter of numerical methods but rather of underlying theory. The numerical system produced in the present study is fully concordant with the theory of the Braun- Blanquet approach in phytosociology and for this reason it could serve as a basic reference system for a Prodrome of European salt marsh communities.

Summary

This work is devoted to a hierarchical classification of European salt marsh vegetation based on numerical methods. A numerical syntaxonomy has been obtained by automatic methods of classification applied to synoptic tables of salt marsh communities. The data have been reduced by a process of species selection based on a weighting procedure which gives high weights to species with high frequency density in Raunkiaer class V. The methods, including both monothetic divisive and polythetic agglomerative procedures, give very similar results at low hierarchical levels. The greater part of clusters defined at the first level above the association, correspond to well-described alliances in phytosociological literature.

The methods give different results at higher hierarchical levels. The hierarchical structure has been evaluated in terms of correspondence between species and clusters of synoptic tables by Concentration Analysis and by the ratio: mutual information/equivocation information. The most efficient structure has been provided by methods based on presence-absence data.

The resulting numerical syntaxonomical scheme encompasses 14 groups of tables at the order level and 4 groups at the class level. The scheme suggests similarity values within alliance, order and class level which are quite comparable with those already reported in literature. A strong relationship between the syntaxonomical scheme and ecological and chorological factors is emphasized. The need to use logical models on which to base syntaxonomical decisions is stressed.

Zusammenfassung

Die vorliegende Arbeit befaßt sich mit der hierarchischen Klassifikation europäischer Pflanzengesellschaften auf Salzböden mit Hilfe numerischer Methoden. Dabei konnte eine numerische Syntaxonomie durch automatische Klassifikationsmethoden, die ihrerseits auf synoptische Tabellen von Salzböden-Vegetation gestützt sind, erzielt werden. Die Angaben wurden durch eine Auswahl von Arten reduziert, indem jenen Arten mehr Bedeutung beigemessen wurde, deren Häufigkeit zu Klasse V nach Raunkiaer gehört. Die Methoden, die sowohl das monothetisch trennende als auch das polythetisch zusammenfassende Verfahren einschließen, ergeben bei niederem hierarchischen Niveau sehr ähnliche Resultate. Der Großteil der Clusters, die sich bei der ersten Stufe oberhalb der Assoziation definieren lassen, stimmt mit gut beschriebenen Verbänden in der bereits bestehenden pflanzensoziologischen Literatur überein.

Auf höherem hierarchischen Niveau zeigen die Methoden verschiedene Ergebnisse. Die hierarchische Struktur wurde nach der Übereinstimmung zwischen Arten und Clusters synoptischer Tabellen durch die Konzentrationsanalyse und den Ratio: mutuelle Information/Equivocation geschätzt. Als wirkungsvollste Struktur erwiesen sich dabei Methoden, die sich auf Angaben über Vorhandensein bzw. Nichtvorhandensein von Arten stützen.

Das sich daraus ergebende numerische syntaxonomische Schema umfaßt 14 Tabellengruppen mit Ordnungs- und 4 Gruppen mit Klassen-Niveau. Aus dem Schema ergeben sich Ähnlichkeitswerte innerhalb des Verbands-, Ordnungs-, und Klassen-Niveaus, die den in der Literatur bestehenden ziemlich vergleichbar sind. Enge verwandtschaftliche Beziehungen zwischen dem syntaxonomischen Schema und ökologischen sowie chorologischen Faktoren lassen sich feststellen. Die Notwendigkeit, syntaxonomische Entscheidungen aufgrund logischer Modelle zu treffen, scheint naheliegend.

References

Adriani, M.J. 1945. Sur la phytosociologie, la synécologie et le bilan d'eau de halophytes. Thèse Amsterdam, SIGMA Comm. 88, Montpellier, 217 p.

Beeftink, W.G. 1968. Die Systematik der Europäischen Salzpflanzengesellschaften. In R. Tüxen (ed.) Pflanzensoziologische Systematik. Ber. Int. Symp. Stolzenau, p. 239–272. Junk, Den Haag.

Beeftink, W.G. 1972. Übersicht über die Anzahl der Aufnahmen europäischer und nordafrikanischer Salzpflanzengesellschaften für das Projekt der Arbeitsgruppe für Datenverarbeitung. In E. van der Maarel & R. Tüxen (eds.), Grundfragen und Methoden der Pflanzensoziologie. Ber. Int. Symp. Rinteln 1970, p. 371–396. Junk, Den Haag.

Bodrogközy, G. 1959. Ökologische Verhältnisse der Standortstypen der Sandweiden von falschem Schafschwingel (Potentillo-Festucetum pseudovinae danubiale) in Süd-Kiskunsag. Acta Biol. N.S. 5: 145–160.

Bodrogközy, G. 1960. Phytozönologische und bodenökologische Untersuchungen an den Sumpfwiesen im Süden des Gebietes Kiskunsag (Klein-Kumanien). Acta Bot. Acad. Scient. Hung. 6: 171–207.

Bodrogközy, G. 1962. Die standortökologischen Verhältnisse der halophilen Pflanzengesellschaften des Pannonicum. I. Untersuchungen an den Solontschak-Szikböden der Südlichen Kiskansag. Acta Bot. Acad. Scient. Hung. 8: 1–37.

Bodrogközy, G. 1965a. Ecology of the halophilic vegetation of the Pannonicum. II. Correlation between Alkali (Szik)plant communities and genetic soil classification in the northern Hortobagy. Acta Bot. Acad. Scient. Hung. 11: 1–51.

Bodrogközy, G. 1965b. Ecology of the halophilic vegetation of the Pannonicum. III. Results of the investigation of the Solonetz of Oroshaza. Acta biol. 11: 3–25.

Bodrogközy, G. 1965c. Ecology of the halophilic vegetation of the Pannonicum. IV. Investigations on the Solonetz meadow soils of Oroshaza. Acta Biol. N.S. 11: 207–227.

Bodrogközy, G. 1966. Ecology of the halophilic vegetation of the Pannonicum. V. Results of the investigation of the 'Fehérto' of Oroshaza. Acta Bot. Acad. Scient. Hung. 12: 9–26.

Bolós, O. de. 1962. El paisaje vegetal barcelonés. Fac. de filos. y let. Univ. Barcelona.

Braun-Blanquet, J. 1974. Die höheren Gesellschaftseinheiten der Vegetation des südeuropäisch-westmediterranen Raumes. SIGMA Comm. 204, Montpellier 7 pp.

Braun-Blanquet, J. & O. de Bolós. 1957. Les groupements végétaux du bassin moyen de l'Ebre et leur dynamisme. An. Estac. Exp. Aula Dei 5: 1–266.

Braun-Blanquet, J. & Cl. de Ramm. 1957. Contribution à la connaissance de la végétation du littoral méditerranéen. Les prés salés du Languedoc Méditerranéen. Bull. Mus. Hist. Nat. Marseille. 17: 5–43.

Braun-Blanquet, J., N. Roussine & R. Nègre. 1952. Les groupements végétaux de la France Méditerranéenne. Centre Nat. Rech. Scient.: 297 pp. Montpellier.

Braun-Blanquet, J. & R. Tüxen. 1952. Irische Pflanzengesellschaften. Veröff. Geobot. Forsch. Inst. Rübel Zürich 25: 224–421.

Gorillion, M.R. 1953. Les halipèdes du Nord de la Bretagne. Revue Gén. Bot. 60: 609–658, 707–775.

Cristofolini, G., D. Lausi & S. Pignatti. 1970. Über statistische Eigenschaften der Charakterarten und deren Verwertung zur Aufstellung einer empirischen Systematik der Pflanzengesellschaften. In R. Tüxen (ed.). Gesellschaftsmorphologie. Ber. Int. Symp. Rinteln, p. 8–15. Junk, Den Haag.

Feoli, E. 1977. A criterion for monothetic classification of phytosociological entities on the basis of species ordination. Vegetatio 33: 147–152.

Feoli, E. & M. Lagonegro. 1978. Intersection analysis in phytosociology: computer program and applications. Paper presented at 'Advanced Study Institute and Research Workshop in Statistical Ecology'. Parma 1978.

Feoli, E. & L. Orlóci. 1979. Concentration analysis and detection of underlying factors in structured tables. Vegetatio (in press).

Feoli Chiapella, L. & E. Feoli. 1977. A numerical phytosociological study of the summits of the Majella massive (Italy). Vegetatio 34: 21–39.

Fewster, P. & L. Orlóci. 1978. Stereograms to aid group recognition and trend identification in vegetation data. Can. J. Bot. 56: 162–165.

Fontes, F.C. 1945. Algumas caracteristicas fitosociologicas dos 'Salgados' de Sacavém. Bol. Soc. Broteriana, Ser. 2 19: 789–813.

Frei, M. 1937. Studi fitosociologici su alcune associazioni litorali in Sicilia. N.G. Bot. Ital. 44: 273–294.

Gillner, V. 1960. Vegetations- und Standortsuntersuchungen in den Strandwiesen der schwedischen Westküste. Acta Phytogeogr. Suec. 43, 198 pp.

Gower, J.C. & G.S. Ross. 1969. Minimum spanning tree and single link age cluster analysis. Appl. Statist. 18: 47–54.

Hocquette, M., J.M. Gehu & M. Fauquet. 1965. Contributions à l'étude phytosociologique de l'estuaire de l'Authie. Bull. Soc. Bot. Nord de la France. 18: 114–143.

Ivimey-Cook, R.B. & M.C.F. Proctor. 1966. The application of association analysis to phytosociology. J. Ecol. 54: 179–192.

Kortekaas, W.M., E. van der Maarel & W.G. Beeftink. 1976. A numerical classification of European Spartina communities. Vegetatio 33: 51–60.

Kullback, S. 1959. Information theory and statistics. New York, Wiley, 395 pp.

Lausi, D. 1972. Die Logik der pflanzensoziologischen Vegetationsanalyse. Ein Deutungsversuch. In E. van der Maarel & R. Tüxen (eds.), Grundfragen und Methoden der Pflanzensoziologie, Ber. Int. Symp. Rinteln. 1970, p. 17–24. Junk, Den Haag.

Lausi, D. 1973. Contributo delle elaborazioni quantitative per analisi e sintesi in fitosociologia. Not. Fitosoc., 7: 1–15.

Maarel, E. van der. 1971. Bericht über die Arbeitsgruppe für Datenverargeitung der Internationalen Vereinigung für Vegetationskunde. Vegetatio 22: 269–273.

Maarel, E. van der. 1974. The working Group for Data-processing of the International Society for Plant Geography and Ecology in 1972–1973. Vegetatio 29: 63–67.

Maarel, E. van der. 1975. The Braun-Blanquet approach in perspective. Vegetatio 30: 213–219.

Maarel, E. van der, J.G.M. Janssen & J.M.W. Louppen. 1978. TABORD, a program for structuring phytosociological

tables. Vegetatio 38: 143–156.
Maarel, E. van der, L. Orlóci & S. Pignatti. 1976. Data-processing in phytosociology, retrospect and anticipation. Vegetatio 32, 2: 65–72.
Molinier, René & G. Tallon. 1965. Etudes botaniques en Camargue II. Rev. Hist. Natur. Paris.
Nègre, R. 1956. Recherches Phytosociologiques sur le Sedd-el-Messjoun. Trav. Inst. Scient. Chérif. Ser. Bot. 10–1–193.
Nordhagen, R. 1954. Studies on the vegetation of salt and brackish marshes in Finmark (Norway). Vegetatio 5/6: 381–394.
Oberdorfer, E. 1952. Beitrag zur Kenntnis der nordägäischen Küstenvegetation. Vegetatio 3: 329–349.
Orlóci, L. 1967. An agglomerative method for classification of plant communities. J. Ecol. 55: 193–206.
Orlóci, L. 1973. Ranking characters by a dispersion criterion. Nature 244: 371–373.
Orlóci, L. 1976. Ranking species by an information criterion. J. Ecol. 64: 417–419.
Orlóci, L. 1978. Multivariate analysis in Vegetation research. 2nd ed. The Hague, Junk, 451 pp.
Parks, J.M. 1970. Fortran IV Program for Q-mode cluster analysis on distance function with printed dendrogram. State Geological survey. The University of Kansas, Lawrence. Computer contribution 46: 1–31.
Pignatti, S. 1966. La vegetazione alofila della Laguna Veneta. Ist. Veneto Sci. Lett. Arti. Classe Sci. Mat. Natur. 33: 1–174
Pignatti, S. 1976. A system for coding plant species for data-processing in phytosociology. Vegetatio 33: 23–32.
Rivas-Goday, S. 1945. Fàcies subhàlofitas del Schoenetum nigricantis: origin y sucesion. Bol. Soc. Broteriana 19, ser. 2: 789–813.
Rivas-Goday, S. 1955. Aportaciones a la fitosociologia hispanica. An. Inst. Bot. A.J. Cavanilles. 13: 394.
Sneath, P.H.A. & R.R. Sokal. 1973. Numerical taxonomy. The principles and practice of numerical classification. Freeman. S. Francisco. 573 pp.
Soó, R. 1947. Conspectus des groupement végétaux dans les Bassins Carpathiques I. Les associations halophiles. Inst. Bot. Univ. Debrecen. 60 pp.
Soó, R. 1968. Neue Übersicht der höheren zönologischen Einheiten der ungarischen Vegetation. Acta Bot. Acad. Scient. Hung. 14: 385–394.
Tüxen, R. & V. Westhoff. 1963. Saginetea maritimae, eine Gesellschaftsgruppe im wechselhalinen Grenzbereich der europäischen Meeresküsten. Mitt. Flor. Soz. Arb. Gemeinsch. N. F. 10: 116–129. Stolzenau/Weser.
Wendelberger, G. 1950. Zur Soziologie der kontinentalen Halophytenvegetation Mitteleuropas. Unter besonderer Berücksichtigung der Salzpflanzengesellschaften am Neusiedler See. Denkschr. Österr. Akad. Wissensch. 108(5). 180 pp.
Westhoff, V. & E. van der Maarel. 1978. The Braun-Blanquet approach. 2nd ed.In: R.H. Whittaker (ed.) Classification of plant communities, pp. 287–399. Junk, The Hague.
Whishart, D. 1969. Clustan 1A. User Manual. St. Andrews Computing Center.
Whittaker, R.H. 1972. Evolution and measurement of species diversity. Taxon 21: 213–251.
Wilkón-Michalska, J. 1963. Halofity Kujaw. Studia Soc. Torunensis, D 7: 1–122.
Wolff, W.J. 1968. The halophilous vegetation of the Lagoons of Mesolonghi, Greece. Vegetatio 16: 1–4.

ANALYSIS OF CONCENTRATION AND DETECTION OF UNDERLYING FACTORS IN STRUCTURED TABLES

Enrico FEOLI[1,2] & László ORLÓCI[1]*

[1] Department of Plant Sciences University of Western Ontario London, Ontario, Canada N6A 5B7
[2] Permanent address: Istituto ed Orto Botanico, Università, Via Valerio 30, I 34100 Trieste, Italy

Keywords:
Classification, Concentration, Data, Table, Test

Introduction

Data management by structuring data tables is one of the methods used by data analysts in many fields of science to describe the group structure in given tables. It involves rearrangements of the rows and the columns of the table to concentrate the non-zero values, as much as possible, in distinct blocks. Table 1 is a typical example. This table is structured and it has 18 blocks.

It is true that the rearrangement of a data table can be an arbitrary step. Yet it may greatly help the interpretation of the data, since by structuring the data table specific properties may come to light suggestive of the existence of trends and classes in the sampled population. Structured phytosociological tables (van der Maarel, Janssen & Louppen 1978) have indeed been used traditionally for the purpose of describing the vegetation and to show trends of variation. Mueller-Dombois & Ellenberg (1974), Westhoff & van der Maarel (1978) and Orlóci (1978) review the state of the art.

The main point to be made in the present paper is that for a reliable interpretation, specific table properties have to be measured and correlated with external variables. It is in this connection that the *analysis of concentration* (*AOC*) is proposed. *AOC* measures the sharpness of the group structure in the table and it helps identify the underlying factors wich influence the compositional differences among the groups.

* This work is part of a research project supported by a National Research Council of Canada Grant (Orlóci) and an Italian C.N.R. grant: project 'Promozione qualità dell'ambiente', Subproject 'Metodologie matematiche e basi di dati' (Feoli).

Block structure

Important as they are the methods of cluster analysis which can impose a block structure on data tables are not considered here. Interested readers are referred to reviews elsewhere (see Orlóci 1978 for references). How well a given method of cluster analysis succeeds is the problem to be addressed. The solution is simple, since it only requires the measurement of the divergence χ^2 of two distributions $\boldsymbol{F}$ and $\boldsymbol{F}^0$. These both have $t \times z$ elements and identical totals ($F_{..}$). The number of rows (t) in $\boldsymbol{F}$ or $\boldsymbol{F}^0$ is equal to the number of row groups recognized in the structured table. The number of columns (z) is equal to the number of column groups. The row and column groups intersect in blocks of which there are $t \times z = 18$ in Table 1. An element F_{ij} of $\boldsymbol{F}$ is given meaning as the density of the non-zero values in the ij block. An element F^0_{ij} of $\boldsymbol{F}^0$ is an expectation of F_{ij}. The actual value of χ^2 will depend on the definition of F^0_{ij}. The more commonly used definitions include,

$$F^0_{ij} = F_{..}/tz \qquad \text{for any } i \text{ or } j \tag{1}$$

$$F^0_{ij} = F_{i.}F_{.j}/F_{..} \tag{2}$$

$F_{..}$ is the total number of non-zero values in the table. $F_{i.}$ and $F_{.j}$ are the counts of non-zero values respectively in the ith row group and in the jth column group.

Since $\boldsymbol{F}$ describes the block structure as observed and $\boldsymbol{F}^0$ describes a table with the same marginal totals as in $\boldsymbol{F}$ but without a real block structure, the quantity

$$\chi^2 = \sum_{i,j} (F_{ij} - F^0_{ij})^2/F^0_{ij}, \qquad i = 1,\ldots,t; \; j = 1,\ldots,z \;. \tag{3}$$

Table 1. The Coquihalla data (following Fewster & Orlóci 1978). The species and relevés (columns) are arranged according to sum of squares clustering. Species names listed twice in the table indicate species occurrence in two strata of the sampled community. The plant names follow the use of nomenclature by Taylor & MacBryde (1977) and Grout (1940).

			a	b	c
1	1	Abies grandis	--------------	--1-----------------	-----------
	22	Philadelphus lewisii	--------------	---------1----------	-----------
	59	Streptopus roseus	--------------	---------2----------	-----------
	10	Alnus rubra	--------------	--------------------	------2----
	48	Lilium columbianum	--------------	--------------------	-----2-----
	56	Cinna latifolia	2-------------	--------------------	-----------
	60	Tolmiea menziesii	-2------------	--------------------	-----------
	64	Veronica officinalis	--------------	--------------------	-------2---
	26	Rubus parviflorus	--------------	--1-------------2--1	-----------
	66	Dicranum scoparium	--------------	--------------------	----------2
	71	Rhacomitrium canescens	--------------	--------------------	----------3
	24	Ribes bracteosum	-22---------2-	------------2-------	-----------
	52	Polypodium hesperium	-----------23-	--------------------	-----------
	18	Menziensia ferruginea	--------------	------2---2---------	-----------
	37	Athyrium filix-femina	-----------1--	------2-------------	-----------
	44	Dryopteris filix-mas	--------2-----	---1--2-------------	-----------
	23	Rhamnus purshianus	--------------	-----22---2-----2---	---------2-
	8	Acer glabrum	------2---2---	-------2--------2---	2--2------2
	19	Mahonia aquifolium	3----22-------	--------------------	-----------
	38	Bromus vulgaris	2----2----2---	--------------------	-----------
	28	Sambucus pubens	----2-2----2--	--------------------	-----------
	63	Urtica dioica	------3----3--	--------------------	-----------
	33	Tsuga heterophylla	--------------	--5-----------------	--3--------
	31	Taxus brevifolia	------------2-	-1---2---2----2222--	-----------
	55	Pyrola grandiflora	--------------	----------------22--	------5-2--
	13	Cornus nuttallii	--------2--2--	-1-2-2-2----------2-	2---22-22--
	58	Streptopus amplexifolius	---------5--3-	--------------------	-----------
	65	Viola glabella	--------23-2--	-------------------2	-----------
	73	Rhytidiopsis robusta	---------3----	--------------3-----	-----------
	12	Betula papyrifera	-------------2	------2-5-----------	--3--22----
	62	Trillium ovatum	---223-2------	2---3---3-----2321-2	--2--------
	3	Betula papyrifera	--------------	332----------3---2-2	-3--3--22--
	14	Corylus cornuta	--1----1--2---	11--2--2---222--2--3	-3--2--2--2
2	11	Amelanchier alnifolia	-3------------	------22-22---------	1-3--2--325
	35	Vaccinium parvifolium	--------------	-----2---3----------	233-333-355
	16	Holodiscus discolor	--------------	--------------------	-3--23---23
	34	Vaccinium membranaceum	--------------	--------------------	-2-23322225
	29	Spiraea douglasii	--------------	--------------------	2-3232-2223
	17	Lonicera ciliosa	--------------	---2-------3--------	-32----3333
	25	Rosa gymnocarpa	2-------------	------22--2-2-------	-3---222222
	32	Thuja plicata	--------------	--------------------	---------8-
	39	Chimaphila umbellata	--------------	--2--------------2--	35-33225533
	49	Linnaea borealis	--------------	-------------------2	--233255355
3	2	Acer macrophyllum	5--333-13-35--	-------3--3---------	-----------
	45	Galium triflorum	-2--22-122223-	2-----2232-32---2---	-----------
	50	Claytonia sibirica	-3-3253-2-323-	--------------------	-----------
	51	Osmorhiza chilensis	222523--23-22-	----2-----22---22---	-----------
	70	Mnium spinulosum	----------78--	---2----------------	-----------
	9	Acer macrophyllum	-25---7--3--33	-----------3--------	3--------2-
	40	Circaea alpina	73---25---5323	--------------------	-----------
	42	Dicentra formosa	33323352222355	--------------------	-----------
	57	Smilacina stellata	5535335---3---	---------3-353------	-----------
4	6	Tsuga heterophylla	--------------	-----55-3---1-888887	-----------
	41	Clintonia uniflora	--------------	355575-573-3-2575555	------3----
	27	Rubus spectabilis	----2---22---2	2222322232222323-223	2-323222---
	46	Goodyera oblongifolia	--------------	23323232222--322222-	---22-22322
	61	Trientalis latifolia	---3---2-2----	2-2232233-32-223322-	2322222-22-
	54	Pteridium aquilinum	--------------	3-52325533-2-3235332	-----------
	72	Rhytidiadelphus loreus	23233--33-3-3-	3---3333333333-33323	33253335331
5	30	Symphoricarpos albus	77535582325728	--12-----23532------	-----2---2-
	53	Polystichum munitum	88878787877877	5--2-35-355375--2--2	1----------
	43	Disporum hookeri	53325753357578	35--22353322552-2213	-----------
	47	Lactuca canadensis	253555-3553383	3-2--235555555322-2-	-----------
	69	Mnium insigne	7858588587----	22-------23-3---2--3	2----------
6	4	Pseudotsuga menziesii	87859898999798	78788585578887332833	78888888898
	7	Acer circinatum	79788778777888	855337755-7577378733	7332225--23
	36	Achlys triphylla	32555553535557	5552555777553-755335	--522--2-33
	5	Thuja plicata	757855-1------	77777555755755775775	557555585-2
	20	Mahonia nervosa	-3733--8235-3-	77777577775753787787	7773-2255-5
	67	Eurhynchium oreganum	22-23327353333	77575355335557233775	53353557537
	68	Hylocomium splendens	----2----32-88	55757775855575887557	7887587779-
	15	Gaultheria shallon	--------------	5552253-333---2--23-	77877877888
	21	Paxistima myrsinites	-2-----2------	353-33355233-525-555	75775857777

is a direct measure of the divergence of $\boldsymbol{F}$ from $\boldsymbol{F}^0$, and thus, a direct measure of the sharpness of the blocks in the table. The quantity,

$$C = \chi^2/[F_{..} \min (t-1, z-1)]$$

is a relative measure of divergence.

The numerical values of F_{ij} and the totals are given in Table 2 for the data in Table 1. The definition of the expectations (F^0_{ij}) could be based on formula (1), but since this assumes an equidistribution the definition would prove to be too restrictive. It would increase the chances of declaring χ^2 significant. It is thus formula (2) on which our definition of the expectation F^0_{ij} and the subsequent results are based:

$$\chi^2 = 165.4$$

$$C = 0.21$$

Considering that the limits of C are 0 and 1, the value obtained for the sample indicates a moderately strong divergence. But to expect that clustering algorithms will produce perfectly structured tables in which no row group has non-zero values in more than one column group, a condition of C to reach the maximum, would be unrealistic.

After the observation that

$$\chi^2 > \chi^2_{.05,10} = 18.81,$$

the block structure in Table 1 appears to indicate a strong type structure in the sampled vegetation. $\chi^2_{.05,10}$ is the 0.05 probability point of the chi square distribution.

Table 2. Joint frequencies of the six species groups and the three relevé groups (a, b, c) in Table 1. The entries in the cells are counts of non-zero values in the corresponding blocks of Table 1 adjusted to minimum block size. The adjustments follow the assignment $F_{ij} = :F_{ij}\, min\ B/B_{ij}$ where *min B* is the size of the smallest block and B_{ij} the size of the *ij* block. E.g., $F_{2a} = :F_{2a}B_{5c}$ $B_{2a} = 2(55/140) = 0.79$.

	a	b	c	Total
1	4.64	5.00	4.39	14.03
2	0.79	4.13	35.50	40.42
3	32.30	6.42	1.11	39.83
4	8.98	44.00	25.00	77.98
5	51.07	31.90	4.00	86.97
6	34.05	51.33	49.44	134.82
Total	131.83	142.78	119.44	394.05

Underlying factors

If it is concluded from the test that the type structure is significant and sharp, it will make sense to undertake a search for the influential factors which underlie the compositional variation among the blocks. The search can be conveniently accomplished in two steps: a. Extract intrinsic (vegetational) variables, known as canonical variates, based on an eigenanalysis. b. Correlate the canonical variates with given environmental variables. To do these standard computer routines are available. The justification will become obvious as the steps are described.

It is known from the statistical theory of contingency tables (cf. Lancaster 1949, 1969, Williams 1952) that when a data table is structured, such as Table 1, two sets of $q \leq \min (t, z)$ internal variables are generated. These are known as the *canonical variates* of the table. We represent these in symbolic terms by

$$\boldsymbol{X}_1, \ldots, \boldsymbol{X}_q$$

$$\boldsymbol{Y}_1, \ldots, \boldsymbol{Y}_q$$

The first set ($\boldsymbol{X}$) comes about after clustering the species (row groups) and the second set ($\boldsymbol{Y}$) after clustering the relevé (column) groups. The canonical variates partition the total χ^2 into q additive components, given by

$$\chi^2 = \chi^2_1 + \ldots + \chi^2_q = R^2_1 F_{..} + \ldots + R^2_q F_{..} \qquad (4)$$

in proportion with the squared *canonical correlations* $R^2_1, \ldots, R^2_q$. The kth of these R^2_k relates $\boldsymbol{X}_k$ to $\boldsymbol{Y}_k$.

It is convenient to use the canonical variates when searching for influential factors, but this is not entirely without pitfalls. Problems may arise on the count of the assumption that variation among the blocks is linear in the sense that the relationships of the rows and columns in the table are linear. And yet, this does not have to be that way. The relationships may indeed be non-linear.

Linear canonical variates can be easily extracted if an eigenanalysis program is available. A cross products matrix

$$\boldsymbol{S} = \boldsymbol{U}\boldsymbol{U}'$$

with elements defined by

$$U_{ij} = \frac{F_{ij}}{(F_{i.}F_{.j})^{\frac{1}{2}}} - \frac{(F_{i.}F_{.j})^{\frac{1}{2}}}{F_{..}}$$

is calculated first. The q non-zero eigenvalues of $\boldsymbol{S}$

$$R^2_1, \ldots, R^2_q$$

are extracted next. The eigenvalues represent the squared canonical correlations. The kth canonical correlation is associated with the pair,

$$\boldsymbol{X}_k = (X_{1k}, \ldots, X_{tk})$$

$$\boldsymbol{Y}_k = (Y_{1k}, \ldots, Y_{zk})$$

whose elements are the canonical scores.

X_{ik} is a canonical score of the ith row group on the kth canonical variate. In terms of the determining algorithm,

$$X_{ik} = \alpha_{ik}(F_{..}/F_{i.})^{\frac{1}{2}}$$

in which α_{ik} is the ith element of the kth eigenvector of $\boldsymbol{S}$. The α_{ik}, $i = 1, \ldots, t$, are such that their sum is zero and their sum of squares is unity. The underlying theory is discussed elsewhere. Orlóci (1978) gives references.

Y_{jk} is the kth canonical score of column group j. It is defined by

$$Y_{jk} = \sum_{i=1}^{t} F_{ij} X_{ik}/(F_{.j} R_k)$$

Based on the pair $\boldsymbol{X}_k$, $\boldsymbol{X}_m$ or $\boldsymbol{Y}_k$, $\boldsymbol{Y}_m$, scatter diagrams can be constructed of the row groups or the column groups (Fig. 1). These can be used in a similar manner as the scatter diagrams in ordinations.

The Coquihalla sample (Table 2) has only two canonical variates. The canonical correlations and scores are given in Table 3.

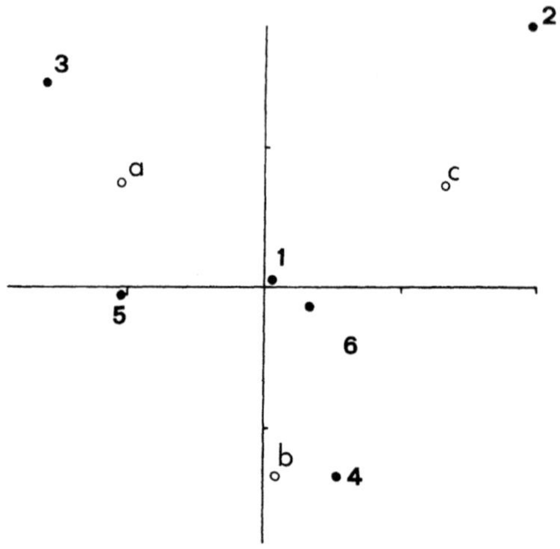

Fig. 1. Joint distribution of canonical variates for relevé groups (empty circles) and species groups (full circles). The method is explained in the main text. Coordinates are given in Table 3.

The indices,

$$Q_1 = 100\, R_1^2 F_{..}/\chi^2 = 80\,\%$$

$$Q_2 = 100\, R_2^2 F_{..}/\chi^2 = 20\%$$

indicate that the first canonical variate accounts for an overwhelming portion of the total chi square.

Having extracted the canonical variates, their corre-

Table 3. Canonical scores and correlations for the species and relevé groups in Table 1. The method of analysis is described in the main text.

Canonical variate	Species group 1	2	3	4	5	6	Relevé group a	b	c	Canonical correlation
1	0.03	1.95	−1.61	0.49	−1.10	0.31	−1.20	0.01	1.31	0.580
2	0.04	1.89	1.44	−1.45	−0.04	−0.13	0.75	−1.33	0.76	0.288

Table 4. Characterization of the Coquihalla flood plain by level

Level*	Height	Vegetation type**	Soil type	Productivity	Scores Y_1	Y_2
a	Low	Polystichum	Acid brown forest	Excellent	−1.20	0.75
b	Intermediate	Eurhynchium – Mahonia	Minimum podsol	Medium	0.01	−1.33
c	High	Mahonia – Gaulteria	Podsol	Poor	1.31	0.76

* Level symbols are those used for relevé groups in Table 1.
** Names of vegetation types follow Horváth (1963).

lation with the external variables have to be determined. The point in doing this in the Coquihalla sample is that such correlations reveal information about the type and the intensity of the environmental influence to which the vegetation responds. In such terms, a clearly defined relationship can be seen between the flood plain level and the first canonical variate (Table 4). This suggests that varying elevation above the river is the most influential factor (or factor complex) accounting for the recognized group structure in the sample (Table 1). The ordering of the relevé groups on the second canonical variate (Table 4) suggests a weak, secondary influence. This appears to be disturbance.

Local influence of the factors

Once the influential factors are identified, further insight can be gained if their influence is measured within the row or column groups and within the blocks. To do this the total χ^2 have to be subdivided into additive components specific to the canonical variates, row groups and blocks:

Source	Level of influence measured
All factors	$\chi^2 = R_1^2 F_{..} + \ldots + R_q^2 F_{..}$
*k*th factor	$\chi_k^2 = R_k^2 F_{..}$
*k*th factor in *i*th row group	$\chi_{ik}^2 = (\sum_{j=1}^{z} F_{ij} Y_{jk})^2 / F_{i.}$*
*k*th factor in *ij* block	$\chi_{ijk} = F_{ij} Y_{jk} / (F_{i.})^{1/2}$

* This is an $F_{..}$ multiple of the ith term in the expression of R_k^2 (cf. Williams, 1952).

** This quantity may be positive or negative depending on the position of the column group, as a point, relative to the origin on the canonical axis.

The components of χ^2 can be expressed as relative quantities:

$$C_k = \chi_k^2 / \chi^2$$

$$C_{ik} = \chi_{ik}^2 / \chi_k^2$$

$$C_{ijk} = |\chi_{ijk} / \chi_{ik}| = |F_{ij} Y_{jk} / \sum_j F_{ij} Y_{jk}|.$$

The Coquihalla results are given in Table 5. Occupancy values ($E_{ij} = F_{ij}/F_{i.}$) of the row groups are listed in Table 6. The same could be done, but has not been done, for the column groups. The row groups are of special interest, for

Table 5. The level of influence of the underlying factors associated with canonical variates in the row groups.

Row group	Factor 1	Factor 2	Total χ^2
1	0.01	0.00	
2	51.58	12.07	
3	34.59	6.83	
4	6.40	13.67	
5	35.56	0.01	
6	4.44	0.20	
Total	132.58	32.78	165.36

Table 6. The level of occupancy of the row groups within the column groups of Table 1 and 2.

Row group	Column group a	b	c
1	0.33	0.36	0.31
2	0.02	0.10	0.88
3	0.81	0.16	0.03
4	0.11	0.56	0.32
5	0.59	0.37	0.04
6	0.25	0.38	0.37

these represent the collections of species which have definite utility in diagnosing the different factors and their level of influence in the column groups.

Discussion

The users of numerical methods frequently rely on cluster analysis to impose a block structure on the data table. The possible number of different block structures is usually very large. Yet the success of a particular outcome, is rarely measured. In the present paper it is assumed that a structure has been imposed on the table and that the remaining problem is to measure its sharpness and to identify the underlying factors.

Analysis of concentration expects data input from a structured table. The input data represent frequencies. The first test is concerned with the sharpness of the block structure. If this test indicates that the block structure is significant, eigenanalysis is performed on the frequencies. Eigenanalysis yields estimates for the canonical correlations and canonical scores. When the canonical scores are correlated with given environmental variables, the factors which influence variation among the blocks can be identified. That the analysis reveals information about structured tables gives the algorithm special significance as an extention to cluster analysis. Only the case of presence/absence type data is treated.

Summary

If the rows and columns of a data table are rearranged in such a way that the non-zero values are concentrated in distinct blocks, the table is said to be structured. In such tables, two specific properties are often subject to interpretations: the sharpness of the block structure and the compositional variation among the blocks. A simple chi square analysis is satisfactory to test the overall sharpness of the block structure. The interpretation of compositional variation is more difficult, since the dimensionality of such variation will almost certainly exceed the dimensionality of the table. A solution can nevertheless be obtained based on a two-step analysis. In the first step, the canonical variates of compositional variation are identified. These are characteristics of the vegetation. In the second step, the environmental variables are identified which are highly correlated with the canonical variates.

The example in the paper uses data from fossil flood plains. The data table is structured into 18 blocks of 6 species groups and 3 relevé groups. The chi square test indicates a sharp block structure. The structuring generated two canonical variates. The dominant of these two signifies the environmental influence of changing elevation on the flood plain. The second seems to be a response of the vegetation to disturbance.

The algorithm which performs analysis of concentration is *AOC*. A program is available to enquirers free of charge. The programming language is BASIC.

References

Fewster, P. & L. Orlóci. 1978. Presentation of ordination results using stereograms. Can. J. Bot. 56: 162–165.

Grout, A.J. 1940. List of mosses of North America and Mexico. Bryologist 43: 117–131.

Horváth, O. 1963. Contributions to nesting ecology of forest birds. M.Sc. Thesis, University of British Columbia.

Lancaster, H.O. 1949. The derivation and partition of χ^2 in certain discrete distributions. Biometrika 36: 117–129.

Lancaster, H.O. 1969. The Chi-squared distribution. Wiley, New York.

Maarel, van der E., J.G.M. Janssen & J.M.W. Louppen. 1978. TABORD, a program for structuring phytosociological tables. Vegetatio 38: 143–156.

Mueller-Dombois, D. & H. Ellenberg. 1974. Aims and methods of vegetation ecology. Wiley, New York

Orlóci, L. 1978. Multivariate Analysis in Vegetation Research. 2nd ed. Junk, The Hague.

Taylor, R.L. & B. MacBryde. 1977. Vascular plants of British Columbia. A descriptive resource inventory. The University of British Columbia Press, Vancouver, Canada.

Westhoff, V. & E. van der Maarel. 1978. The Braun-Blanquet approach. 2nd ed. In R.H. Whittaker (ed.). Classification of plant communities. p. 287–399. Junk, The Hague.

Williams, E.J. 1952. Use of scores for the analysis of association in contingency tables. Biometrika 39: 274–289.

INTERSECTION ANALYSIS IN PHYTOSOCIOLOGY: COMPUTER PROGRAM AND APPLICATION*

Enrico FEOLI[1] & Mario LAGONEGRO[2]

[1] Istituto ed Orto Botanico, Università degli Studi, Trieste 34100, Italy
[2] Istituto di Fisica, Università degli Studi, Trieste, 34100 Italy

Keywords:
Classification, Gradient, Intersection, Seriation, Table rearrangement

Introduction

A central problem of phytosociology is the discovery of environmental gradients which influence vegetational variation. For this purpose numerical ordinations (e.g. Orlóci 1978, 1979, Prentice 1977) and table rearrangement methods (e.g., Lieth & Moore 1971, Dale & Quadraccia 1973, Van der Maarel, Janssen & Louppen 1978, Feoli & Orlóci 1979) have been proposed.

Intersection analysis, *IA*, is offered as an alternative method to analyse large samples of relevés for which the traditional methods of ordination or classification would not be suitable. Although *IA* can produce structured tables by use of a monothetic, divisive clustering (Feoli 1977), it is presented in a more general context.

Intersection analysis as a classificatory method

IA assumes that each species can give rise to two sets of relevés, one which presents it and another which does not: *IA* can incorporate different classificatory criteria. One is the coincidence of sets measured as the maximal intersection. For two sets, this may be formulated as follows:

if $\mathbf{x} \in \{\mathbf{A}\}$ then $x \in \{\mathbf{B}\} \equiv \{\mathbf{A}\}\ \{\mathbf{B}\}$.

* Contribution from the Working Group for Data-Processing in Phytosociology, International Society for Vegetation Science. This paper has been presented at the Workshop of the International Statistical Ecology Program, Parma 1978. A computer program, written by M. Lagonegro, is available from the authors on request free of charge. The Italian CNR funded the study (project 'Promozione qualità dell'ambiente', subproject 'Metodologie matematiche e basi di dati'). We thank Mr. A. Zampar for technical assistance and Dr. E. van der Maarel and Prof. L. Orlóci for comments on the manuscript.

The intersection is maximized for $\{\mathbf{A}\}$ or for $\mathbf{B}\}$ which are sets defined respectively by species *A* and *B*, and **x** is a relevé. The intersection can be measured based on any symmetric resemblance function (Orlóci 1972, Goodall 1973). Feoli (1977) suggests the Jaccard coefficient which is easily interpreted in set theoretical terms. The clustering is applied iteratively to a matrix of intersections between sets defined by single species to give rise to a minimal-connected graph (minimum spanning tree, Gower & Ross 1969) for species.

To obtain groups of relevés, the minimum spanning tree (*MST*) is split at increasing values of intersection to obtain groups of species to which the relevés are assigned based on species composition. Since the relevés can have species of different species groups, *IA* produces overlapping groups of relevés (cf. Jardine & Sibson 1971, Williams 1971; Yarranton, Beasleigh, Morrison & Shafi 1972). The classification of a sample based on the maximum intersection criterion is presented in Fig. 1a. *IA* can be revised to produce an exclusive classification by incorporating ranking for the species groups (Fig. 1b). The species groups can be ordered analogously to single characters, following different criteria (Orlóci 1978). Only one will be considered in this paper for relevé seriation to obtain diagonal tables.

Intersection analysis as an ordination method

In order to obtain a diagonal table the blocks with the

if $\mathbf{x} \in \{\mathbf{A}\}$ then $\mathbf{x} \in \{\mathbf{B}\} \Rightarrow \{\mathbf{A}\} \cap \{\mathbf{B}\}$ maximal for $\{\mathbf{A}\}$ or for $\{\mathbf{B}\}$.

highest concentration have to be moved as near to the diagonally positions as possible. By concentration we mean the rate of occupancy of the blocks by non-zero values. Each block is a submatrix in the data matrix of p (species) rows and z (relevés) columns.

IA can produce a diagonal block structure by arranging the species in a linear order starting with those which have the highest intersection, followed by species which share the highest intersection with one of them and so forth. Following this, the relevés are arranged into groups: the first group includes the relevés containing the first of the rearranged species, the second group includes the relevés containing the second species, excluding the relevés of the first group, third group includes the relevés containing the third species, excluding the relevés in the first and second group, and so on.

If two or more groups of species have identical values of intersection then either may serve as the next group in the linear order. Such a criterion of ordering may be the basis for recognizing species groups based on the nearest neighbour criterion. The species sequence so derived can dismember the species groups defined by the maximal intersection criterion, since it could happen that within a species group some intersection values are lower than between the species of different groups. Low intersection values may be due to species with low probability, even if in a favourable environment, or with a tendency to be equally probable in different environments. In either of these cases the species are poor sources for conveying ecological information.

When applying a nearest neighbour criterion in species seriation, a distinction should be made between two groups of species:

a) Species of the first type have intersections greater than the intersections of the species groups.

b) Species of the second type have intersections less than the intersections of the species groups.

In a structured table only the species of the first type should be used.

The seriation of relevés, according to the seriation of species based on the nearest neighbour criterion, relies on two assumptions:

1. The species which give rise to the sets of relevés with the maximal intersection are confined to a portion of the dominant environmental gradient, where the influence is strongest.

2. That portion is located near an extreme point.

Neither of these assumptions have yet been proved; they are to be considered only as working hypotheses.

The computer program

IA (*MLTAX1*) has been programmed in FORTRAN IV for large data matrices. In the latter the rows are the relevés and the columns are the species. A file containing the names, code numbers (cf. Pignatti 1976), and an index for the column numbers of the species must be stored. The column index allows the user to base the analysis only on a selected subset of species according to some ranking procedure (Orlóci 1978). To run the program the user completes a parameter card. On this the number of species (*NSPEC*), the number of relevés (*NQUAD*), and options are specified. If *IFPR* is given a value greater than zero the similarity matrix of the species based on the Jaccard coefficient is printed. If *LASOLP* is chosen to be greater than zero, similarities are computed using presence (1) absence (0) data. If *LASOLP* is set equal to zero, quantitative data are expected for input. If *INSEQ* is given a value greater than zero the similarity matrix is not computed and the relevés are written in an output table according to an externally specified sequence (file *ISPOC*). The external sequence could be output from another method of ordination, or it may come from interpolations of species points in an ordination scatter diagram (Feoli & Feoli Chiapella 1976, Philipps 1978).

The principal steps of the program include:

1. Read and print the numeric and alphanumeric labels of species.
2. Read data using species specified by column indices. (Species not specified are skipped.)
3. Compute a similarity matrix between species (*INSEQ* = 0).
4. Print the similarity matrix (*IFPR* > 0).
5. Construct a minimum spanning tree based on the similarity matrix. (The highest rank is given to the species group with the highest similarity value. The species sequence is defined according to a nearest neighbour criterion.)
6. Print the species sequence and the maximum similarities between species and the species groups in the form of a dendrogram table.
7. Define the relevé sequence.
8. Print the output table.

MLTAX1 can be used to cluster relevés. For this, it is sufficient to transpose the data matrix and to create a file, including codes for relevés and column indices.

The program will allow a certain flexibility regarding the choice of species that could be used in the ecological interpretation of the data. Option *INSEQ* > 0 permits to

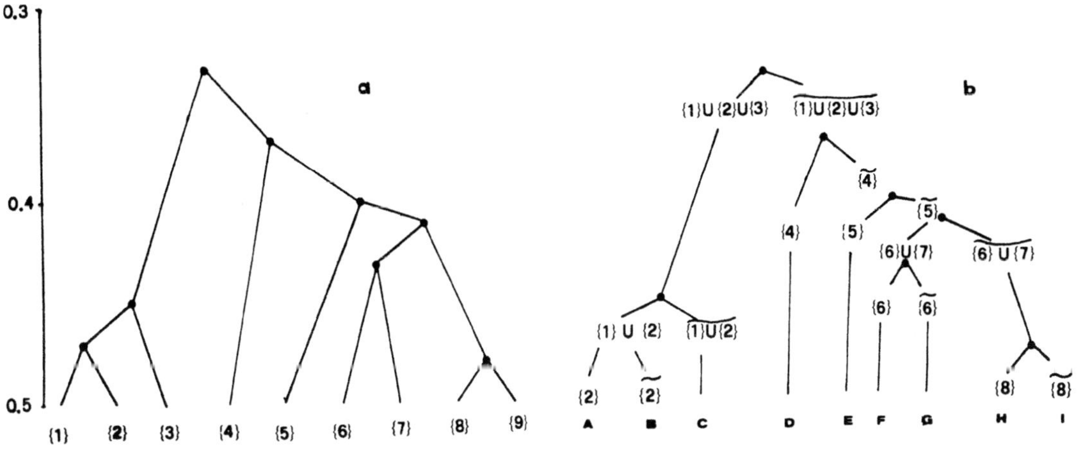

Fig. 1. a) Hierarchial classification of species groups (see Fig. 2) yielding the overlapping relevé groups ({1}, {2}, etc.). The dichotomies follow the principle of maximal intersection between species within sets.
b) Hierarchical classification of relevés in non-overlapping groups based on the presence/ absence of species. The highest ranking species group is 1; it has the highest intersection values between its species. The symbol {1} signifies a set of relevés delimited by species group 1. The symbol $\{\overline{1}\}$ represents the exclusion set of relevés without species of group 1. $\{1\} \cup \{2\} \cup \{3\}$ means a common set of relevés based on species of group 1, 2 and 3. $\overline{\{1\} \cup \{2\} \cup \{3\}}$ means the exclusion set.

obtain several structured tables on which to base further analyses.

*MLTAX*1 has been written to run on a 48 Kw *CDC* 6200 machine. The word size is 60 bits. 10 bytes of 6 bits each are allowed. In each byte an integer number is stored with value between 0 and 63. Via an appropriate routine *MOVC*, the relevé data are packed 10 per word into a vector (*DATA*) of 10 Kw. This allows the user to handle up to 100 species and 1000 relevés at a time. The subroutine *MOVC* (*N*, *A*, *M*, *B*) picks up the *N*-th byte of vector (or word) *A* and puts it into the *M*-th byte of vector (or word) *B*. This subroutine uses the *COMPASS* language which is machine dependent. There are, however, corresponding routines at most computer installations.

Since in *MLTAX*1 a great deal of iterative scanning of the similarity matrix is requested, it may be better to use a simple squared symmetric matrix instead of a triangular one even though it is more memory consuming. The whole program, including a large blank *COMMON* and several other vectors, needs about 30 k on a *CDC* 6200 machine.

Example

With the aim of illustrating the performance of *IA*, *MLTAX*1 has been run with 90 relevés selected at random from a set of 450 relevés of the beech woods in the Apennines (Italy). This sample constitutes part of a data bank for the Italian vegetation (Anzaldi & Mirri 1977). In the sample, 90 species have been selected at random from 200 available species which have shown a certain level of discriminatory power between the different phytosociological tables in the data bank. Only presence/absence data have been considered (*LASOLP* > 0).

The hierarchical classification of the relevés is presented in two ways in Fig. 1: (a) is the hierarchical classification of the species groups. The associated classification of the relevé group ({1}, {2}, {3} . . . etc.) is of course overlapping since a relevé can correspond with more than one species group. The noda in (b) show sets or union sets based on the species groups. The relevé groups are defined through the noda from top to bottom. Obviously, the classification is an exclusive one.

Fig. 2 has been rearranged according to the classification in Fig. 1(b). It is the final output of *MLTAX*1. It is one of the many outputs of the program which could be obtained. Fig. 2 contains only the species of the first type in the set of 51 species capable to represent the whole set of 90 relevés. The relevés with only species of the second type have been omitted so that the output table is considerably reduced.

In fig. 2 an indicator value is given for each species. This is according to Landolt (1977). The indicator values help to determine if the sequence of relevés obtained by IA can indeed reflect the influence of a specific, major environmental factor. Conscious of the fact that the

Index of column	Species group	Species	Indicator values	A	B	C	D	E	F	G	H	I
					111111111	12222	222222333	33333334	444444444	5	5555555556	666
			FRNHDLTK	1234567	89012345678	90123	456789012	34567890	123456789	0	1234567890	123
24	1	Euphorbia dulcis	34334232	1322222	22232223							
148		Geranium nodosum	23333444	3773592	225322 225							
68	2	Cardamine eptaphylla	34444232	33731								
144		Galium aristatum	33344232	2232223								
130	3	Daphne mezereum	3433X233	23 5	22332 2	2222						
65		Actaea spicata	43443133	332 3	2	222 2						
124	4	Corallorhiza trifida	32254223		3		232					
17		Melittis melissophyllum	24233344		2 2	2	32					
95		Anemone trifolia	23343243		32 2		322					
126		Coronilla emerus	24233344		8		2					
153		Helleborus viridis	34334232	5	12 2		35 2					
125		Cornus sanguinea	34334343		3		23 222					
44		Fraxinus ornus	24233352	52	2	2 3	23 22					
10	5	Tamus communis	34334342		2		22	2222				
7		Acer campestre	34334342		3		3	222 2223				
85	6	Festuca altissima	33344232				7	732 22	3257			
53		Arum italicum	43434252				3	32 3				
42		Anemone apennina	23343243				33	533 55	35 32335			
26	7	Carpinus betulus	33344243				5	132 2				
9		Euonymus europaea	34335332				2 22	22 2		3		
61	8	Cyclamen hederyfolium	34333252		3	2	2	122 22	3323	3	232333	
12		Melica uniflora	34234233	5	3	2 2	3 3	553 335	2 32	3	37232 3323	
38		Ruscus aculeatus	23234252	2			7	223 322	3352	3	5 33	
39	9	Hedera helix	33334242	5	33		335	53325522	733 2	3	722 22	22
45		Ilex aquifolium	33344242	32			72	323 5 3	3 32 3	2	7322 3232	2 2

Fig. 2. Output of *MLTAX*1 (explained in the text). The nomenclature of species follows Pignatti (1979). The relevés are coded from 1 to 63 (digits vertically stacked). The code for indicator value accords with Landolt (1977): F = humidity value, R = reaction value, N = nutrient value, H = humus value, D = dispersion value, L = light value, T = temperature value, K = continentality value.

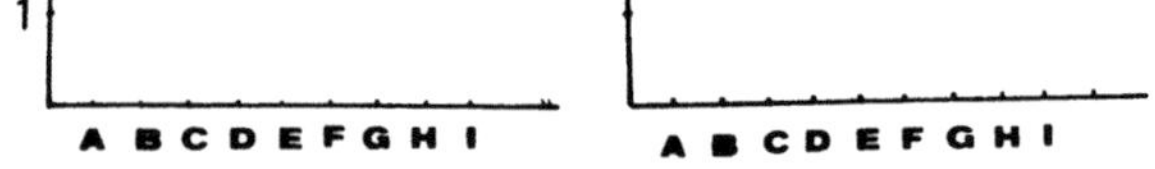

Fig. 3. a) Mean values for temperature within sets of relevés (Fig. 2) based on the first type indicator species (see the text for explanation).
b) Mean values for temperature within sets of relevés of (Fig. 2) based on first and second type indicator species.

indicator values have been established on the basis of observations and experiences in Switzerland, their use here may still be justified considering that they fit closely our experience in the mountainous regions of northern Italy. For each set of relevés the mean indicator value for each ecological factor has been calculated. Among these, the mean indicator values of temperature show a slight but steady rise to group *E*, after which they remain constant in the other groups (Fig. 3a). By considering the species of the second type, and all relevés, the mean indicator value of temperature increases steadily from the first group (*A*) to the last (*I*) as seen in Fig. 3(b). The mean indicator values of the other factors undergo oscillatory variation throughout the relevé groups. The temperature gradient seems to be the most important influence on variation in the beech wood vegetation of the Apennines.

Discussion

According to the terminology of Orlóci (1978) *IA* can be considered an *R*-clustering method. As presented here, it can help in classification problems as well as in ordinations where the seriation of relevés is considered more suitable than the determination of co-ordinates on gradient axis. *IA* gives structured tables which can serve to analyze variation in the vegetation based on compositional gradients. The method suggested in this work ranks the species groups and then the species define the assignments of the relevés to sets on the basis of the presence/absence. Just as in the case of relevés 14, 23 and 28 of Fig. 2, misclassifications can occur. But these are really common to monothetic classifications. Methods have been suggested to avoid misclassifications (see Orlóci 1978, pp. 224–233 for references). Some of these (e.g. Hill, Bunce & Shaw 1975) involve eigen analysis and other intricate procedures. It is however our opinion that to force polythetic criteria after a monothetic procedure has been chosen is not logical.

We believe that misclassifications in monothetic classifications as well as in polythetic classifications should be identified based on analyses of the dependence between sets of relevés and environmental data. *MLTAX*1 offers a table in which the likely misclassifications become apparent because the species groups have a logical sequence. If these suggest reallocations they can be done with a method of cutting and rearranging which is familiar to phytosociologists who have used the tabular methods.

Summary

Intersection analysis is described as a method for structuring phytosociological tables. It is based on the analysis of intersections of sets of releves defined in the sample on the basis of single species. The nearest neighbour criterion for such sets is proposed for seriation of the relevés. The method is illustrated based on an example using data from the beech woods of the Apennines (Italy).

References

Anzaldi, C. & L. Mirri. 1977. Un esperimento di strutturazione di dati floristici e vegetazionali. Quaderni IAC. Roma.

Dale, M.B. & L. Quadraccia. 1973. Computer assisted tabular sorting of phytosociological data. Vegetatio 28: 77–83.

Feoli, E. 1977. A criterion for monothetic classification of phytosociological entities on the basis of species ordination. Vegetatio 33: 147–152.

Feoli, E. & L. Feoli Chiapella. 1976. Due associazioni rupicole della Majella. Not. Fitosoc. 12: 67–75.

Feoli, E. & L. Orlóci. 1979. Analysis of concentration and detection of underlying factors in structured tables. Vegetatio 40/1: 49–54.

Goodall, D.W. 1973. Sample similarity and species correlation. In: R.H. Whittaker(ed), Handbook of Vegetation Science, Vol. 5, pp. 105–156. The Hague, Junk.

Gower, J.C. & G.J.S. Ross. 1969. Minimum spanning tree and single linkage cluster analysis. Appl. Stat. 18: 54–64.

Hill, M.O., R.G.H. Bunce & M.W. Shaw. 1975. Indicator species analysis, a divisive polythetic method of classification, and its application to a survey of native pinewoods in Scotland. J. Ecol. 63: 597–613.

Jardine, N. & R. Sibson. 1971. Mathematical taxonomy. New York, Wiley. 268 pp.

Landolt, E. 1977. Ökologische Zeigerwerte zur Schweizer Flora. Ber. Geobot. Inst. ETH. 64: 64–207.

Lieth, H. & G.W. Moore. 1971. Computerized clustering of species in phytosociological tables and its utilization for field work. In: G.P. Patil, E.C. Pielou & W.E. Waters (eds.), Statistical Ecology, Vol. 1, pp. 403–422. London, Penn. State Univ. Press.

Maarel, E. van der, J.G.M. Janssen & J.M.W. Louppen. 1978. TABORD, a program for structuring phytosociological tables. Vegetatio 38: 143–156.

Orlóci, L. 1972. On objective functions of phytosociological resemblance. Am. Midl. Nat. 88: 28–55.

Orlóci, L. 1978. Multivariate analysis in vegetation research. 2nd ed. Junk, The Hague.451 pp.

Orlóci, L. 1979. Non-linear data structures and their description. International Statistical Ecology Program. Parma, August 1978.

Pignatti, S. 1976. A system for coding plant species for data-processing in Phytosociology. Vegetatio 33: 23–32.

Pignatti, S. 1979. Flora d'Italia. Calderini, Bologna. (In press)

Prentice, I.C. 1977. Non-metric ordination methods in ecology. J. Ecol. 65: 85–94.

Phillips, D.L. 1978. Non-linear ordination. Field and computer simulation testing of a new method. Vegetatio 37: 129–140.

Williams, W.T. 1971. Principles of clustering. Ann. Rev. Ecol. Syst. 2: 303–326.

Yarranton, G.A., W.J. Beasleigh, R.G. Morrison & M.I. Shafi. 1972. On the classification of phytosociological data into non-exclusive groups with a conjecture about determining the optimum number of groups in a classification. Vegetatio 24: 1–12.

CLUSLA: A COMPUTER PROGRAM FOR THE CLUSTERING OF LARGE PHYTOSOCIOLOGICAL DATA SETS*

J. M. W. LOUPPEN & E. van der MAAREL**

Division of Geobotany, University of Nijmegen, Toernooiveld, 6525 ED Nijmegen, The Netherlands

Keywords:
Classification, Clustering, Data-processing, Numerical phytosociology, Salt marsh vegetation

Introduction

The aim of the program is to subdivide a large set of phytosociological relevés into a number of clusters, each of which is characterized by one or more constant species and can be handled subsequently with more sophisticated techniques. CLUSLA is an elaboration of Janssen's (1975) simple procedure for preliminary classification of very large phytosociological data sets. The essence of the method is that each relevé is considered separately and only in relation to the relevés considered before. The procedure starts with taking the first relevé and designating it as a cluster. The second relevé then is compared to the first based on a (dis)similarity measure. If the (dis)similarity between relevé 1 and 2 exceeds a threshold value, relevé 2 is added to the cluster of relevé 1, otherwise it acts as a separate cluster.

Two extensions of the original program have been developed:

1. After all relevés have been treated, each relevé is reallocated to the cluster to which it appears to fit best.
2. The significance of occurrence of each species in the clusters is calculated on the basis of a chi-square test.

Both extensions have been adopted from our program TABORD for rearranging phytosociological tables (Van der Maarel, Janssen & Louppen 1978), the first one in a simplified form. It should be noted that CLUSLA is primarily intended for sets of relevés which are stored according to the phytosociological table to which they were previously assigned in vegetation descriptions. (See Van der Maarel, Orlóci & Pignatti 1976 and Pignatti 1976 for the storage system developed by the Working Group). As Janssen (1975) already indicated the sequence of relevés obtained in this way may effect the result of the clustering. We will briefly report on this effect in this paper.

Material and methods

Phytosociological data collection

The program was applied to all phytosociological tables of selection *g* of European coastal and continental salt marsh communities (see Van der Maarel et al. 1976), as well as tables from mainly Baltic and (Sub-) Arctic Europe, which were added to fill regional gaps still present in selection *g*. The data set consists of 6072 relevés and 889 species, which are stored table-wise on magnetic tape. Each species entry involves a species code, a table code and the scores of this species in the table. The scores are Braun-Blanquet – (and in few cases Hult-Sernander – and Domin-) values, all transformed according to Van der Maarel (see 1979a, b).

Outline of the program

The method comprises 5 independent steps, which operate in successive runs (see also Fig. 1.).

* Contribution from the Working Group for Data-Processing in Phytosociology, International Society for Vegetation Science. Nomenclature follows the Trieste system, which will be published later.

** The authors are very grateful to Drs. Jan Janssen, Mike Dale, László Orlóci and Mike Austin for their comments on drafts of the program, and to Wil Kortekaas for her help in the interpretation of the tables.

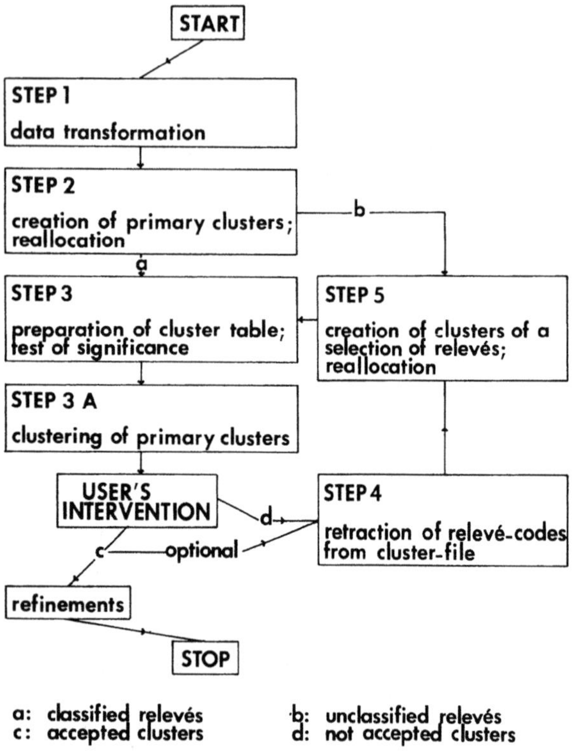

Fig. 1. Flow-chart of CLUSLA.

Step 1: Data transformation

In view of the possible subsequent use of more elaborate techniques, we decided to simplify the initial procedure by transforming the data into binary ones. The tables of the original file are read one by one, the number of relevés and species in each table is calculated and all species scores unequal to 0 are transformed into a logical 'T' (true), while all zero scores are replaced by a logical 'F' (false). This transformation has two advantages, viz. (1) by using presence-absence data the comparison between relevés and clusters is faster, and (2) the use of LOGICAL *1 arrays saves computer-space. 'The transformed data are put table-wise on a new file, each table being headed by the table code and its number of species and relevés.

Step 2: Creation of primary clusters; reallocation

The first relevé of the first table in the file is taken as the first cluster nucleus. All successive relevés are compared to it using Bray and Curtis' dissimilarity measure (cf. Clifford & Stephenson 1975), which reads in its general form:

$$d(C_j, R_k) = \frac{\sum_{i=1}^{m} |C_{ij} - r_{ik}|}{\sum_{i=1}^{m} (C_{ij} + r_{ik})}$$

were $d(C_j, R_k)$ is the dissimilarity between cluster C_j ($j = 1, \ldots, n$) and relevé $R_k (k = 1, \ldots, s)$ c_{ij} and r_{ik} are the scores of the i-th species ($i = 1, \ldots, m$) in C_j and R_k which can be 1 or 0 depending on whether the species is present ('T') or absent ('F'); s is the number of relevés, m is the number of species and n is the number of clusters ($n_{max} = 100$). The value of d ranges from 0 to 1.

If for the second relevé the value of $d(C_1, R_2)$ is lower than or equal to a given threshold value R_2 will join C_1. Otherwise R_2 is taken as a new cluster nucleus C_2. This procedure is continued until all relevés are compared, or the maximum number of clusters is reached. Joining a relevé with a cluster implies, that the scores of those species of the relevé not already occurring in the cluster are introduced as present, i.e. set 'T'.

Then a simple reallocation procedure is applied in which each relevé is compared with the centroids of each cluster formed so far. A relevé is either put in a remainder group if no single d(C, R) value exceeds the threshold value, or placed in the cluster with which the highest $d(C, R)$ value is found. After this reallocation the program gives information on the definite position of the relevés and the size of the clusters. Reallocation can have the following effects:
1) A relevé changes cluster because it now fits best into a cluster which did not yet exist at the moment of the first comparison.
2) A relevé may leave its original cluster because the centroid of this cluster has changed in the course of the procedure.
3. A relevé which was not yet assigned to a cluster may now be placed because of the fusion criterion is satisfied for a newly formed cluster or a cluster with a changed centroid.
4. Because of 1 and 2 a cluster created in the first round may be emptied during the reallocation.

The relevés are coded; the code is obtained by multiplying the table code by 1000 and adding to that value the position value of that relevé in the table, for example the 61-th relevé of table 2237 will be coded as 2237061. The relevés which have been assigned to a cluster are entered on a third file with their relevé code and the code of the cluster to which they have been assigned. The codes of unplaced relevés are entered in a remainder file.

Step 3: Preparation of cluster-table and test of significance

The classified relevés are taken from the file, containing the 0–1 transformed data. A simple cluster-table is produced with clusters in sequence of their creation and the species in sequence of the entrance in the clusters. This table provides information on the frequency of the species in the clusters. It therefore cannot be stored in a LOGICAL *1-array, but needs an INTEGER*2-array. In addition the number of relevés per cluster and the total frequency of the species in the cluster-table are calculated. The significance of occurrence is tested with a chi-square test (Sokal & Rohlf 1969), which compares the frequency of a species in any one cluster with that in all other clusters. A species occurrence is called 'significant' if a chi-square value of 6.64 is reached, i.e. for $P < 0.01$.

The cluster-table is stored in a disc-file, and we may obtain an output on paper or on a television display. Species occurrence is indicated per cluster with separate symbols for

- frequency $\geqq 60\%$;
- significance of occurrence
- both significance of occurrence and frequency $\geqq 60\%$.

Species which do not attain any of these values in any one cluster are not printed.

Step 3A: Clustering of primary clusters

To obtain an idea of the relation between the clusters, Ward's clustering method as derived from the CLUSTAN-package (Wishart 1975) may be applied to the primary clusters, with species scores standardized over the cluster: $x_{ij} = c_{ij}/c_{.j}$

It my be useful to interpret the resulting dendrogram with a criterion to find an optimum splitting-level e.g. Hogeweg's (1976) criterion, or the information criterion of Lausi & Feoli (1979), but we did not yet include such criteria in our program.

At this point in the procedure we may decide which clusters are sufficiently characterized by frequent and/or significant species. The relevés of the latter category of clusters can be removed to the residual group of unclassified relevés formed during step 2. It can also be decided whether clusters need to be split up before they can be treated by more elaborate procedures.

Step 4: Retraction of relevé codes from the cluster file

In this step the relevé codes of clusters which are considered unsatisfactory from a phytosociological viewpoint (according to the procedure described in step 3) are removed from the classification file and added to the file of unclassified relevés.

Step 5: Creation of clusters of a selection of relevés; reallocation

In this step relevés are collected which are either still unclassified, or belonging to clusters which have to be split and then subjected to the procedure described in step 2.

Steps 4 and 5 should be repeated until all relevés are classified and the clusters are of a size that can be handled by more elaborate methods as HIERAR and MODE (Wishart 1975), indicator species analysis (Hill, Bunce & Shaw 1975) or TABORD (Van der Maarel et al. 1978, see also Van der Maarel 1979b).

Technical information on the program

All calculations were performed on the IBM 370/158 of the Computer Centre of the University of Nijmegen, which is operational under the VMS-system and has a virtual storage campacity of 4 M bytes. CLUSLA is written in FORTRAN IV; a listing is available at the address of the authors.

Regarding the required computor storage capacity it can be noticed, that this is mainly determined by the total number of species in the data set and the number of primary clusters. For a storage of 512 Kbytes, which we can use for six hours CPU without special arrangements, the maximum size is ca. 140.000 cells. With a maximum of 100 primary clusters, which we adopted, the dataset can contain up to ca. 1400 species. The total number of relevés in the data set is of little importance, because the relevés are considered one at a time.

This 100×1400 size seems sufficient for any very large phytosociological data set, unless this would be a very heterogeneous one, so heterogeneous that it would not be realistic to treat it as one set. If, in an exceptional case, the number of species would increase above 1400, we can always reduce the maximum number of clusters, which can be created per cycle. As we will show, with our data set the final result is not essentially influenced by such a reduction.

Survey of treatments

As Janssen (1975) pointed out the result of CLUSLA may

be influenced by the sequence of the relevés as they are clustered. To test this influence in some runs the relevés were treated in the reverse sequence. In addition various threshold values were tried. This resulted in the following treatments:

1. Classification of the complete data set with the relevés in the order of collection (cf. Van der Maarel et al. 1976) and with a threshold d-value of 0.95, which is rather close to the maximum d-value of 1.00. Previous runs showed that decreasing the threshold value to 0.85 would result in considerable growth of computer-time, because the maximum number of 100 clusters is reached soon and many remaining relevés have to be compared with all 100 clusters. An additional disadvantage is that the number of unclassified relevés is rather high (35 %). For these reasons the fusion level during the first run should be at a high d-value.
2. Classification of the complete data set with the relevés in a reversed order, while using the same threshold value of 0.95.
3. From both classification runs 1 and 2 one large cluster was analysed anew at two different lower threshold values, viz. 0.90 and 0.80.

The various treatments are referred to as CLUSLA95 (1), ALSULC95 (2) and CLUSLA4490, CLUSLA4480, ALSULC4290 and ALSULC4280 (3), where 44 and 42 are the codes of the respective clusters in CLUSLA95 and ALSULC95.

For the interpretation of the tables we consulted Westhoff & den Held (1969), Beeftink (1977), Vicherek (1973) and the survey of the salt-marsh data by Kortekaas, Lausi & Beeftink (1979), in which the origin of the phytosociological tables, their character species and the geographical distribution of the originally published syntaxa are included.

Results

CLUSLA95 (Fig. 2, Table 1)

This run resulted in 44 clusters, comprising 96 % of the relevés. Despite the large number of allocated relevés the number of clusters remained low as compared to the maximal possible number of 100. Most of the clusters are well characterized by frequent and/or significant species. Two clusters, nrs. 42 and 44, are very large and apparently heterogeneous.

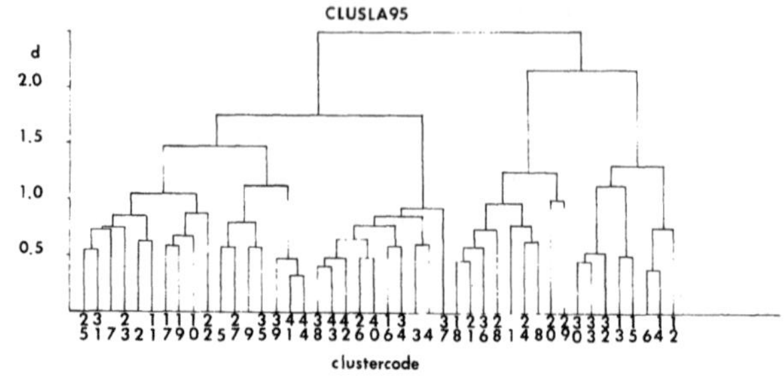

Fig. 2. Agglomerative clustering of the standardized clustertable of CLUSLA95 using Ward's method (Wishart 1975).

The syntaxonomical interpretation is as follows: At the 4 group-level, clusters 30 to 12 are characterized by continental halophytic species from the Pontic-South-Siberian Region. Communities of the *Artemisio-Festucetalia pseudovinae* (with *Petrosimonia diandra*, *Halimione verrucifera*), *Crypsidetalia* (*Crypsis aculeata*) and *Thero-Suadetalia* (with *Camphorosma annua* as character species) are found in cluster 33. Clusters 15 to 12 comprise *Artemisio-Festucetalia pseudovinae* communities, but they belong to different syntaxa of that order (*Puccinellion limosae* in cluster 15, *Festucion pseudovinae* in 6 and 12, *Artemisio-Festucetum pseudovinae* in 14). Cluster 13 relates to the *Beckmannion* from the Pannonic Region. Opposite to the continental communities we find (Sub-) Atlantic *Zosterion* communities in 29, the *Frankenieto-Koelpinetum* from Northern Africa in 20 and the Mediterranean *Salicornietalia fruticosae* and *Limonietalia* in 8 up to 18.

While these groups are clearly distinguished geographically, for the other ones ecological differences are more important. The typical salt-marsh communities of the *Salicornietalia*, the *Spartinetalia* and the *Glauco-Puccinellietalia* are found in clusters 39, 41 and 44. Clusters 31 and 25 may be considered as more halic communities of the Pontic-South-Siberion Region of the *Scorzonero-Juncetalia gerardii* (31) and the *Festuco-Puccinellietalia* (25). Similarly we find clusters 11, 2 and 23 here with elements from the (West-) Mediterranean Region (*Juncion maritimae* with *Tetragonolobus maritimus* and *Carex distans* in 11, *Limonietalia* in 2 and *Plantaginion crassifoliae* in 23).

A development towards fresh conditions is shown in cluster 35 with species of the *Centaurio-Saginetum*. The *Cakiletalia maritimae* are represented in cluster 22.

Clusters 38 up to 37 are characterized by species of *Armerion*-communities of contact-zones between salt and fresh conditions. In this group the (Sub-) Arctic and the Boreal Regions are represented by relevés from Eastern Sweden and Estland with species of the *Agropyro-Rumicion crispi* (*Trifolium fragiferum*, *Potentilla anserina*, *Leontodon autumnalis*). Cluster 42 contains species of the Arctic *Puccinellion phryganodes*, *Eleocharetum uniglumis*, *Juncetum gerardii* and *Caricetum pulchelli*. This group shows a decrease in salinity from cluster 38 to 37. Cluster 38 is characterized by the general salt marsh species *Plantago maritima*, *Juncus maritima*, *Juncus gerardii* and *Glaux maritima* in addition to the regional indicators *Puccinellia phryganodes*, *Stellaria humifusa* and *Carex glareosa*.

The clusters which are not mentioned here, are insufficiently characterized. It is up to the user to decide

Table 1. Cluster table of CLUSLA95.
Cluster sequence based on the hierarchical classification of the standardized cluster table by Ward's method (Wishart 1975). Species in sequence of entrance in the clusters. + species-frequency at least 60%; x species is significant; o species is significant and has a frequency of at least 60%.

	23 2 11112 2 3344	3442413 3	1232 2 22	33311 11
CLUSTER NUMBERS	51732179025795914	8326064347	816814809	03235642
	00000000000000011	0000000000	000000000	00000000
NUMBER OF RELEVES	01000000000000302	1040000001	011000100	06101000
	41523131311121179	6871500013	914415152	25153362
	87022162076880602	7064141811	102426382	18607057
Aster tripolium	xx------------xxo	----------	---------	-x------
Plantago maritima	x------------+xxx	xxxo------	---------	-x------
Juncus gerardii	xxx----+------xxx	xoxo--+---	--o------	-x------
Agrostis stolonifera	ox---oxoo----+xxx	ooooo--++x	---------	---ox---
Trifolium fragiferum	xx-----+o--------	----------	---------	--------
Lotus tenuis	o-o+----o--------	----------	---------	--------
Potentilla anserina	x-------o--------	xoxo--+---	---------	--------
Puccinellia limosa	x----------------	----------	---------	----x---
Puccinellia distans	-x------------xx-	----------	---------	xxx-----
Carex distans	-x---o--o--------	---+------	---------	--------
Taraxacum bessarabicum	-x---------------	----------	---------	-x------
Carex divisa	-ox--------------	----------	---------	--------
Juncus maritimus	--x-------------x	----------	-----xx--	--------
Sonchus arvensis	--x--------------	----------	---------	--------
Alopecurus bulbosus	--x--------------	----------	---------	--------
Scorzonera parviflora	--x--------------	----------	---------	--------
Glaux maritima	---o----x-----xxx	xxxo--++--	---------	--------
Plantago crassifolia	---o-------------	----------	---------	--------
Hordeum marinum	---+-------------	----------	x--------	--------
Leontodon saxatilis	---o-------------	----------	---------	--------
Limonium vulgare	----x--------+--o	----------	xox--x---	--------
Sueda maritima	----x---------xxx	----------	x-x------	-x------
Puccinellia palustris	----o------------	----------	-x---x---	--------
Arthrocnemum fruticosum	----o------------	----------	xoox-x---	--------
Artemisia caerulescens	----x------------	----------	-----x---	--------
Tetragonolobus maritimus	-----o-----------	----------	---------	--------
Plantago major	------x-x--------	----------	---------	--------
Festuca rubra	-------o-----+xxx	xoxo-++--x	---------	--------
Eleocharis uniglumis	-------o---------	--x-------	---------	----x---
Triglochin palustre	-------o---------	--x-------	---------	--------
Poa pratensis	-------o---------	-o-o--+---	---------	--------
Trifolium repens	-------o---------	-o-o--+---	---------	--------
Bellis perennis	-------o---------	----------	---------	--------
Poa trivialis	-------o---------	----------	---------	--------
Holcus lanatus	-------+---------	----------	---------	--------
Phragmites australis	--------x------xx	--x-------	------x--	-x------
Festuca arundinacea	--------x--------	----------	---------	--------
Potentilla reptans	--------x--------	----------	---------	--------
Trifolium pratense	--------x--------	----------	---------	--------
Drepanocladus aduncus	--------o--------	----------	---------	--------
Odontites verna	--------x----+---	----------	---------	--------
Leontodon autumnalis	--------x--------	xoxo--+---	---------	--------
Agropyron repens	--------x--------	---+------	---------	--------
Festuca pratensis	--------x--------	----------	---------	--------
Juncus compressus	--------x--------	----------	---------	--------
Plantago media	--------o--------	----------	---------	--------
Inula brittanica	--------o--------	----------	---------	------x-
Achillea millefolium	--------x--------	---+------	---------	--------
Atriplex littoralis	---------o-------	----------	---------	--------
Atriplex hastata	---------o----x-x	----------	---------	--------
Cakile maritima	---------o-------	----------	---------	--------
Parapholis strigosa	----------+--+x-x	----------	x--------	--------
Agropyron pycnanthum	-------------+x-x	----------	---------	--------
Armeria maritima	-------------+xxx	--xo------	---------	--------
Halimione pedunculata	-------------+-x-	----------	---------	--------
Plantago coronopus	-------------+x--	----------	xx-------	--------
Sagina nodosa	-------------+---	----------	---------	--------
Centaurium littorale	-------------+---	--x-------	---------	--------
Cochlearia danica	-------------+x--	----------	---------	--------
Sedum acre	-------------+---	----------	---------	--------
Sagina maritima	-------------+x--	----------	---------	--------
Leontodon taraxacoides	-------------+---	-------+--	---------	--------
Cerastium diffusum	-------------+---	----------	---------	--------
Hippophae rhamnoides	-------------+---	----------	---------	--------
Cladonia spec.	-------------+---	----------	---------	--------
Salicornia europaea	--------------xxx	----------	-xx------	-x------
Halimione portulacoides	--------------x-x	----------	xxxx-x---	--------
Puccinellia maritima	--------------xxx	----------	---------	--------
Triglochin maritimum	--------------xxx	oxo---+--x	---------	--------
Spergularia media	--------------xxx	----------	---------	-x------
Spergularia marina	--------------xx-	----------	---------	--------
Spartina townsendii	----------------x	----------	---------	--------
Artemisia maritima	----------------x	----------	---------	-x------
Scirpus maritimus	----------------x	--x-------	---------	--------
Carex extensa	----------------x	--x-------	---------	--------
Puccinellia phryganodes	-----------------	x--------x	---------	--------
Stellaria humifusa	-----------------	x--------x	---------	--------
Carex glareosa	-----------------	x--------x	---------	--------
Drepanocladus uncinatus	-----------------	---------x	---------	--------
Carex fusca	-----------------	-x----+---	---------	--------
Scirpus tabernaemontani	-----------------	--x-------	---------	--------
Centaurium pulchellum	-----------------	--x-------	---------	--------
Taraxacum officinale	-----------------	---+------	---------	--------
Cerastium fontanum	-----------------	---+------	---------	--------
Lotus corniculatus	-----------------	------+---	---------	-------x
Linum catharticum	-----------------	------+---	---------	--------
Calliergonella cuspidatum	-----------------	------+---	---------	--------
Phleum pratense	-----------------	------+---	---------	--------
Galium palustre	-----------------	------+---	---------	--------
Campylium polygamum	-----------------	------x--x	---------	--------
Carex oederi	-----------------	------+---	---------	--------
Carex panicea	-----------------	------+---	---------	--------
Galium uliginosum	-----------------	------+---	---------	--------
Campylium stellatum	-----------------	------+---	---------	--------
Musci	-----------------	---------x	---------	--------
Carex subspathacea	-----------------	---------x	---------	--------
Arthrocnemum glaucum	-----------------	----------	xx-------	--------
Arthrocnemum perenne	-----------------	----------	-xx------	--------
Inula crithmoides	-----------------	----------	-----xx--	--------
Schoenus nigricans	-----------------	----------	------x--	--------
Suaeda pruinosa	-----------------	----------	------xx-	--------
Sphenopus divaricatus	-----------------	----------	------x--	--------
Spergularia diandra	-----------------	----------	-------x-	--------
Mesembryanthemum nodiflorum	-----------------	----------	-------o-	--------
Beta macrocarpa	-----------------	----------	-------x-	--------
Frankenia corymbosa	-----------------	----------	-------x-	--------
Zostera noltii	-----------------	----------	--------o	--------
Atriplex tatarica	-----------------	----------	---------	-x------
Polygonum aviculare	-----------------	----------	---------	-x--x-x-
Cynodon dactylon	-----------------	----------	---------	-x---x-o
Crypsis aculeata	-----------------	----------	---------	-x------
Camphorosma annua	-----------------	----------	---------	-x------
Limonium gmelinii	-----------------	----------	---------	-x--xxx-
Petrosimonia triandra	-----------------	----------	---------	-x------
Halimione verrucifera	-----------------	----------	---------	-x------
Pholiurus pannonicus	-----------------	----------	---------	--x-x---
Chamomilla recutita	-----------------	----------	---------	--x-x---
Alopecurus pratensis	-----------------	----------	---------	---xxx--
Beckmannia eruciformis	-----------------	----------	---------	---x----
Festuca pseudovina	-----------------	----------	---------	----xooo
Arachnospermum canum	-----------------	----------	---------	----xox-
Plantago tenuiflora	-----------------	----------	---------	----x---
Rorippa sylvestris	-----------------	----------	---------	----x---
Mentha pulegium	-----------------	----------	---------	----x---
Plantago lanceolatum	-----------------	----------	---------	-----x--
Erophila verna	-----------------	----------	---------	-----xx-
Achillea colina	-----------------	----------	---------	-----xx-
Poa bulbosa	-----------------	----------	---------	------x-
Artemisia santonicum	-----------------	----------	---------	------x-
Ononis spinosa	-----------------	----------	---------	-------x
Potentilla cinerea	-----------------	----------	---------	-------o
Achillea pannonica	-----------------	----------	---------	-------x
Euphorbia cyparissias	-----------------	----------	---------	-------o
Bryum salinum	-----------------	x---------	---------	--------
Ceratodon purpureus	-----------------	----------	---------	------x-

whether to reject such clusters or to test their contents first by means of TABORD (Van der Maarel et al. 1979) or a similar method.

ALSULC95 (Fig. 3)

The general structure of the dendogram resulting from this reversed treatment is rather similar. Again two very large clusters, now the numbers 41 and 44, were formed and they are equally characterized by many common salt marsh species. The syntaxonomical interpretation is also similar, although there are differences in the separation of certain clusters.

At the 6 group-level we find a clear gradient from haline to fresh conditions in the group of clusters 43 and 34. From the *Salicornietalia* communities with *Salicornia europaea* and

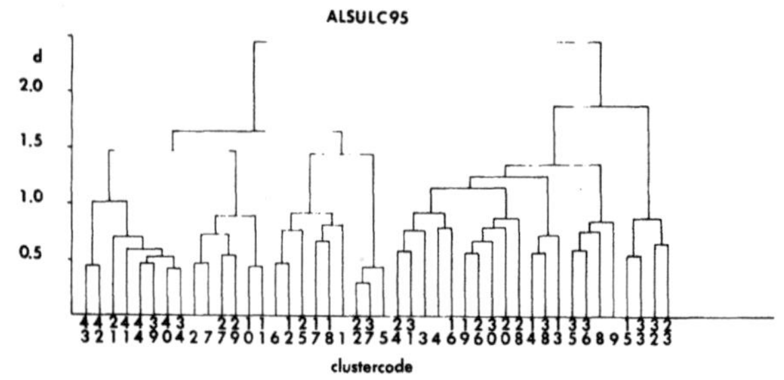

Fig. 3. Agglomerative clustering of the standardized cluster-table of ALSULC95.

Suaeda maritima in cluster 43, via the *Armerion* we proceed towards clusters 40 and 34, which show *Agropyro-Rumicion crispi* characteristics (*Phragmites australis*, *Leontodon autumnalis*, *Potentilla anserina*, *Trifolium repens*).

As in CLUSLA95 clusters 39, 40 and 44, 42 and 43 represent the typical salt marsh communities.

Clusters 2 to 11 show a transition within the more fresh, Northern European relevés from the Baltic *Armerion* communities with *Carex fusca* to the (Sub-) Arctic *Puccinellion phryganodes* vegetation with *Carex glareosa* and *Puccinellia phryganodes*. Clusters 6 to 1 include some splinter-groups of the *Agropyro-Rumicion crispi* and even the *Arrhenatherion* (with *Bellis perennis*, *Festuca pratensis*, *Ranunculus acris*) in 6 and 17.

Clusters 22, 37 and 5 with *Sagina maritima* and *Cochlearia danica* belong to the Baltic and Arctic *Saginion* vegetation. This part of the table shows a gradient from haline to fresh, from 43 to 5, as well as a preponderance of species from Northern Europe, although a large number of Mediterranean relevés are included (for example in cluster 43 with its *Arthrocneum*-species).

Probably because of a lack of typical salt marsh species, the relevés of Southern and Continental Europe could not join the clusters of the first part of the table. Among the clusters which can easily be defined, we recognize from the Pontic-South-Siberion Region the *Thero-Camphorosmion* in clusters 19 and 38, the *Festucion pseudovinae* in 14 and *Artemisio-Festucetalia pseudovinae*, the *Inulo-Festucion pseudovinae* and *Festucion pseudovinae* in 33, *Beckmannion* and *Festuco-Puccinellietalia* in 32 and in 23 the *Festucion pseudovinae*. From the Mediterranean cluster 28 with *Saginio-Limonietalia* communities and the *Frankenieto-Koelpinetum* (Morocco) and 9 with *Spartina patens* communities.

If we compare the two tables with regard to the characterizing species we find that in CLUSLA 95 136 species are used and in ALSULC95 125, of which 100 are the same

This again shows the overall similarity in results. 30 of the 44 species selected by Lausi & Feoli (1979) to characterize the 2605 relevés of selection *c* of the Working-Group for Data Processing (Van der Maarel, Orlóci & Pignatti 1976) contribute to the differentiation of the clusters of CLUSLA 95 and ALSULC95.

CLUSLA4490 and CLUSLA4480 (Fig. 4)

In CLUSLA4480, at the 3 group-level, clusters 1 to 12 may be considered as *Armerion* vegetation with a transition to the *Saginion*. Clusters 9 to 15 comprise *Puccinellion*, *Salicornion* and *Salicornion fruticosae* communities. As in the first group clusters 2 and 18 belong to the *Armerion*, but they are not very distinct. However, because of their size they can be treated by finer methods.

The threshold value in CLUSLA4490, resulting in only 17 clusters, was obviously too feeble to enable a more detailed description as in CLUSLA95.

ALSULC4290 (Fig. 5) and ALSULC4280 (Fig. 6).

As compared with the preceding tables the two threshold values result in more differentiated clusters here. Remarkable is the occurrence in both tables of a cluster, characterized by *Pottia heimii*, *Parapholis strigosa* and *Sagina maritima*, which was called a transition to the *Saginion* in CLUSLA4480. This is a group of 6 respectively 11 relevés, which appear in a more or less pure composition.

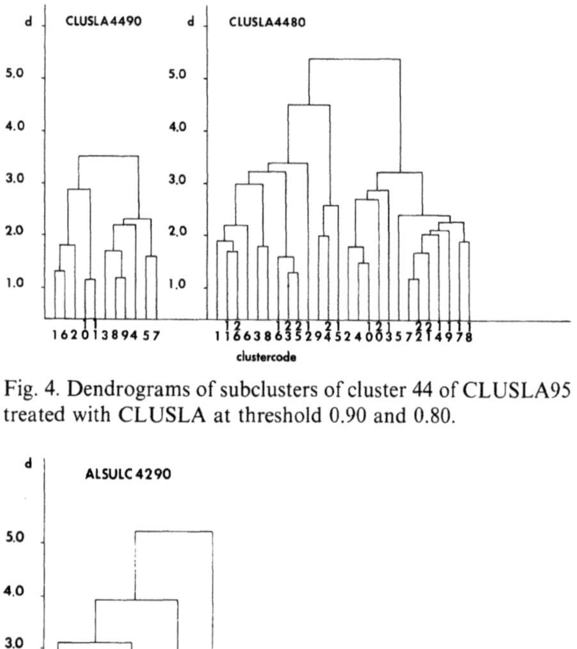

Fig. 4. Dendrograms of subclusters of cluster 44 of CLUSLA95 treated with CLUSLA at threshold 0.90 and 0.80.

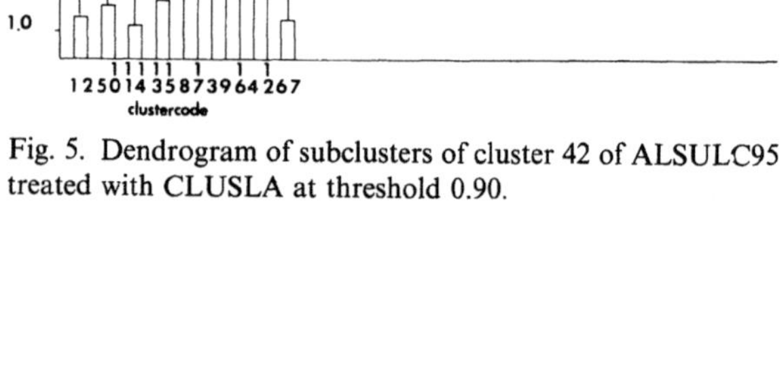

Fig. 5. Dendrogram of subclusters of cluster 42 of ALSULC95 treated with CLUSLA at threshold 0.90.

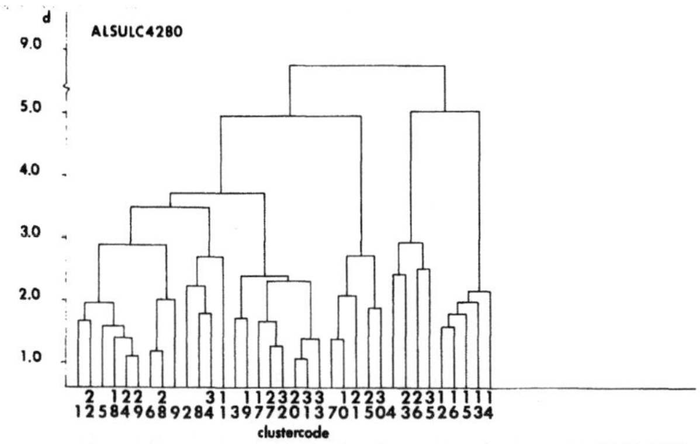

Fig. 6. Dendrogram of subclusters of cluster 42 from ALSULC95 treated with CLUSLA at threshold 0.80.

Of the group of clusters 5 to 17 in ALSULC4290 cluster 15 is characterized by species of the *Glauco-Puccinellietalia*. The *Junco-Caricetum extensae* is represented in 11 and the *Juncetum gerardii* in 14. *Scirpus maritimus* vegetation is found in 17.
Of the clusters 3 to 12 the former is recognized as the *Frankenio-Limonietum lychnidifolii*, an eu-Atlantic community from Western France. Species of the *Puccinellio-Spergularion salinae* and the *Puccinellion maritimae* characterize cluster 16. Cluster 6 and 7 exist of disturbance communities of the *Puccinellio-Spergularion salinae* with elements of the low salt-marshes (*Salicornia europaea*, *Sueada maritima*). The same cluster, with *Spergularia media*, *Artemisia maritima*, *Halimione pedunculata* and *Puccinellia distans*, is established as 16 in ALSULC4280. In this table the large cluster of the *Glauco-Puccinellietalia* (15 in ALSULC4290) is divided into three (20, 31, and 33). The cluster of *Scirpus maritimus* communities is split into a group with and one without species such as *Plantago maritima*, *Triglochin maritima* (34 and 35). In cluster 7 up to 30 the *Puccinellio-Spergularion salinae*, the *Puccinellion* and the *Frankenio-Limonietum* as distinguished in ALSULC4290 are represented.

Discussion

Generally the results obtained with CLUSLA are satisfactory. A huge material can be arranged rather fast (1 to 2 hours CPU-time) in reasonably interpretable homogeneous clusters. If large clusters, obtained with CLUSLA in a first run, are treated separately with the same program again an interpretable substructure arises. With the help of more elaborate programs the smaller groups can be split up further. An important check (which we did not yet perform) would be the comparison of all final low-level clusters in order to discover similarities between lower units from different major groups as resulting from the first run.

Two aspects of the program need further discussion, viz. the choice of the resemblance function and the order in which the relevés are entered. The resemblance functions considered for use in CLUSLA were the similarity ratio as used in TABORD and some distance measures (cf Orlóci 1978, Van der Maarel 1979b). Mainly for reasons of computational simplicity we chose the simple dissimilarity measure of Bray & Curtis, which is a standardized, symmetric distance measure on presence-absence basis for calculating the dissimilarity between relevés and clusters. In an earlier version we used the asymmetric form of this formula by calculating $d(C, R)$ if $r \neq 0$, to reduce the 'absorbing' influence of a species-rich cluster on a species – poor relevé. This resulted in a classification with only a few clusters, one of them containing 97 % of the relevés, the others containing some small and isolated groups. This can be understood from the special character of the relevés of this set of salt-marsh data, which are composed of a relatively small number of typical and common salt-marsh species accompanied by groups of regionally differentiating species. Although the frequency of species never reaches high values (Kortekaas et al. in prep.) the overlap between the relevés is still too large to profit from this formula. It could however well be applied to more heterogeneous data sets containing well-distinguished species groups.

The effect of changing the order of relevés was only studied by taking exactly the reverse order. This ALSULC-treatment succeeded in a similar way as the CLUSLA-treatment did. Although the main divisions in both treatments are similar, there are some clear differences, as is demonstrated in CLUSLA95 (Fig. 2, table 1) and ALSULC 95 (Fig. 3). One should realize that the collection first contains data from the Netherlands, then from Atlantic, Continental and Mediterranean Europe and finally from Sweden. CLUSLA95 reflects this sequence to some extent through a number of continental (splinter-) groups, while in ALSULC95 the Baltic-Arctic communities are more distinctly represented. In the latter run these clusters were created already at an early stage.

Evidently such effects may again be diminished by adopting a stronger threshold. Let us consider a community with strong local or regional characteristics, still some typical salt marsh species. It can be predicted that such communities will become separate clusters if they enter the treatment at an early stage. If a general salt-marsh cluster already exists, however, relevés of the local-characteristics type will join the general cluster, because of the share in common salt marsh species. If we would strengthen the threshold, i.e. adopting a lower d-value, they will still form clusters of their own. However, in view of the increase in computer-time, the threshold should not be too strong in a first run. Locally characteristic communities can then be extracted in subsequent runs. In this connection it can be doubted whether it is really necessary to adopt such a high maximum of 100 clusters in the first run.

The gain in information after the subsequent clustering of large primary clusters at two different threshold values was not large. However, in accordance with the aim of CLUSLA, the secondary clusters still became better accessible to other methods.

Summary

CLUSLA, a computer program for the clustering of very large phytosociological data sets is described. It is an elaboration of Janssen's (1975) simple procedure. The essence of the program is the creation of clusters, each starting with one relevé, as the relevés are entered in the program. Each new relevé that is sufficiently distinct from already existing clusters is considered a new cluster. The fusion criterion is the attainment of a certain level of (dis-) similarity between relevé and cluster. Bray and Curtis' dissimilarity measure with presence-absence data was used.

The program, written in FORTRAN for an IBM 370–158 system, can deal with practically unlimited numbers of relevés, provided the product of the number of primary clusters and the number of species does not exceed 140.000. We adopted maxima of 100 and 1400 respectively.

After the primary clustering round a reallocation is performed. Then a simple table is printed with information on the significance of occurrence of species in clusters according to a chi-square approach. The primary clusters can be treated again with a higher fusion threshold; or approached with more elaborate methods, in our case particularly the TABORD program.

The program is demonstrated with a collection of 6072 relevés with 889 species of salt marsh vegetation from the Working-Group for Data-Processing.

References

Beeftink, W.G. 1977. The coastal salt marshes of Western and Northern Europe: an ecological and phytosociological approach. In: V.J. Chapman (ed): Ecosystems of the world Vol. 1 p. 109–155. Elsevier Scientific Publishing Company. Amsterdam, Oxford, New York.

Clifford, H.T. & W. Stephenson. 1975. An introduction to numerical classification. Academic Press. New York, San Francisco, London.

Hill, M.O., R.G.H, Bunce & M.W. Shaw. 1975. Indicator species analysis, a divisive polythetic method of classification, and its application to a survey of native pinewoods in Scotland. J. Ecol. 63: 597–615.

Hogeweg, P. 1976. Iterative character weighing in numerical taxonomy. Comput. Biol. Med. 6: 199–211.

Janssen, J.G.M. 1975. A simple clustering procedure for preliminary classification of very large sets of phytosociological relevés. Vegetatio 30: 67–71.

Kortekaas, W.M., D. Lausi & W.G. Beeftink. 1979. Description of the salt marsh data set of the Working-Group for Data-Processing. Phytosociology. In E. van der Maarel, L. Orlóci & S. Pignatti (eds): Data-Processing in Phytosociology, Junk, The Hague (in press).

Lausi, D. & E. Feoli. 1979. Hierarchical classification of European salt marsh vegetation based on numerical methods. Vegetatio 40 (in press).

Maarel, E. van der. 1979a. Transformation of cover-abundance values in phytosociology and its effect on community similarity. Vegetatio 39: 97–114.

Maarel, E. van der. 1979b. Multivariate methods in phytosociology, with reference to the Netherlands. In M. J. A. Werger (ed.) The study of vegetation, p. 161–225. Junk, The Hague.

Maarel, E. van der, J.G.M. Janssen & J.M.W. Louppen. 1978. TABORD. A program for structuring phytosociological tables. Vegetatio 38: 143–156.

Maarel, E. van der, L. Orlóci & S. Pignatti. 1976. Data-processing in phytosociology, retrospect and anticipation. Vegetatio 32: 65–72.

Orlóci, L. 1978. Multivariate analysis in vegetation research. 2nd Ed. Junk, The Hague – Boston.

Pignatti, S. 1976. A system for coding plant species for data processing in phytosociology. Vegetatio 33: 23–32.

Sokal, R.R. & F.J. Rohlf. 1969. Biometry. W.H. Freeman and Company, San Francisco.

Vicherek, J. 1973. Die Pflanzengesellschaften der Halophyten- und Subhalophytenvegetation der Tschechoslowakei. Academia Verlag, Prague.

Westhoff, V. & A.J. den Held. 1969. Plantengemeenschappen in Nederland. W.J. Thieme & Cie., Zutphen.

Wishart, D. 1975. Clustan 1c. User manual. University College. London.

ELEMENTS FOR A SYSTEM OF DATA PROCESSING IN PHYTOSOCIOLOGY AND ECOLOGY*

P. DAVID[1], J. LEPART[2] & F. ROMANE[1]**

[1] Centre d'études phytosociologiques et écologiques L. Emberger (CEPE – L. Emberger, CNRS), B.P. 5051, 34033 Montpellier Cedex, France
[2] Ecothèque méditerranéenne (CNRS), B.P. 5051, 34033 Montpellier Cedex, France

Keywords:
Data-processing, Ecothèque, Factor analysis of correspondences, Phytosociology, Salt marsh

Introduction

This contribution presents a system of data-processing which has gradually been developed at our institute to solve concrete problems arising in the various vegetation studies carried out here. The system comprises a data storage and retrieval system as well as a series of statistical tests and multivariate methods for use in 'phytosociology and ecology'. (cf. the name of our institute). According to our usage these terms have the following meaning. Phytosociology is considered as the study of relations between plants, sensu Krylov (1898), or the study of plant associations, 'associations végétales', sensu Guinochet (1973). Plant (or vegetation) ecology is considered as the study of the relations between vegetation and its environment, including studies of indicator species and ecological groups, and subsequent analyses of vegetation structure and distribution ecology of animal species (e.g. birds and molluscs). This definition of phytosociology deviates from definitions such as of Braun-Blanquet (1932), Cain (1944) and Becking (1957) which include environmental relations (Cf. Westhoff & van der Maarel 1978 for a review of the Braun-Blanquet approach which emphasizes synecology as an essential approach in phytosociology).

The standardisation of phytosociological and ecological observations and their codification

In order to have ecological studies fit in a coherent system of data-processing, it is necessary that the observers use field methods as objective as possible and moreover a definite language, before synthesizing their results. The standardization of the observations and their statistical interpretation dictate the use of a uniform, unambiguous code.

To establish such a code requires carefull consideration of the significance of the terms used. It is in this connection that L. Emberger (in Godron et al. 1968) presented the 'Code for systematic surveying of the vegetation and the environment', a document widely used in the CEPE-L. Emberger for studies within France and in other temperate regions. We will only briefly discuss the essentials of the code.

The main item to which the codification is applied, is the *relevé*, which is defined as a set of phytosociological and ecological observations from a well defined site. Some attempts have been made to codify descriptions on higher levels of perception, such as ecological sector and ecological region (Anon. 1971, Son 1975) but this is rather difficult and does not always respond to the objectives for which the relevés are made.

The nature of the observations in a relevé may be of two different types: 'raw' or 'elaborated'. The raw observations are those needing no further interpretation, e.g., altitude, slope, presence or absence of given species, etc. In cases where the observations are quantitative, it may be convenient to use scales to describe specific caracteristics. But the

* Contribution from the Working Group for Data-Processing in Phytosociology, International Society for Vegetation Science.
** We appreciate the help of Miss Gautier for a first translation of our French manuscript, of Drs. L. Orlóci and E. van der Maarel for improving the English, and of Dr. E. van der Maarel and Dr. W.G. Beeftink for commenting on the results of our treatment of salt marsh data.

Numéro .... (2 3 4)
Auteur .... (5 6)
Année .... (7 8)
Elément .... (9)

RELEVÉ DE RECONNAISSANCE

Nom de la carte .... Date : Jour (12 13) Mois (10 11)
Latitude .... Numéro de la carte (15 16) (17 18)
Longitude .... Auteur (s) ....
Département .... Commune ....
Région naturelle .... (21 22 23 24)

Surface de l'élément .... % (14)
Photographies noir ☐
Photographies couleur ☐ (19)
Pas de photographie ☐
Surface du relevé
.... m² (20)

Degrés d'ouverture des strates principales

ligneux hauts (> 2m) ( )
ligneux bas (≤ 2m) (52)
herbacées (53)

1. Fermé (> 90 %)
2. Peu ouvert (75 à 90 %)
3. Assez ouvert (50 à 75 %)
4. Ouvert (25 à 50 %)
5. Très ouvert (10 à 25 %)
6. Extrêmement ouvert (0 à 10 %)
7. Totalement ouvert (0 %)

Formation (55)
0. Zone à végétation très claire ou nulle
1. Formation ligneuse haute dense
2. Formation ligneuse haute assez claire
3. Formation ligneuse haute claire
4. Formation ligneuse basse
5. Formation herbacée
6. Formation complexe ligneuse
7. Formation complexe herbacées - ligneux hauts
8. Formation complexe herbacées - ligneux bas
9. Formation complexe herbacées - ligneux bas et hauts

Degré d'artificialisation (56)
1. Végétation climacique
2. Artificialisation faible
3. Artificialisation assez faible
4. Artificialisation moyenne
5. Artificialisation assez forte
6. Artificialisation forte
7. Milieux artificiellement dépourvus de végétation

1ère Espèce dominante .... (57 58 59 60 61)
2ème Espèce dominante .... (62 63 64 65 66)

Altitude .... m (10' 11')

Exposition (12')
0. Terrain plat ou sons exposition définie
1. N 5. S
2. NE 6. SW
3. E 7. W
4. SE 8. NW

Situation topographique (13')
0. Terrain plat
1. Sommet vif (pic, arête, éperon)
2. Escarpement (corniche)
3. Sommet arrondi (butte, mamelon, crête, croupe)
4. Haut de versant (talus)
5. Mi - versant
6. Replat
7. Bas de versant
8. Dépression ouverte
9. Dépression fermée

Pente *(noter en clair la valeur observée)* (14')

0. 0 à 0,9 %	6. 36 à 48 %
1. 1 à 3,9 %	7. 49 à 63 %
2. 4 à 8,9 %	8. 64 à 80 %
3. 9 à 15 %	9. 81 à 99 %
4. 16 à 24 %	11. 100 à 275 %
5. 25 à 35 %	12. plus de 275 %

Age de la roche *(voir Code)* .... (15' 16' 17')

Réaction à ClH de la roche affleurante (19')
0. La roche n'affleure pas
1. La roche affleurante ne fait pas effervescence
2. Effervescence très faible
3. Effervescence faible
4. Effervescence forte
5. Effervescence très forte
6. Effervescence nettement localisée

Nature de la roche *(voir Code)* (20' 21' 22')
....

Surface couverte par :

la roche dure et les blocs ..... % (25' 26')
les pierrailles ..... % (27' 28')
la terre fine ..... % (29' 30')
la végétation ..... % (31' 32')
(recouvrement basal)
la litière ..... % (33' 34')

Humidité apparente de la station (36')
0. Cas particuliers
1. Station très sèche
2. Station sèche
3. Station assez sèche
4. Station moyenne
5. Station assez humide
6. Station humide
7. Station très humide (sol saturé)
8. Station extrêmement humide (sol sursaturé)

Submersion (37')
1. Station apparemment jamais inondée
2. Station inondable accidentellement
3. Station submergée périodiquement (moins de 6 mois)
4. Station submergée périodiquement (plus de 6 mois)
5. Station toujours submergée en eau peu profonde
6. Station toujours submergée en eau profonde
11. Eau circulante oxygénée
12. Eau stagnante

Hauteur de la couche d'eau de submersion
0. Pas d'eau de submersion
11. Profondeur non déterminée
12. Profondeur très variable d'un point à un autre
.... cm

Profondeur de la nappe au moment du relevé
(Echelle II)
11. Nappe non observée
.... cm

Fig. 1. Standard sheet for collecting and codifying information from the field.

use of scales may result in the loss of information. Still, at the phase of data-processing scaling is useful since quantitative and qualitative variables can be taken into account at the same time.

Elaborate observations are the ones which take into account previous knowledge. The risk of subjectivity is obvious in elaborate observations, but it appears very difficult to do entirely without these sorts of observations, e.g., type of soil. In the case of utilization of elaborated variables, we must pay special attention not to introduce in the observations the very relations we are just trying to demonstrate. The choice of variables as described in the proposed Code (Godron et al. 1968) is founded on the actual knowledge of the scientist. Only such variables are retained which seem able to explain variation in the vegetation.

Every user will have to find from the list the variables which seem to be appropriate to solve the problems involved. We see advantages to keep the proposed set of descriptions or at least compatible ones, to allow comparisons with observations made by others. For example, in the case where the description of a variable is too fine, it would be better to group the existent classes than to create new classes of equivalent amplitude but with different bounds. When the proposed list is found incomplete and new variables have to be described, it would be useful for the user to think of the more general aspects of his problem in order to reach a decision useful to those who, in future, may have to use the same set of variables. By such cooperation, different sets of observations will enrich the same data base and the comparison between different sets will be more fruitful. Using a set of relevés from the field, the chosen variables, their description as well as their codification will be considered in an example (Figure 1).

Data management system

The effort of coding the data on cards and the management of the data constitutes normalization management. The two phases constitute the beginning of setting up a data base. In addition to the coding of field data a codification must be devised for the transcription of the observations on cards and for the subsequent data management.

Relevé data are entered on punched cards in two files:

1. Card-file CDR for the description of vegetation, with:
   - number of the relevé
   - a list of the code of species found in the relevé, with, if possible, a note on abundance, dominance, cover, etc. and indication of the strata where they were found. There may be several cards for one relevé.
2. Card-file CMR for the description of environmental factors with:
   - number of the relevé
   - one column field is assigned to a variable with the code of the state assumed by the variable in the relevé.

   Several cards may have to be used for one relevé.

Options for data-management are kept separate from those for data-processing. This gives the system greater flexibility, particularly, in the setting up of new processing programmes. The management programmes to run require only the definition of parameters such as number of relevés, number of variables, etc., and format statements. The data files are transferred to magnetic disks. The access to the data is then quick and the data format is constant. In cases where processing occurs outside the Institute or data are stored for a longer time data are transferred to magnetic tapes.

Data management involves three steps:

(1) Loading on disk

This is done with a recording structure which is similar to the one of the punched cards. This step allows to detect punching errors.

(2) Inventory of data

This is done by two programmes. The first, PROFIL, produces the list of the variables and the numbers of relevés containing the species. The second, INVERT, reverses the card file of the floristic observations and produces for every species a list of relevés where it is present. During this step the code number of the species involved are decoded with a floristic catalogue and species names are written in full in a file, for use in subsequent treatments.

(3) Preparation of the data tables

Various programmes can be used to create matrices which will be then subjected to analysis. These include presence or absence of species in the relevé (or, reverse), table of the frequencies of occurrence of the species for the classes of the variables, distances (or similarities) between relevés, distances between species, etc. These programmes allow to sort species, relevés or environmental variables on different criteria.

Statistical data processing

Organization

We will not discuss the value of given methods of processing here, but only mention the organizational aspects. Some methods will not be considered here, not that they are considered of less value, but simply because programmes are not available in Montpellier or are not yet adapted to the hardware to which we have access (Romane et al., 1973). The organization of our data-processing system is outlined in Fig. 2 and specified in Figs. 3–5. For our present purpose the following elements may be mentioned separately:
– The data processing methods related to the description of the environment (Fig. 3)
– The methods for the study of the relations between the vegetation and its environment, either by an analytical approach (Fig. 4a) or by a global approach (Fig. 4b).
–The data processing methods for floristical observations only (Fig. 5a) and techniques related to the study of vegetation pattern (Fig. 5b).

Such a stratification of the processing allows to present the various analytical methods we have at our disposal and to compare the results obtained with various approaches (phytosociological and plant ecological).

Relation between environmental variables

A first possible step, at least for quantitative variables, is the comparison with a normal distribution (AKURTO

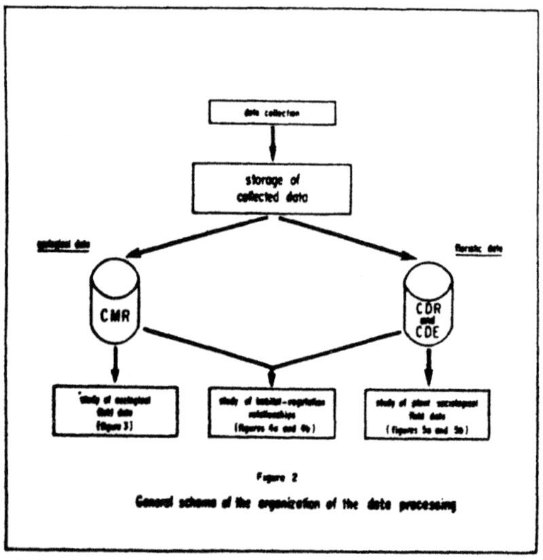

Fig. 2. Organization of data processing.

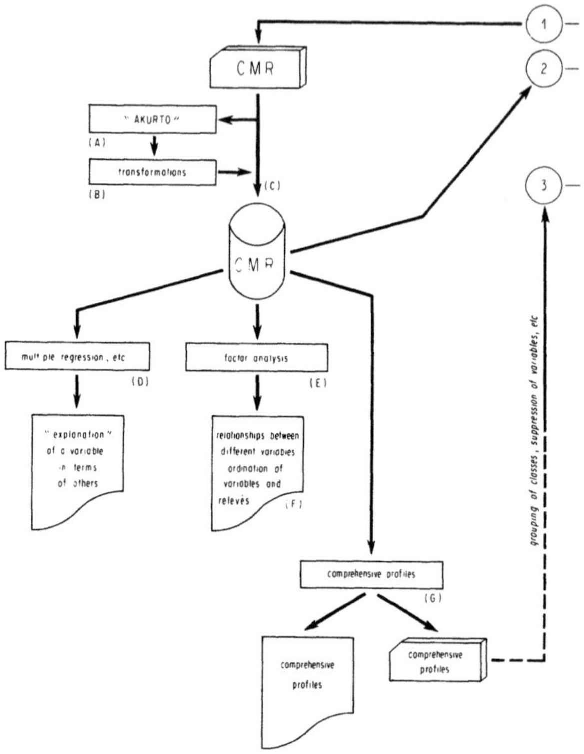

Fig. 3. Study of ecological field data.

programme, followed by transformation (logarithm, square root, arcsinus, etc.) of the variables if this appears useful (cf Dagnelie 1965, Decourt et al. 1969). The analysis of the relationships between the environmental variables may be done next with the help of analytical techniques relying on mutual information between variables, or multivariate analyses such as multiple regression analysis, factor analysis, etc. Another possibility is the erection of comprehensive profiles, including the distribution of the frequencies of the relevés in each class of a variable (Fig. 3).

Relations between environmental and floristical observations

Two approaches may be followed here:
(1) *Analytical approach*
This involves corrected profiles (Godron 1968) of species along environmental variables, i.e. the distribution of species response parameters (e.g. simply the number of occurrences related to the total frequency of the species) in the various classes devised for each environmental variable. To avoid excessive numbers of results we may select the species according to the occurrence of correlations with environmental variables (criterion of mutual

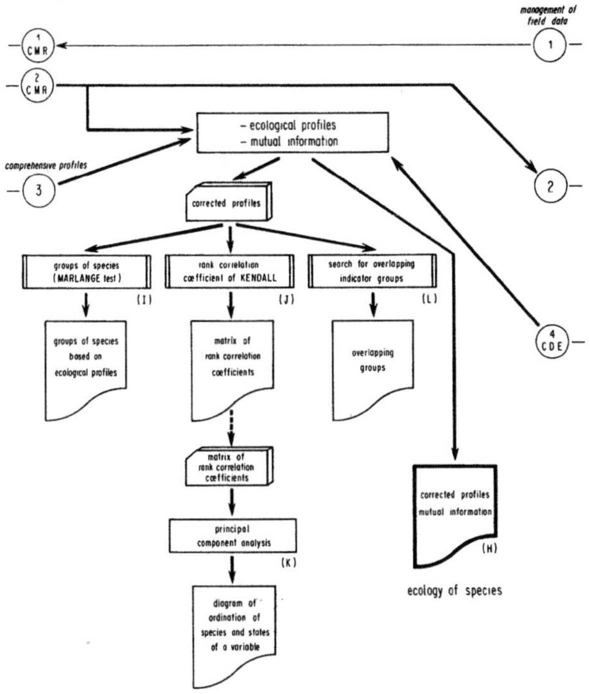

Fig. 4a. The analytical approach: Study of the relationships between plant sociological variables and ecological variables.

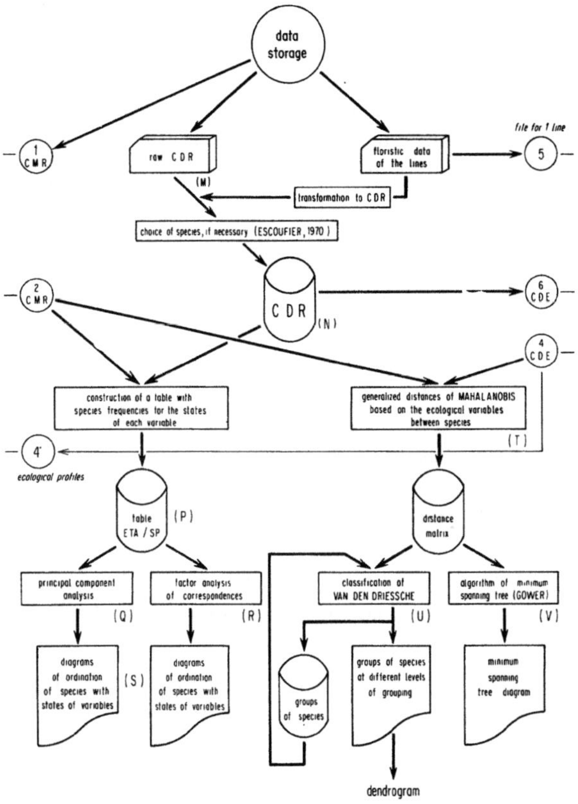

Fig. 4b. The comprehensive approach: Study of the relationships between plant sociological variables and ecological variables.

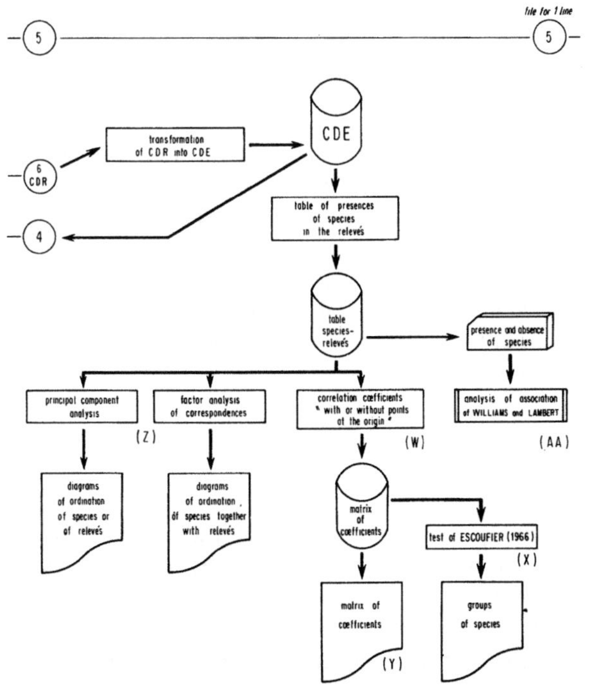

Fig. 5a. Relationships among species. Study of plant sociological field data.

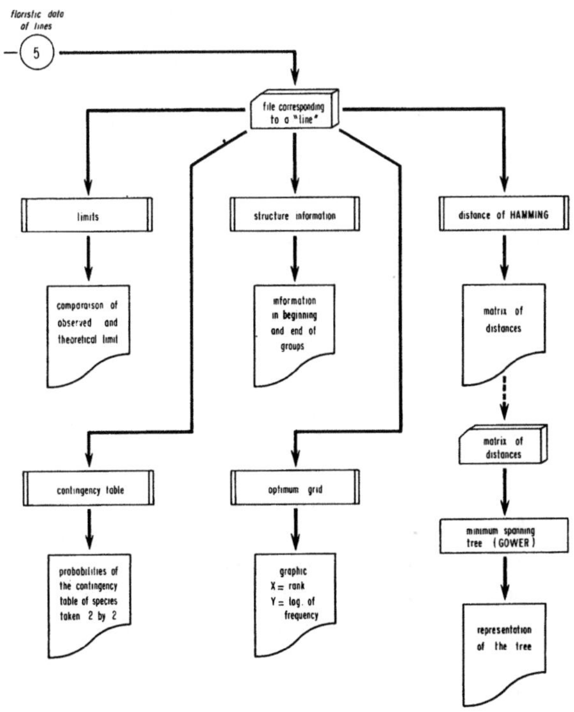

Fig. 5b. Horizontal structure of vegetation.

information) with others, and the environmental variables themselves, according to the impact they have on species distribution (criterion of average mutual information).
(2) *Synthetical approach*
This is global in character and it allows a more general perception of the observations with the help of techniques including factorial analysis, analysis of generalized distances of Mahalanobis, polar ordination, etc. (Goodall, 1954, van der Maarel 1969, Romane 1972, see Orlóci 1978 and van der Maarel 1979 for recent reviews).

Floristical observations

The study of sociological relations between species is an aspect which has been widely considered since the advance of computers, either by utilizing the tabular methods used before the computer era, or with the help of methods which have been developed more recently for computers. Westhoff & van der Maarel (1978, also van der Maarel 1979) reviewed the different possibilities of computer analyses to learn about these relationships within the framework of the Braun-Blanquet approach. Our system does not yet include tabular sorting programs such as TABORD (van der Maarel et al. 1978), but we incorporated some simple related techniques such as the use of matrices of fourfold point correlation coefficients. The factorial analysis of correspondences (as elucidated by Guinochet 1973) which we also incorporated, give results concordant with the tabular method. An example with a selection of salt marsh relevés is presented below. Other methods particularly for cluster analysis (single linkage, average linkage, etc.) are now also available, but they are not yet included in the scheme.

As to the analysis of horizontal structure, we use observations on the presence or absence of species in the segments of a transect laid down across the site).

If, by convention, the heterogeneity is estimated by the information measure $\log_2 C_N^F$ (F = frequency, N = nr. of segments) it is possible to study (Godron 1971):

- The location of an 'optimal limit', dividing the transect into two portions as different as possible.
- The occurrence of tendencies towards a macro- or microheterogeneity in the transect.
- The independance of species in their distributing along the transect, as well as the information linked to the first appearance, the vanishing and the contagiousness of species.

Software-hardware relations

The development of this set of programmes has taken several years during which the available computer systems changed (IBM 1620, 64 K; IBM 360, 512 K; CII MITRA 15, 64 K). Moreover some important elements were presented to other research institutions. Thus the set is not linked to a given hardware. All in all, only a few programmes are really original. Our main effort has been in the development of a system for the organization and managing of observations for easy use by the ecologist.

Treatment of salt marsh data, an example

Selection *e* of the Working-Group for Data-Processing in Phytosociology (Van der Maarel, Orlóci & Pignatti 1976) comprising 58 relevés with 89 species of *Glauco- Puccinellietalia* communities was inserted in our data-processing system and a factor analysis of correspondences was carried out. This technique enables a representation of relevés and species in the same space. Fig. 6 presents the result for the axis 1 and 2. The lines of variation present in the material are very well reflected and in general accordance with the results obtained with the same selection by Feoli (1977a, b) and van der Maarel et al. (1978). In fact the diagram presented here shows somewhat more variation, mainly because of the extra information provided by the species. On the basis of the general ecological knowledge on the species as summarized by the authors of the original publications from which the relevés were taken (cf. Feoli 1977a, b for a survey) and discussed at previous meetings with W.G. Beeftink and E. van der Maarel (pers. comm.), we have drawn some lines reflecting the probable environmental gradients involved. (Fig. 6) The example just shows the usefulness of such a principal analysis for a first clarification of relationships in a new material understudy.

Conclusion

We have presented here only those aspects of data organization that are linked with phytosociology and plant-ecology. However, the scope of the application in Montpellier has been progressively extended to neighbouring fields: such as successive mapping, study of regional climates, zooecology and ecosystem modelling.

In this way our data-processing system may keep its

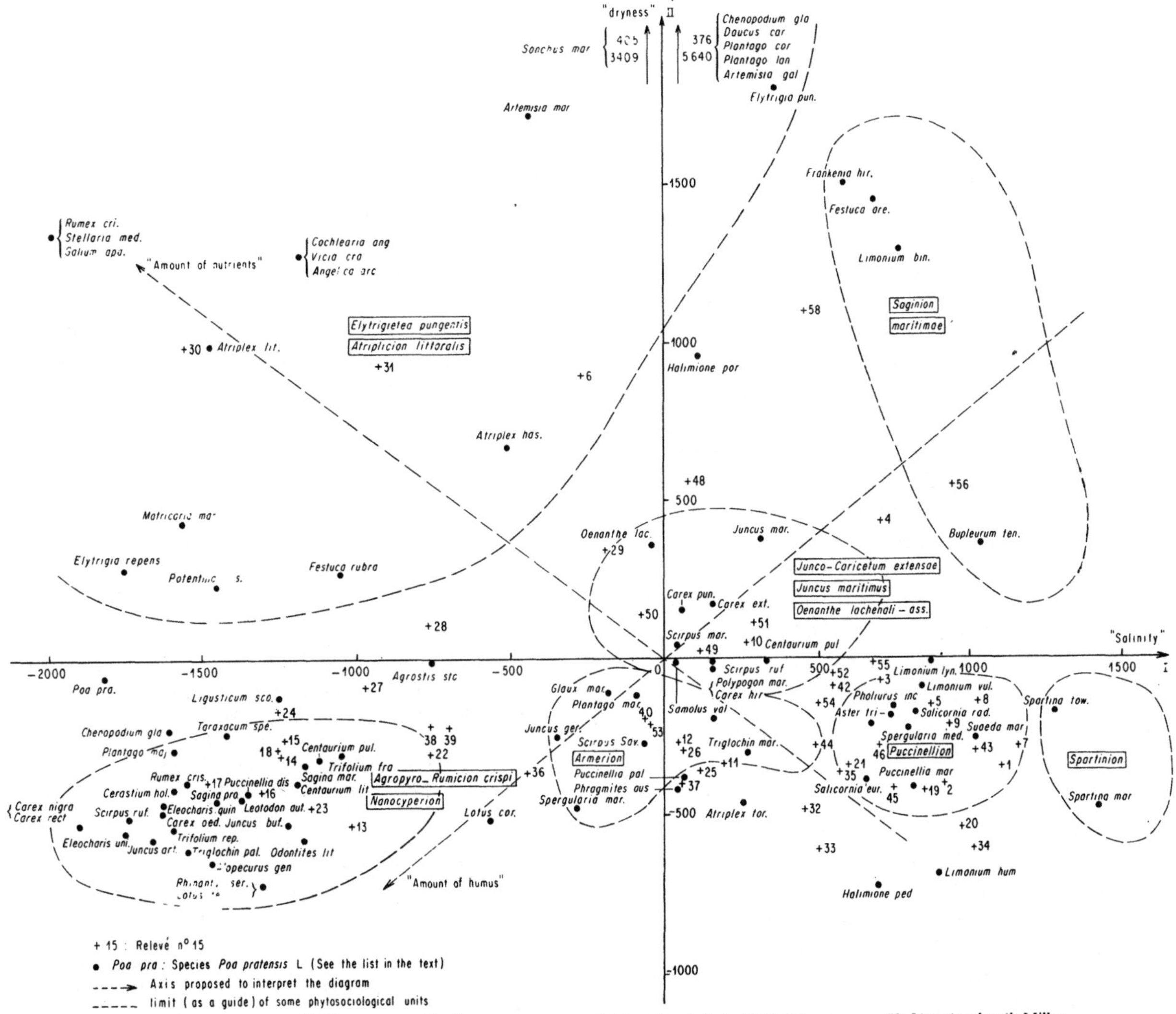

1 *Triglochin maritimum* L.
2 *Triglochin palustre* L.
3 *Pholiurus incurvatus* Hitchk.
4 *Alopecurus geniculatus* L.
5 *Polypogon ma.:timus* Willd.
6 *Agrostis stolonifera* L.
7 *Spartina maritima* (Curt.) Fernald.
8 *Spartina townsendi* Groves
9 *Phragmites australis* Trin. *ex* Steud.
10 *Poa pratensis* L.
11 *Puccinellia distans* Parlat.
12 *Puccinellia maritima* (Huds.) Parl.
13 *Puccinellia palustris* Hayek
14 *Festuca arenaria* Osbeck
15 *Festuca rubba* L.
16 *Elytrigia pungens*
17 *Elytrigia repens* Devs.
18 *Scirpus rufus* (Huds.) Schrad.
19 *Scripus maritimus* L.
20 *Scirpus savii* Seb. et Maur.
21 *Scirpus tabernaemontani* Gmel.
22 *Eleocharis quinqueflora* Schwartz
23 *Eleocharis uniglumis* Schult.
24 *Carex extensa* Good.
25 *Carex nigra* All.
26 *Carex recta* Boott.
27 *Carex oederi* Ehrh.
28 *Carex punctata* Gaud.
29 *Carex hirta* L.
30 *Juncus articulatus* L.
31 *Juncus bufonius* L.
32 *Juncus gerardi* Lois.
33 *Juncus maritimus* Lamk.
34 *Rumex crispus* L.
35 *Polygonum aviculare* L.
36 *Chenopodium glaucum* L.
37 *Beta maritima* L.
38 *Atriplex tornabeni* Tin.
39 *Atriplex littoralis* L.
40 *Atriplex hastata* L.
41 *Halimione portulacoides* (L) Aellen
42 *Halimione pedunculata* (L.) Aellen
43 *Salicornia radicans* Sm.
44 *Salicornia europaea* L.
45 *Suaeda maritima* Dum.
46 *Stellaria media* (L.) Vill.
47 *Cerastium holosteoides* Fries
48 *Spergularia media* (L.) C. Presl.
49 *Spergularia marina* (L.) Griseb.
50 *Sagina maritima* G. Don
51 *Sagina procumbens* L.
52 *Cochlearia officinalis* L.
53 *Cochlearia anglica* L.
54 *Potentilla anserina* L.
55 *Trifolium repens* L.
56 *Trifolium fragiferum* L.
57 *Lotus tenuifolius* L.
58 *Lotus corniculatus* L.
59 *Vicia cracca* L.
60 *Frankenia hirsuta* L.
61 *Bupleurum tenuissimum* L.
62 *Oenanthe lachenali* Gmelin
63 *Ligusticum scoticum* L.
64 *Angelica archangelica* L.
65 *Daucus carota* L.
66 *Samolus valerandi* L.
77 *Glaux maritima* L.
68 *Armeria maritima* Willd.
69 *Limonium vulgare* Miller
70 *Limonium humile* Miller
71 *Limonium binervosum* (Sm.) Kuntze
72 *Limonium lychnidifolium* (Gir.) Kuntze
73 *Centaurium littorale* (D. Turner) Gilmour
74 *Centaurium pulchellum* (Swartz) Druce
75 *Centaurium tenuiflorum* (Hoffg. et L.)
76 *Odontites litoralis* Fries
77 *Rhinantus serotinus* Schoenh.
78 *Plantago coronopus* L.
79 *Plantago lanceolata* L.
80 *Plantago major* L.
81 *Plantago maritima* L.
82 *Aster tripolium* L.
83 *Galium aparine* L.
84 *Matricaria maritima* L.
85 *Artemisia gallica* Willd.
86 *Artemisia maritima* L.
87 *Leontodon autumnalis* L.
88 *Taraxacum speciosum* Ramk.
89 *Sonchus maritimus* L.

Fig. 6. Example of analysis of correspondences; applied to 58 relevés of *Glauco-Puccinellietalia* (See text and species list).

own position in phytosociological and plant ecological circles besides other systems within and outside the Working Group, such as the systems in Nijmegen (Van der Maarel et al. 1978) and London Ont. (Wildi, Fewster & Orlóci 1978).

Realising all possible applications of the system would involve personal and material assistance beyond the possibilities of the CEPE-Louis Emberger. This fact together with the unique position of Montpellier in the Mediterranean Zone led to the establishment of the 'Ecothèque méditerranéenne' (Romane et al. 1973). In collaboration with the CEPE-Louis Emberger, which continues its activities, the Ecothèque méditerranéenne will concentrate on data processing activities, notably codification of floristical and environmental variables, compatability of data files, acquisition and management of a data base, programme libraries, etc.) with special emphasis on the mediterranean isoclimatic zone. In addition a program for training researchers and technicians is planned for the Ecothèque méditerranéenne, which it will realize together with the University of Montpellier and the CEPE-Louis Emberger.

Résumé

Les auteurs proposent ici une organisation du traitement des données en phytosociologie et phyto-écologie. L'organisation repose sur un système bien développé pour rassembler et codifier l'information des relèves. (fig. 1). Un ordinogramme (Fig. 2–5) présente l'agencement et l'enchaînement des différentes méthodes d'analyse des données (profils écologiques, analyse factorielle, . . .) qui permettent d'étudier les relations entre variables du milieu et variables floristiques. La conception de l'enchaînement des fichiers et des programmes de traitement permet d'aboutir assez rapidement aux résultats, facilitant ainsi la tâche du phytosociologue et du phyto-écologue. Ce travail a été réalisé en commun entre le C.E.P.E. Louis Emberger et l'Ecothèque méditerranéenne conduisant à une coopération nécessaire et utile. Une exemple est traité utilissant la sélection '*e*' de 58 relevés de milieux salés, sélection retenue par le groupe de travail 'traitement des données en phytosociologie' de l'Association internationale de phytosociologie.

Summary

An organization for data-processing in phyto-sociology and plant ecology is described. A good preparation by means of standard sheets for collecting and codifying information from the field is emphasized (Fig. 1). Fig. 2–5 present the arrangement and the logical sequence of the different methods of data analysis (ecological profiles, factor analysis,. . .) which allows the study of the relations between environmental and floristical variables, in a quick and efficient way. Cooperation with the Ecothèque méditerranéene at Montpellier is both necessary and fruitful. One example is presented using the selection '*e*' of 58 relevés of salt marsh vegetation retained by the Working Group for Data-Processing of the International Society for Vegetation Science.

References

Anon, 1971. Glossaire de pédologie. Description de l'environnement en vue du traitement informatique. Association 'Informatique et Biosphère', Paris, 173 pp.

Becking, R.W. 1957. The Zürich-Montpellier School of Phytosociology. Bot. Rev. 23: 411–488.

Braun-Blanquet, J. 1932. Plant Sociology. The study of Plant Communities (Authorized English Translation of 'Pflanzensoziologie' by Fuller G.D. and Conrad H.S.) Mc Graw Hill Book New York and London. 439 pp.

Cain, S. 1944. Foundations of Plant Geography. Harper and Brothers, New York and London, 556 pp.

Dagnélie, P. 1965. A propos des transformations de variables. Biométrie-Praximétrie 6: 59–78.

Decourt, N., Godron, M., Romane, F., Tomassone, R. 1969. Comparaison de diverses méthodes d'interprétation statistique de liaison entre le milieu et la production du Pin sylvestre en Sologne. Ann. Sci. For. 26: 413–443.

Escoufier, Y., 1966. Analyse des composantes principales. Utilisation de groupes de variables dans la recherche de la solution. Thèse troisième cycle, Faculté des Sciences. Montpellier, 53 pp.

Escoufier, Y., 1970. Echantillonnage dans une population de variables aléatoires réelles. Thèse doctorat d'état, Faculté des Sciences, Montpellier, N^0 CNRS AO 4508, 49 pp.

Feoli, E. 1977. On the resolving power of principal component analysis in plant community ordination. Vegetatio 33: 119–125.

Feoli, E. 1977b. A criterion for monothetic classification of phytosociological entities on the basis of species ordination. Vegetatio 33: 147–152.

Godron, M. 1968. Quelques applications de la notion de frequence en écologie végétale. Oecol. Plant. 3: 185–212.

Godron, M., Daget, Ph., Emberger, L., Long, G., Le Floc'h, E., Poissonet, J., Sauvage, Ch., Wacquant, J. P. 1968. Code pour le relevé méthodique de la végétation et du milieu. Principes et transcription sur cartes perforées. Editions du Centre national de la recherche scientifique, Paris, 292 pp.

Godron, M. 1971. Essai sur une approche probabiliste de l'écologie des végétaux. Thèse, Univ. Sci. Techn. Languedoc, Montpellier, 247 pp.

Goodall, D.W. 1954. Objective methods for the classification of vegetation. III. An essay in the use of factor analysis. Aust. J. Bot. 2: 304–324.
Gower, J.C. & G.J.S. Ross. 1969. Minimum spanning trees and single linkage cluster analysis. Appl. Statistics 18: 54–64.
Guinochet, M. 1973. Phytosociologie. Collection d'écologie. I. Masson et Cie. Paris, 227 pp.
Krylov, P.N. 1898. La végétation du gouvernement de Tomsk (Russian), in R.W. Becking 1957.
Maarel, E. van der 1969. On the use of ordination models in phytosociology. Vegetatio 19: 21–46.
Maarel, E. van der 1979. Multivariate methods in phytosociology with reference to the Netherlands. In M.J.A. Werger (ed.). The study of vegetation. Junk, The Hague, p. 161–225.
Maarel, E. van der, J.G.M. Janssen & J.M.W. Louppen. 1978. TABORD, a program for structuring phytosociological tables. Vegetatio 38: 143–156.
Maarel, E. van der, L. Orlòci & S. Pignatti. 1976. Data processing in phytosociology, retrospect and anticipation. Vegetatio 32: 65–72.
Marlange, M. 1973. Ordination et groupement des profils écologiques. Recueil de méthodes phyto-écologiques de l'école d'Emberger. CEPE/CNRS, Montpellier, Fasc. 3, 18 pp.
Orlóci, L. 1978. Multivariate analysis in vegetation research. 2nd. ed. Junk, The Hague, IX + 451 pp.
Romane, F. 1972. Quelques aspects de l'analyse multivariable des données phyto-écologiques. Invest. Pesq. 36: 131–139.
Romane, F., A.M. Bacou, P. David, P. David, M. Godron, & J. Lepart. 1973. An example of the organisation of plant ecology data processing. Compendium on plant ecological methods, school of L. Emberger. CEPE L. Emberger, Montpellier, Fasc. 6, 24 pp.
Son, Y. 1975. Essai sur la codification des relevés géomorphologiques pour les études d'écologie et d'aménagement. Diplôme d'études approfondies en écologie. Univ. Sc. Techn. Languedoc. Montpellier, 58 pp. + ann.
Van den Driessche, R. 1965. La recherche des constellations de groupes à partir de distances généralisées D^2 de Mahalanobis. Biométrie-Praximétrie 6: 36–47.
Westhoff, V. & E. van der Maarel. 1978. The Braun-Blanquet approach. 2nd ed. In R.H. Whittaker (ed.) Classification of plant communities, p. 287–399. Junk, The Hague.
Wildi, O., P.H. Fewster & L. Orlóci. 1978. Management and multivariate analysis of large data sets in vegetation research. Dept. Plant Sciences. Univ. Western Ontario, London W. Ont., 99 pp. + pp.

DATA-PROCESSING IN PHYTOSOCIOLOGY: CONCLUSIONS AND PERSPECTIVE

Eddy van der MAAREL, László ORLÓCI & Sandro PIGNATTI

Introduction

This series of contributions was opened with the paper: Data-processing in Phytosociology: retrospect and anticipation (van der Maarel, Orlóci & Pignatti 1976), with a look back on preparations and the subsequent realisation of the scientific program and a look forward on the results that the Working-Group was expected to produce. Although we may not have achieved all, we are satisfied with the result we present: in this volume of over twenty contributions. It is accurate to say that the group represents three different, though closely related branches of vegetation science, which we can broadly describe as theoretical vegetation ecology, quantitative plant ecology and floristic-geographical phytosociology. The authors of this concluding paper are representatives of these branches of vegetation science. They hope that through this volume and the common efforts which produced it, the mutual understanding of approaches and syntheses is enhanced!

The authors summarise the main results reached and problems left unsolved and offer their analysis of the future of phytosociology.

Data-processing as a common practice in phytosociology

Many phytosociological centres now use data-processing in the normal phytosociological practice, of type recognition through elaboration of relevé tables (cluster analysis), and in the clarification of the compositional relationships between types (ordination). Relatively simple programs with the familiar output table such as in the program of Spatz (Spatz & Siegmund 1973, see also Mueller- Dombois & Ellenberg 1974, Streng & Schönfelder 1978) certainly helped phytosociologists to overcome their initial hesitation to use a computer. Many colleagues learn simple computer programming very quickly, once they realize the potentials in these techniques.

As van der Maarel remarked in the Introduction to this book the rapid gain of data-processing methods in many phytosociological centres is a direct consequence of the activities of the Working-Group. At special courses, notably at Lund, Parma and Rome, and through visits by young scientists and students, many phytosociologists and plant ecologists were initiated in the art of data-processing. Our own centres continued to develop new programs and progress has been made in establishing an integrated data-processing package. The London, Ontario packages are the most complete and they are easily accessible (Orlóci, Feoli & Fewster 1978, Wildi, Fewster & Orlóci 1978, see also Orlóci 1978). Another many-sided system is the 'écothèque' at the Centre d'Etudes Phytosociologiques et Ecologiques, Montpellier (Godron et al 1968, David, Lepart & Romane 1979). Satellite packages are developed in other centres, e.g. at Lund (Persson 1977, 1978). Amongst other packages being developed in the world of plant ecology that of Cornell is the most important one (Gauch 1978, 1979, Hill 1979b).
Hill 1979b).

We should point out that typological and ecological studies such as by Bouxin (1976), Adam (1977), van Gils & Kovács (1977), Meyer (1977), Eysink et al (1978), Wiegleb (1978), Jensén (1979) and Tyler (1979) are representative examples for the effective use of automated data-processing in phytosociology. Notwithstanding the fact that the majority of the leading authorities in syntaxonomy have not fully accepted the numerical approach, the cooperation between the users of classical and numerical approach which started with the *Spartinetea* project has been continuing ever since.

Data storage and coding

The 'Trieste' system of coding and storing phytosociological data (Cristofolini, Lausi & Pignatti 1969, Pignatti 1976) has proven its usefulness in various treatments. The coding did not cause serious difficulties. The coding of genera according to the system of Dalla Torre & Harms is satisfactory, despite the changes which occur in the delimitation and nomenclature of some genera. Still, it is a handicap that the Flora Europaea on which the species code is based has not appeared in full. The recoding of provisionally coded species after a new volume has appeared, takes much time and attention.

The storage of relevés in the form of the relevé tables in which they are available through publications or manuscripts has proven to be very effective. One major problem in the treatment of collections of tables was the composition of entire species - relevé matrices. This problem has been solved now by inserting information on the number of relevés and the number of species at the beginning of each table. Although in most local data banks we know the data are stored per relevé rather than per table (e.g. in the Montpellier system (David, Lepart & Romane 1979)), we think that for broad phytosociological purposes the Trieste system is preferable.

The botanical institutes at Trieste and Nijmegen have extended their data banks beyond the large collection of salt marsh relevés. In Trieste ca. 500 relevés of alpine grasslands, 700 of Italian woodlands and 200 of riverbank communities are stored, in Nijmegen ca. 3500 relevés of dune grasslands as well as a first selection of 60 published tables concerning Dutch vegetation types with ca. 1500 relevés are also stored. Unfortunately, neither centre has the needed technical assistance for further extension of their data banks.

Table rearrangement

The formation of clusters with the help of numerical methods (Orlóci 1978) has become a common practice among many phytosociologists. Table rearrangement programs combine a clustering device with a comprehensible tabular output. Table rearrangement programs were reviewed by Westhoff & van der Maarel (1978).

Program TABORD, developed within the Working Group (van der Maarel, Janssen & Louppen 1978) is a widely used table structuring program. The main characteristic of TABORD is that it starts with clusters in stead of single relevés. Most users prefer to base their initial clusters on personal judgement rather than on automatical clustering procedures. In most applications with the Nijmegen group this judgement stems directly from field experience. In Lund (e.g. Persson 1977, Jensén 1978) and elsewhere, numerical classification programs, e.g. DIVINF. are used for pre-classification.

Clusters are arranged in TABORD in a final table according to the occurrence of constant species. The minimum frequency a species has to reach to be considered constant can be chosen by the user. This minimum constancy level is usually 60–80 %. This implies that the average TABORD user only accepts clusters for further interpretation which contain at least one species with a high degree of constancy, i.e. high diagnostic value. This appears to imply a shift in preference to the constancy aspect of the characteristic species combination in a syntaxon over the character-and differential-species (cf Westhoff & van der Maarel 1978). The emphasis is on the *characteristic combination of constant species*. Naturally a constant species can be characteristic for one or more types, or character- and differential species can be constant, but in principle, these features should be considered independently. Within our Working-Group Moravec (1978) paid special attention to the use of differentiating groups of constant species (see also Lausi & Feoli 1979). The importance of character-and differential species appears steadily diminishing, in most cases of application. Their distinction is almost completely blurred: the character-species being an exceptional case of the differential species. Yet, (non-constant) differentiating species can still be useful as indicators. The TABORD-program provides a simple chi square test to indicate both positive and negative deviations from random occurrence. In many cases positively differentiating species are added to the block of constant species in order to strengthen the diagnosis of types.

The arrangement of groups in the structured table is best obtained according to the 'most efficient starting group' option. This means in TABORD that the first cluster in the table is the one with the lowest value of the number of constant species / number of relevés ratio. In this way relatively many potential constant species are left for arranging the following clusters, while relatively few relevés are still to be arranged.

Various improvements or at least additions to simple table rearrangement programs such as TABORD can be considered. They integrate the clustering of relevés and of species. Some of them are mentioned here: The block structure in an ordered table can be improved by arranging the species into groups. This has in fact been done in nodal (two-way) analysis by Lambert & Williams (1962), but it has not been incorporated in the table rearrangement procedures of phytosociology. Rearrangement of rows and columns may of course be based on different criteria. Van der Maarel (1979) suggested reciprocal averaging, earlier known as correspondance analysis (Hill 1973, Orlóci 1978) for this purpose. Correspondance analysis has been used by the Montpellier group for many years (Romane 1972, Romane, Guillerm & Waksman 1977, David, Lepart & Romane 1979). The Montpellier work has been mainly devoted to plant - environment relations, but others have applied the method to phytosociological problems, e.g. Lacoste (1975,1976) and Klein (1979). TABORD- clustering and reciprocal averaging have been compared by van Groenendael (1978). This, and also a comparison with indicator species analysis (Hill, Bunce & Shaw 1975) confirmed that these methods are mutually complementary.

A recently developed package of Wildi & Orlóci (1979) combines efficiency in handling large data sets with flexibility, which few other packages can offer in the choice of different criteria, subjective or objective, on which two way table arrangements may be based. This package also offers many peripheral devices for table printing and other data manipulation, which makes it one of the most versatile packages available to group members.

Outside the group Hill's (1979b) table rearrangement program TWINSPAN is mentioned. In this program two-way ordering of data is based on his earlier methods. TWINSPAN, to which the ideas behind TABORD have contributed, has yet to be tested on phytosociological data sets.

A similar approach has been followed by group members Feoli & Lagonegro (1979) who grouped species on the basis of a minimum spanning tree based on a nearest neighbour criterion. The relevés are grouped with the help of intersection analysis and arranged according to the representation of the respective species groups. This approach is monothetic (cf Feoli 1977), in fact oligothetic (as the method of Hill, Bunce & Shaw 1975), implying a risk of misclassifications. These methods and other such R-clustering methods (Orlóci 1978) have the attractive features that new relevés can be easily entered into an existing classification. (cf Hill 1977,1979b). Another two-parametric approach has been developed by Dale (1971, Dale & Anderson 1973, Dale & Webb 1975), but not yet applied to large phytosociological data sets.

A test on the sharpness of block structure and the existence of compositional gradients between the groups arranged in a table has been developed by Feoli & Orlóci (1979) and applied by Orlóci & Stanek (1980) on a large data set with good results.

Classification of very large sets of relevés

With the growing salt marsh relevé collection the group's need for classification programs for very large data sets - in the order of thousands of relevés and hundreds of species - increased. The respectives centres of the group developed special programs. The first such program developed as part of the activities of the group was the classification program of Stockinger & Holzner (1972, 1973) to establish groups of correlated species. This program is applicable to phytosociological by tables as shown by Holzner, Werger & Ellenbroek (1978), but it was little used for the construction of structured tables. Other packages developed by group members include TABORD and a set of programs by Wildi, Fewster & Orlóci (1979). These have already been characterized. These both can handle up to 1000 relevés, or even more with some of the options.

CLUSLA (Janssen 1975, Louppen & van der Maarel 1979) can handle thousands of relevés. This program has the characteristic (which may be felt a drawback) of the clustering results being dependent on the sequence in which the relevés are presented for analysis. The program was developed for data sets stored in the Trieste system as series of already classified relevés (hence the name serial clustering, van der Maarel 1979).

There are other programs from outside the group with good potential when the given clustering criteria and expected low level of precision are satisfactory. TWINSPAN has been mentioned. Another is COMPCLUS (Gauch 1979). This program starts with random sets of relevés. It stores the data more effectively than CLUSLA and it seems to be faster. Another new Cornell program DECORANA (Hill 1979) is a very fast reciprocal averaging procedure (coping with the problem of axis dependance) which arranges thousands of relevés.

It seems that with the present range of programs we can manage almost any feasible very large phytosociological data set. Two important aspects may be kept in mind here. Firstly the treatment of such very large data sets will always be suboptimal and of a provisional kind. Secondly, since it is virtually impossible to synthesize thousands of relevés within one table, the main aim of such programs is to obtain manageable smaller more homogeneous sets, which can then be classified and ordinated with more refined programs. As Louppen & van der Maarel (1979) remarked, the clusters resulting from detailed analysis of the various subsets obtained should be compared with each other, while treating the clusters as new individuals on the basis of synoptic tables.

Numerical ordination

The efforts of the Working Group have been concentrated on the structuring of large data sets, which can be understood from the background and the proper aim of the group. Still, ordination techniques (see Orlóci 1978 for a review) were regularly applied and contributions were made to the understanding of ordination methods in phytosociological and ecological contexts:

Ordination have been used as a tool in the classification of relevés. Fresco (1971) arrived at a system of overlapping species clusters by means of sequential R-type ordinations. Feoli (1977, Feoli & Lagonegro 1979) developed a similar approach and incorporated the minimum spanning tree clustering approach, as did Romane, Guillerm & Waksman (1977). Indicator species analysis (Hill, Bunce & Shaw 1975) is another example of using ordination (in this case a Q-type) for finding species groups. Ordination and subsequent grouping of species may not be of the same technical importance as it was in the time that our computer programs could not deal with all the relevés of large sets but could handle all the species. However, they are of great importance in the establishment and diagnosis of vegetation types.

As was explained in the foregoing section ordination is also part of some table structuring programs. Clearly ordination is used here in either of the broader senses distinguished by Dale (1975): reduction of dimensionality and path seeking and not in the narrower sense of Whittaker (e.g. 1978, Noy-Meir & Whittaker 1978) as arranging relevés or species in relation to environmental gradients. As was expressed repeatedly by group members (e.g. Orlóci 1966, 1978a,b, 1979, Westhoff & van der Maarel 1978) and in that of many others (see Whittaker 1978) it is both theoretically sound and ecologically effective to adopt the model of a phytosociological space (van der Maarel 1979) defined by the species vectors or the relevé vectors of a phytosociological data matrix. Both ordination and classification can be considered as methods to discover main trends of variation within the phytosociological space. Such trends can usually be interpreted, in general terms without exact measurements of the evironmental factors involved. Naturally, it is always an improvement if good environmental data were available. However, the difficulty will always remain since it is hardly possible to measure irregular environmental fluctuations, effects of present and former management and historical factors.

A serious problem in the interpretation of such a model, but not only in that case, is the non-linearity of species response to environmental gradients (see reviews by Austin 1976, Phillips 1977 and Orlóci 1978a), a problem inevitably faced if the phytosociological space is replaced by an ecological space: We are aware of the problem (cf. Orlóci 1978a), but not much we could do since in our phytosociological data sets systematic environmental measurements are not available. Evidently one of the urgent tasks for phytosociology in the near future will be the performance of integrated vegetation- environmental analyses to test hypotheses derived from phytosociological studies.

The first symposium of our broader Working Group has brought to surface many new theories and methods on ordination, including aspects of non-linearity. The proceedings of this symposium (van der Maarel 1980a) will supply references for further work.

Perspectives of data-processing for vegetation science

From phytosociology to vegetation science

Within the framework of the activities of the Working Group phytosociology was the obvious term for the discipline in the context of which we considered data-processing. Phytosociology is the direct translation of the classical German Pflanzensoziologie and French phytosociologie. In the broad sense of this word, which was certainly given to it by Braun-Blanquet himself, phytosociology is almost identical to vegetation science. However, by many anglo-american ecologists phytosociology is mainly seen as synonymous with one of its branches, viz. the systematic description of plant communities. Another source for confusion is the meaning of ecology. For European phytosociologists, (syn)ecology is a subdiscipline of their science, but for anglo-american ecologists the description of plant communities has the same subordinate position within their science. Since Egler (1954) drew attention to the confusion, the situation has not improved much. Even within the Working Group the meaning of the term phytosociology is sligthly ambiguous. (David, Lepart & Romane 1979).

Mueller-Dombois & Ellenberg (1974) tried to resolve the problem by suggesting the term vegetation ecology. However they use three different descriptions: 'the study of plant communities, or vegetation' (on p. 4), 'synecology' (on p. 7) and: 'the study of structure, systematics variation in the spatial and geographical sense, development and environmental relations' (on p. 9), i.e. all aspects of phytosociology minus synphysiology.

All in all, we think the term vegetation science covers best what the students of vegetation in different parts of the world are really concerned with and we would propose to use this name as the broadest possible equivalent for phytosociology. In fact, it was the Working Group which suggested this term for use in the English name of the Society and this usage has now been accepted in almost all segments of the Society.

We think that at the moment of concluding the group's activities on data-processing in phytosociology we should emphasize the perspectives of numerical approaches for all branches of vegetation science. We will mention only a few now but express the wish that the members of the extending Working Group will suggest many more!

Numerical syntaxonomy

The formal description of plant communities will remain a corner stone in vegetation science and it is satisfactory to conclude that numerical methods have shown to be very helpful in syntaxonomy. It may be regretted that the results obtained with the *Spartinetea* (Kortekaas, van der Maarel & Beeftink 1976) have not been used in the preparation of the Prodrome of European plant communities (see Tüxen 1972 for an outline of this project). We may only hope, as was stated in the Introduction, that more syntaxonomists will take up the challenge to contribute to numerically based systems of syntaxa within recognized classes. The study by Pignatti & Pignatti (1976) on the syntaxonomy of *Seslerietea* communities is an example and stimulus. From these studies it becomes clear how geographical and local edaphic-ecological conditions determine part of the floristic-sociological variation and this interrelationship is one of the fundamental issues of syntaxonomy (cf. Westhoff & van der Maarel 1978, Pignatti 1975, Tüxen e.g. 1970).

The wider perspectives of a numerical synthesis of a variety of classes over the whole of Europe are shown in the work of Lausi & Feoli (1979). They show hierarchical and reticulate relations of the higher salt marsh syntaxa of Europe. The implications are manyfold:

1. The possibility of using numerical similarity levels for the establishment of syntaxonomical levels is confirmed. The general parallel between the two series of levels as summarized by Westhoff & van der Maarel (1978), appears justified. Neuhäusl (1977) pointed out that at least for his much more complicated material, i.e. *Carpinion* woodlands, both the floristic homogeneity of syntaxa (measured on their average internal similarity) and the similarity between syntaxa of a certain level can vary. Such variation can be caused by (a). the amount of deviant or transitional relevés in the table, (b). the level of disturbance of the vegetation types under study and (c). the total geographical area taken into account. The work of Lausi & Feoli (1979) should stimulate others to perform similar calculations on large data sets.
2. The importance of using constant species as diagnostics of types based on homogeneous sets of relevés is stressed. This idea is in line with general thinking expressed by others within the Working Group. A consequent testing of table homogeneity, which is also stressed by Tüxen (1970, 1977), could lead to still more striking parallels between the syntaxonomical and numerical levels of hierarchies.
3. Although any numerical syntaxonomy should preferably be based on individual relevés, it appears to be feasible in some cases to start with synoptic tables. Obviously, the best possible procedure is to arrive at locally defined homogeneous plant community types, which then can be subject to a synthetical treatment such as presented here with the European salt marshes.

In addition, we may remark that in areas where the basis for a syntaxonomical system is still lacking the essentials of the numerical syntaxonomical approach can very well be adopted as was recently demonstrated in a large scale vegetation survey in the Yukon (Orlóci & Stanek 1980)

Numerical synecology

As was remarked earlier the amount of environmental data taken up in the treatments, and in fact the availability of such data in general, is very scarce. Concrete results in this are consequently almost none. From a methodological point-of-view we can nevertheless make some suggestions:

1. Reciprocal averaging (e.g. Romane 1972, David, Lepart & Romane 1979) and analysis of concentration (Feoli & Orlóci 1979) are promising methods for simultaneously studying variation in floristic and environmental vectors.
2. Canonical correlation analysis with its related techniques (e.g. Kercher & Goldstein 1977), and other contingency table methods (Orlóci 1978) are the obvious tools of comparing sets of biotic and abiotic variables, based on established phytosociological classifications (see also Gauch & Wentworth 1976, Feoli & Orlóci 1979).
3. In the absence of environmental data indirect ways of

interpretation are available. We can use general knowledge on species and species groups which we can insert in ordination diagrams in the form of isolines ('isocenes', van der Maarel 1969). Recently indicator values for species, (Ellenberg 1968, 1979) have been used both in ordination diagrams and in comparison of types (see e.g. Feoli & Lagonegro 1979).

Numerical succession research

The application of numerical methods to succession data was suggested already at the beginning of our work (e.g. van der Maarel 1969) and was emphasized later by Austin (1977) and others. It is becoming standard practice now as could be seen at the last Symposium of the Working Group. With a new Symposium on succession ahead in Montpellier (September 1980), this line of numerical research will certainly be continued.

Closing remarks

Since Pignatti, Cristofolini & Lausi (1968) presented their paper on the use of data-processing in phytosociology, over ten years have passed, and many new contributions appeared in print through which ,gradually, a new approach emerged. It took much time and effort to produce the contributions and to organize them in a volume. One only need to consider the appendices, largely prepared by W.M. Kortekaas, to have an idea of what is involved. And it also takes time to have such an approach absorbed by those new to numerical methodologies. May be ten years were not an unduly long time for our project after all!

We feel satisfied with the results and optimistic about the use of numerical methods in vegetation science. Some of the colleagues speak of a level of saturation and they wish to concentrate now on the application of the methods rather than to continue on the theoretical backgrounds and methodological aspects. We agree, but we also realize that we are in a phase of development, as Greig-Smith (1980) described very well in his opening lecture at the Nijmegen Symposium, which requires the improvement of our techniques. To be more specific, we refer to the contributions to that Symposium. It is quite clear that among the many problems the most pressing include a better environmental theory (van der Maarel 1975), a better predictive approach to understand vegetation (Orlóci 1978a, 1980) and a better epistemological basis for vegetation science (Pignatti 1975,1980).

References

Adam, P., 1977. On the phytosociological status of Juncus maritimus on British saltmarshes. Vegetatio 35:81-94.

Austin, M.P., 1976. On non-linear species response models in ordination. Vegetatio 33:33-41.

Austin, M.P., 1977. Use of ordination and other multivariate descriptive methods to study succession. Vegetatio 35:165-175.

Austin, M.P. & L. Orlóci, 1966. Geometric models in ecology. II. An evaluation of some ordination techniques. J. Ecol. 54:217-227.

Bouxin, G., 1976. Ordination and classification in the upland Rugege forest. Vegetatio 32:97-115.

Clifford, M.T., 1976. Dendrograms and their interpretation. In W.T. Williams (ed.) Pattern analysis in agricultural science, p. 96-101. CSIRO Melbourne, Elsevier, Amsterdam.

Cristofolini, G., D. Lausi & S. Pignatti, 1969. Survey of the system for coding of plantsociological record used by the 'Trieste Group'. Paper Trieste Conference Working Group for Data-Processing in Phytosociology, 13 pp.

Dale, M.B., 1975. On objectives of methods of ordination. Vegetatio 30:15-32.

Dale, M.B., 1977. Planning an adaptive numerical classification. Vegetatio 35:131-136.

Dale, M.B. & D.J. Andersson, 1973. Inosculate analysis of vegetation data. Austr. J. Bot. 21:253–276.

Dale, M.B. & L.J. Webb, 1975. Numerical methods for the establishment of associations. Vegetatio 30:77-87.

Davids, P., J. Lepart & F. Romane, 1979. Elements for a system of data processing in phytosociology and ecology. Vegetatio 49:115-123.

Ellenberg, H., 1979. Zeigerwerte der Gefässpflanzen Mitteleuropas. 2 Aufl. Scripta Geobot. Göttingen, 9, 122 pp.

Egler, F.W., 1954. Philosophical and practical considerations of the Braun-Blanquet system of plant sociology. Castanea 19:45-60.

Eijsink, J., G. Ellenbroek, W. Holzner & M.J.A. Werger, 1978. Dry and semidry grasslands in the Weinviertel, Lower Austria. Vegetatio 36:126-148.

Feoli, E., 1977. A criterion for monothetic classification of phytosociological entities on the basis of a species ordination. Vegetatio 33:147-152.

Feoli, E. & N. Lagonegro, 1979. Intersection analysis in phytosociology: computer program and application. Vegetatio 40:55-59.

Feoli, E. & L. Orlóci, 1979. Analysis of concentration and detection of underlying factors in structure tables. Vegetatio 40:49-54.

Fresco, L.F.M., 1971. Compound analysis: a preliminary report on a new numerical approach in phytosociology. Acta Bot Neerl. 20:589-599.

Gauch, Jr. H.G., 1977. Ordiflex. Release B. Ecology and systematics Cornell University, Ithaca N.Y. 185 pp.

Gauch, Jr. H.G., 1979. COMPCLUS, A Fortran program for rapid initial clustering of large data sets. Ecology and systematics. Cornell University, Ithaca N.Y., 59 pp.

Gauch, Jr. H.G. & T.R. Wentworth, 1976. Canonical correlation analysis as an ordination technique. Vegetatio 33:17-22.

Gils, H. van & A.J. Kovács, 1977. Geranion sanguinei communities in Transsylvania. Vegetatio 33:175-186.

Godron, M. et al. 1968. Code pour le relevé méthodique de la végétation et du milieu. Principes et transcription sur cartes perforées. CNRS, Paris, 292 pp.

Greig-Smith, P., 1980. The development of numerical classification and ordination. In E. van der Maarel (ed.) Advances in vegetation science: Classification and ordination. Vegetatio 42

Groenendael, J.M. van, 1978. The vegetation of Cors Goch, in relation to some environmental factors. Rep. Afd. Geobotanie, Nijmegen, 32 pp + app.

Hill, M.O., 1973. Reciprocal averaging: an eigenvector method of ordination. J. Ecol. 61:237-249.

Hill, M.O., 1977. Use of simple discriminant functions to classify quantitative phytosociological data. In E. Diday, L. Lebart, J.P. Pagas & R. Tomassone (eds.). Proc. 1st. Symp. Data analysis and information, Versailles, p. 181-199. Inst. Rech. Inform. Autom. le Chesnay.

Hill, M.O., 1979. DECORANA - A FORTRAN program for detrended correspondence analysis and reciprocal averaging. Ecology and systematics, Cornell University, Ithaca N.Y., 52 pp.

Hill, M.O. 1979. TWINSPAN, A FORTRAN program for arranging multivariate data in an ordered two-way table by classification of the individuals and attributes. Ecology and systematics, Cornell University Ithaca N.Y. 90 pp.

Hill, M.O., R.G.H. Bunce & M.W. Shaw, 1975. Indicator species analysis, a divisive polythetic method of classification, and its application to a survey of native pinewoods in Scotland. J. Ecol. 63:597-613.

Hogeweg, P., 1976. Topics in biological pattern analysis. Thesis Utrecht 208 pp.

Holzner, W., M.J.A. Werger & G.A. Ellenbroek, 1978. Automatic classification of phytosociological data on the basis of species groeps. Vegetatio 38:157–164.

Jensén, S., 1978. Influences of transformation of cover values on classification and ordination of lake vegetation. Vegetatio 37:19-31.

Janssen, J.G.M., 1975. A simple clustering procedure for preliminary classification of very large sets of phytosociological relevés. Vegetatio 30:67-71.

Kercher, J.R. & R.A. Goldstein, 1977. Analysis of an East Tennessee oak hickory forest by canonical correlation of species and environmental parameters. Vegetatio 35:153-163.

Klein, J.-C., 1979. Application de l'analyse factorielle des correspondances à l'étude phytosociologique de l'étage alpin des Pyrénées centrales. Phytocoenologia 5:125-188.

Kortekaas, W.M., E. van der Maarel & W.G. Beeftink, 1976. A numerical classification of European Spartina communities. Vegetatio 33:51-60.

Lacoste, A., 1975. La végétation de l'étage subalpin du bassin supérieur de la Tinée (Alpes - Maritimes). Phytocoenologia 3:83-345.

Lacoste, A., 1976. Relations floristiques entre les groupements prairiaux du Triseto-Polygonion et les Megaphorbiaies (Adenostylion) dans les Alpes occidentales. Vegetatio 31:161-176.

Lambert, J.M. & W.T. Williams, 1962. Multivariate methods in plant ecology. IV. Nodal analysis. J. Ecol. 50:775-802.

Lausi, D. & E. Feoli, 1979. Hierarchical classification of European salt marsh vegetation based on numerical methods. Vegetatio 39:171-184.

Louppen, J.M.W. & E. van der Maarel, 1979. CLUSLA: a computer program for the clustering of large phytosociological data sets. Vegetatio 40:107–114.

Maarel, E. van der, 1969. On the use of ordination models in phytosociology. Vegetatio 19:21-46.

Maarel, E. van der, 1975. The Braun-Blanquet approach in perspective. Vegetatio 30:213-219.

Maarel, E. van der, 1979. Multivariate methods in phytosociology, with reference to the Netherlands. In M.J.A. Werger (ed) The study of vegetation, p. 161-225. Junk, The Hague.

Maarel, E. van der, J.G.M. Janssen & J.M.W. Louppen, 1978. TABORD, a program for structuring phytosociological tables. Vegetatio 38:143-156.

Maarel, E. van der, L. Orlóci & S. Pignatti, 1976. Data-processing in phytosociology, retrospect and anticipation. Vegetatio 32:65-72.

Meyer, M. Vergleich verschiedener Chrysopogon gryllus -reicher Trockenwiesen des insubrischen Klimabereiches und angrenzender Gebiete. Vegetatio 35:107-114.

Moravec, J. 1978. Application of constancy-species groups for numerical ordering of phytosociological tables - the synoptic table version. Vegetatio 37:33-42.

Mueller-Dombois, D. & M. Ellenberg, 1974. Aims and methods of vegetation ecology. Wiley, New York, XX + 547 pp.

Neuhäusl, R., 1977. Delimination and ranking of floristic - sociological units on the basis of relevé similarity. Vegetatio 35:115-122.

Noy-Meir, I. & R.H. Whittaker, 1978. Recent developments in continuous multivariate techniques. In R.H. Whittaker (ed.) Ordination of plant communities, p. 337-378. Junk, The Hague, 388 pp.

Orlóci, L., 1966. Geometric models in ecology. I. The theory and application of some ordination methods. J. Ecol. 54: 193-215.

Orlóci, L., 1978. Multivariate analysis in vegetation research. 2nd ed. Junk, The Hague, Boston. IX + 451 pp.

Orlóci, L., 1978. Ordination by resemblance functions 2nd ed. In R.H. Whittaker (ed.): Ordination of plant communities, p. 239-275. Junk, The Hague, 388 pp.

Orlóci, L., 1980. An algorithm for predictive ordination. In E. van der Maarel (ed.). Advances in vegetation science: Classification and ordination. Vegetatio 42.

Orlóci, L., E. Feoli & P. Fewster, 1978. Multivariate analysis of vegetation data. Report Dept. Plant Sciences 553. Univ. Western Ontario, London, W.O. 179 pp.

Orlóci, L. & W. Stanek, 1980. Vegetation survey of the Alaska Highway, Yukon Territory: Types and Gradients. Vegetatio 41: 1-56.

Persson, S., 1977. Datorprogram för bearbetning av vegetationsdata. 1. Klassifikations-program -dokumentation och handhavande. Meddn. Avd. Ekol. Bot. Lunds Univ. 33: 1-68.

Persson, S., 1978. Datorprogram för bearbetning av vegetationsdata. 2. Ordinationsprogram. Dokumentation och handhavande. Meddn. Växtekol. Inst. Lunds. Univ. 34:1-64.

Phillips, D.L., 1978. Polynomial ordination: field and computer simulation testing of a new method. Vegetatio 37:129-140.

Pignatti, S., 1975. Pflanzensoziologie am Scheideweg. Vegetatio 30:149-152.
Pignatti, S., 1976. A system for coding plant species for data-processing in phytosociology. Vegetatio 33:23-32.
Pignatti, S., 1980. Reflection on the phytosociological approach and the epistemological basis of vegetation science. In E. van der Maarel (ed.) Advances in vegetation science: Classification and ordination. Vegetatio 42.
Pignatti, S., G. Cristofolini & D. Lausi, 1968. Verwendungsmöglichkeiten einer elektronischen Datenverarbeitungsanlage für die Pflanzensoziologische Dokumentation. Discussion paper 12th Symposium Int. Soc. Vegetation Sc. Rinteln. 6 pp.
Pignatti, E. & S. Pignatti, 1975. Syntaxonomy of the Sesleria varia - grasslands of the calcareous Alps. Vegetatio 30:5-14.
Romane, F., 1972. Un exemple d'utilisation de l'analyse factorielle des correspondances en écologie végétale. In E. van der Maarel & R. Tüxen (eds.) Grundfragen und Methoden in der Pflanzensoziologie. Ber. Int. Symp. Rinteln. 1970:151-162. Junk, The Hague.
Romane, F., J.-L. Guillerm & G. Waksman, 1977. Une utilisation possible de l'arbre de portée minimale en phyto-écologie. Vegetatio 33:99-106.
Spatz, G. & J. Siegmund, 1973. Eine Methode zur tabellarischen Ordination, Klassification und ökologischen Auswertung von pflanzensoziologischen Bestandsaufnahmen durch den Computer. Vegetatio 28:1-17.
Stockinger, F.J. & W.F. Holzner, 1972. Rationelle Methode zur Auswertung pflanzensoziologischer Aufnahmen mittels Elektronenrechner. In E. van der Maarel & R. Tüxen (eds.). Grundfragen und Methoden in der Pflanzensoziologie. Ber. Int. Symp. Rinteln. 1970:239-248, Junk, The Hague.
Stockinger, F.J. & W. Holzner, 1973. Prinzipien einer Implementierung des pflanzensoziologischen Zerlegungsproblems. Mitt. Bot. Linz. 4:87-106.
Streng, R. & P. Schönfelder, 1978. Ein heuristisches Computer-Programm zur Ordnung pflanzensoziologischer Tabellen. Hoppea, Denkschr. Regensb. Bot. Ges. 37:407-433.
Tüxen, R., 1970. Entwicklung, Stand und Ziele der pflanzensoziologischen Systematik (Syntaxonomie). Ber. Dtsch. Bot. Ges. 83:633-639.
Tüxen, R., 1972. Richtlinien für die Aufstellung eines Prodromus der Europaïschen Pflanzengesellschaften. Vegetatio 24:23-29.
Tüxen, R., 1977. Zum Problem der Homogenität von Assoziations-Tabellen. Doc. Phytosoc. NS 1:305-320.
Tyler, C., 1979. Studies on Schoenus vegetation in South and Southeast Sweden. Thesis Lund.
Westhoff, V. & E. van der Maarel, 1978. The Braun-Blanquet approach 2nd ed. In: R.H. Whittaker (ed.). Classification of plant communities. p. 287-399. Junk, The Hague, 408 pp.
Whittaker, R.H., (ed.), 1978. Ordination of plant communities. Junk, The Hague, 388 pp.
Wiegleb, G., 1978. Der soziologische Konnex der 47 häufigsten Makrofyten der Gewässer Mitteleuropas. Vegetatio 38:165-174.
Wildi, O., P.H. Fewster & L. Orlóci, 1978. Management and multivariate analysis of large data sets. Rep. Dept. Plant Sciences. London, Ontario. 99 pp + app.
Williams, W.T., Lambert, J.M. Lance, G.N., 1966. Multivariate methods in plant ecology. V. Similarity analyses and information-analysis. J. Ecol. 52:427–445.

SURVEY OF SALT MARSH RELEVÉS INCLUDED IN THE DATA-BANK OF THE WORKING-GROUP FOR DATA-PROCESSING

W.M. KORTEKAAS[1]*, D. LAUSI[2]**, W.G. BEEFTINK[3]*** & E. van der MAAREL[1]****

[1] Division of Geobotany, University of Nijmegen, Toernooiveld, 6525 ED Nijmegen, The Netherlands
[2] Istituto ed Orto Botanico, Università di Trieste, Via Valerio 30, Trieste I 34100, Italy
[3] Delta Institute for Hydrobiological Research, Yerseke, The Netherlands

Introduction

This survey was announced in some of the papers included in this book, under different names and with different sequences of authors. It should have been ready in 1976. However, the technical work involved in the additional selection of the relevés to form the selection *g* (see Van der Maarel, Orlóci & Pignatti 1976 for a survey of selections) and particularly the coding of the additional species took much more time than was foreseen. Moreover selection *g* was extended with some 20 tables, which mainly represent boreal and subarctic communities to form selection *h*.

Kortekaas did most of the additional species coding for selections *c*, *g* and *h*, and composed the three appendices; Lausi suggested the first big selection *a* of our salt marsh data set, and did the species coding for selections *a* and *c;* Beeftink suggested the later selections (*c*, *g* and *h*); Van der Maarel helped in preparing the appendices and wrote this text.

Survey of tables and relevés

The basis for the selection of relevés is formed by the broad survey of the distribution of higher salt marsh syntaxa composed by Beeftink (1972). As was explained earlier (e.g. Cristofolini, Lausi & Pignatti 1969, Van der Maarel, Orlóci & Pignatti 1976, Pignatti 1976) the relevés were taken table for table from both published accounts on salt marsh vegetation and from manuscripts. Appendix 1 presents the entire list of publications and manuscripts from which tables were taken (in chronological order of their inclusion in the system) and the tables with the numbers they have in the original publication. The names of the plant communities represented in the tables are in English, unless the author has given a name to the table (a syntaxon name or an ad hoc name). Otherwise Kortekaas and Beeftink gave an ad hoc name. The number of relevés per table is also indicated. In cases of heterogeneous tables representing zonations or mosaics, the indication 'various communities' is given.

Two publications need special attention: the tables from Korczagin (publ. 009601) are summarized from the original tables because the same relevés occur in different tables! The relevés included from Rebassoo (publ. 009902) represent samples from the original tables which contain many relevés per table!

Appendix 2 presents the bibliographic details of the publications in appendix 1.

Selections of relevés

Selection *a* includes the relevés from publications 0001-0020. Selection *c* includes selection *a* + the relevés from publications 0022-0042. Selection *g* includes selections *a*, *c*, and *b* (see below) + the relevés from publications 0043-0081.

Selection *h* includes selection *g* + the relevés from publications 0082–0110 and 0120.

Selection *b* includes *Spartinetea* relevés from publications 0082–0120. As was mentioned earlier the selections *d*, *e* and *f* are small selections, *d* is a random selection from *a*, *e* is a selection from *Glauco – Puccinellietalia*- relevés chosen from selection *a*; *f* is a selection of relevés of *Spartina* communities taken from selections *a* and *b*.

*The work done for this survey by the first author formed part of a research project supported by the Netherlands Organization for the Advancement of Pure Research.
**The work done by the second author was supported by several grants from the Italian C.N.R.
***Communication Nr. 195 of the Delta Institute for Hydrobiological Research, Yerseke, The Netherlands.
****We thank José Broekmans for considerable help in the preparation of the appendices.

With the first large selection *a* the Mediterranean salt steppe communities were reasonably well represented, while the other areas were underrepresented.
In selection *c* the Atlantic and continental salt marshes became better represented, while in selection *g* all collections except those of N Europe were completed. Selection *h* completed the total selection with boreal and subarctic material.

Although these nearly 7000 relevés are still only a small part of the total set of relevés which we could possibly collect further – the total number may be in the order of 20 000 – we think that with the present selection we must be able to find the large majority of salt marsh communities ever to be expected. With the system derived from this selection – the work of Lausi & Feoli (1979) forms the first step here – as a framework of reference, further local selections can be treated and assigned.

Nomenclature

The coding of all occurring plant species - mosses, liverworts and lichens included – was a painstaking undertaking. Although the guidelines developed by Lausi and Pignatti (see Pignatti, 1976) were sufficiently clear, a great number of taxa caused difficulties. Many species names used in local studies were difficult to identify as synonyms of current names and some names could not be found at all in the available taxonomical literature. In the latter case we replaced the name by the name most probably belonging to the taxon in view of the entire species composition of the relevé, or we applied an ad hoc coding in cases the identity of taxa could not be checked. The main Flora's and taxonomical-geographical works used as reference works are: Willkomm & Lange for Spain, Fiory for Italy, Hegi for Central Europe, Fournier for France, Hayek & Markgraf for the Balkan, Hultén for Scandinavia, and the Flora Europaea edited by Tutin et al. In principle the latter work should serve as the main basis. A serious problem with this Flora was the incompleteness of the work by the time we started the coding project. For taxa treated in Vols. 1 and 2 we could directly follow nomenclature and species number from Flora Europaea. All other taxa received a provisional number, with the idea that those should be replaced by the later Flora Europaea-numbers. At the moment we have changed numbers up to Vol. 4, but for taxa to be treated in Vol. 5 the numbering is still provisional.

In appendix 3 which is meant as a practical help, we list all taxon names we found in the tables included in the entire selection, with nomenclature, author's name(s) and number according to Flora Europaea Vols. 1-4 and with provisional numbers for the remaining taxa. As explained by Pignatti (1976) the first four digits are the genus code taken from Dalla Torre & Harms (1908) and the species code has three digits. For mosses, liverworts and lichens an ad hoc genus code was adopted. The synonyms we found are also included. As was explained by Pignatti (1976) we did not yet give separate numbers to infraspecific taxa. Unknown species identified down to the genus have generally been given the species nr. 999. Unknown bryophtes have received the nr..9899.999, unknown lichens the nr. 9700.999 (Pignatti also mentions reference works for the nomenclature of cryptogams).

Perspective

We think that the coding system adopted by us is useful, not only for our salt marsh project, but also for further syntaxonomical work on a European level. Through the work of Pignatti for his Flora of Italy, which is in print, and through the work of Kortekaas, Rijken-Kępczyńska and van der Maarel concerning a representative selection of plant communities from the Netherlands and the coasts of W Europe we have a coding system for ca 6000 European species now. It would be worth while trying to complete this system for at least the whole North and Central European regions!

References

Beeftink, W.G., Übersicht über die Anzahl der Aufnahmen europaïscher und nordafrikanischer Salzpflanzengesellschaften für das Projekt der Arbeitsgruppe für Datenverarbeitung. In: E. van der Maarel & R. Tüxen (ed.). Grundfragen und Methoden der Pflanzensoziologie. Ber. Int. Symp. Rinteln 1970, p. 371–396.

Cristofolini, G., D. Lausi & S. Pignatti, 1969. Survey of the system for coding of plantsociological records used by the 'Trieste Group'. Paper Trieste Conference, Working Group for Data-Processing in Phytosociology, 13 pp.

Dalla Torre, G.G. & H. Harms, 1908. Register zu Genera Siphonogamarum ad systema Englerianum conscripta. Engelmann, Leipzig. Reprint Engelmann (J. Cramer) Weinheim, 568 pp.

Lausi, D. & E. Feoli, 1979. Hierarchical classification of European salt marsh vegetation based on numerical methods. Vegetatio 39: 171–184.

Maarel, E. van der, L. Orlóci & S. Pignatti, 1976. Data-processing in phytosociology, retrospect and anticipation. Vegetatio 32: 65–72.

Pignatti, S., 1976. A system for coding plant species for data-processing in phytosociology. Vegetatio 33: 23–32.

Appendix 1

List of publications and manuscripts from which phytosociological tables have been taken up in the selection of salt marsh relevés; no further information is presented on tables from unpublished manuscripts, which are indicated (n.p.) in the list of tables.

The areas to which the relevés refer are briefly indicated; the international postal code for countries is used.

The tables are numbered in the order of their inclusion in the system.

The number of relevés per table is also indicated.

The names of the plant communities are taken from the author or devised ad hoc by Kortekaas and Beeftink, and given in the English language.

Publ.nr. and table nr. PIT	Number of relevés	Author(s)	Location	Plant community
000102	11	Adriani (1945)	SW Netherlands, some relevés from the W Frisian islands, NL	Salicornieto - Spartinetum
000103	24			Puccinellietum maritimae
000104	22			Artemisietum maritimae
000106	10			Artemisietum maritimae (facies Statice limonium)
000201	8	Pignatti (1966)	Lagoon of Venice, I	Scirpetum maritimi - litoralis
000202	11			Phragmitetum
000203	13			Juncetum maritimi
000204	23			Spartinetum strictae
000205	21			Salicornietum venetae
000206	19			Suaedetum maritimae - Kochietum
000207	7			Salsoletum sodae
000208	40			Limonietum (1°)
000209	18			Salicornietum fruticosae
000210	17			Staticeto - Artemisietum coerulescentis
000215	52			Pholiureto - Spergularietum marginatae
000216	16			Cariceto - Festucetum arundinaceae
000217	13			Plantagineto - Tetragonolobetum
000218	11			Juncetum gerardii - Obionetum
000219	7			Junceto - Triglochinetum
000220	9			Agropyreto - Inuletum crithmoidis
000221	9			Salicornietum strictae
000222	31			Limonietum (2°)
000223	30			Limonietum (3°)
000304	21	Gillner (1960)	Bohuslän, Halland, Vastergötland, S	Scirpetum maritimi
000305	12			Caricetum paleaceae
000306	12			Eleocharis uniglumis - Agrostis stolonifera comm.
000307	36			Juncetum gerardii
000309	11			Blysmetum rufi
000312	35			Salicornietum strictissimae
000313	35			Puccinellietum maritimae, typicum
000314	11			Puccinellia distans ass.
000315	42			Juncetum gerardii
000316	9			Artemisietum maritimae
000317	12			Sagineto maritimae - Cochlearietum danicae
000319	23			Salicornietum europaeae
000320	30			Puccinellietum maritimae, subass. Salicornia europaea
000321	42			Juncetum gerardii
000401	12	Wendelberger (1950)	Lake of Neusiedl, A	Crypsis aculeata - Chenopodium glaucum veg.
000402	5			Cyperetum pannonici
000404	47			Puccinellia salinaria - Aster pannonicus ass.
000405	34			Puccinellia salinaria -.Lepidium cartilagineum ass.
000406	11		also Hortobágy, H	Pholiurus pannonicus - Plantago tenuiflora ass.
000408	25		also R, CS	Camphorosmetum annuae
000501	7	Corillion (1953)	Brittanny, F	Spartinetum strictae
000502	6			Spartinetum alterniflorae
000503	8			Salicornietum strictae
000504	5			Salicornietum radicantis
000505	6			Juncus maritimus veg.
000506	16			Puccinellietum maritimae
000507	15			Junceto - Caricetum extensae
000508	14			Juncetum gerardii
000509	13			Festucetum arenariae
000702	8	Tüxen & Westhoff (1963)	SW Netherlands, NL	Sagino maritimae - Cochlearietum danicae subass. Agropyron pungens
001101	1	De Bolos & Molinier (1958)	Majorca, ES	Salicornietum fruticosae
001102	1			Arthrocnemetum
001103	1			Schoeneto - Plantaginetum crassifoliae
001104	1			Juncus maritimus veg.
001106	1			Suaeda maritima veg.
001107	1			Salicornietum fruticosae, subass. Juncus subulatus
001205	14	De Soó (1947)	Carpathian Basin, H	Salicornietum europaeae hungaricum
001405	37	Braun-Blanquet (1957)	Languedoc, F	Junceto - Triglochinetum
001406	22			Caricetum divisae
001524	13	Rivas-Goday (1955)	Valladolid, ES	Puccinellia convoluta ass. subass. Lactuca & Spergularia, Puccinellia convoluta - Crypsis aculeata ass.

Publ.nr. and table nr.	Number of relevés	Author(s)	Location	Plant community
001601	10	Rivas-Goday(1945)	Valencia, ES	Schoenetum nigricantis halophytum
001701	10	Carvalho-Fontes(1945)	Sacavém,P	Salicornietum radicantis atroposum
001702	12			Salicornietum radicantis spartinosum
001703	8			Salicornietum europaeae
001704	11			Arthrocnemetum glauci
001804	12	Molinier & Tallon(1965)	Camargue, F	Salicornietum fruticosae
001953	12	De Bolos(1962)	Barcelona, ES	Schoeno-Plantaginetum crassifoliae
001954	5			Spartino - Juncetum maritimi
001955	2			Salicornietum arabicae
001956	2			Pholiuro - Atriplicetum portulacoidis
001957	4			Salicornietum emerici
002018	33	Braun-Blanquet & De	Zaragosa, ES	Suaedetum brevifoliae
002019	23	Bolos(1957)		Limonietum ovalifolii
002022	10			Salicornia herbacea - Suaeda brevifolia ass.
002023	11			Scnoeneto - Plantaginetum crassifoliae
002024	18			Soncho - Juncetum maritimi
002026	5			Suaedetum brevifoliae frankenietosum reuteri
002205	10	Wilkon-Michalska	Kujawi, P	Salicornietum patulae
002208	15	Wilkon-Michalska	Kujawi, PL	Puccinellia distans - Spergularia salina ass.
002209	20	(1963)		Triglochin maritimum - Glaux maritima ass.
002211	10			Scirpetum maritimi , subass. Puccinellia distans
002212	11			Festuca arundinacea - Potentilla anserina - ass.
002213	7			Blysmo - Juncetum compressi
002214	10			Arrhenatheretum elatioris lotetosum tenuifolii
002301	10	Hocquette, Géhu &	Authie, F	Spartinetum townsendii
002302	15	Fauquet(1965)		Salicornietum europaeae
002303	26			Suaedeto - maritimae Asteretum - tripolii
002304	34			Obionetum portulacoidis
002305	40			Puccinellietum maritimae
002306	20			Artemisietum maritimae
002307	9			Armerieto - Festucetum litoralis, Junceto - Caricetum extensae (3 resp. 6 rél.)
002308	2			Atripliceto - Elytrigietum pungentis
002401	7	Bodrogközy(1959)	Kiskunság,H	Potentillo - Festucetum pseudovinae euphorbietosum
002402	20			Potentillo - Festucetum pseudovinae normale
002403	5			Potentillo - Festucetum pseudovinae staticetosum
002501	16	Bodrogközy(1965)	Orosháza,H	Achilleo - Festucetum pseudovinae
002502	20			Artemisio - Festucetum pseudovinae
002503	16			Camphorosmetum annuae
002504	20			Puccinellietum limosae
002601	7	Bodrogközy(1960)	Kiskunság,H	Agrosti - Caricetum distantis molinietosum
002602	20			Agrosti - Caricetum distantis typicum
002701	13	Bodrogközy(1962)	Kiskunság,H	Lepidio - Camphorosmetum annuae
002702	10			Lepidio - Puccinellietum limosae
002703	5			Suaedetum maritimae hungaricum
002704	5			Lepidio - Puccinellietum normale
002705	3			Acorelletum pannonici
002706	1			Bolboschoenetum maritimi agrostidetosum
002801	10	Bodrogközy(1965)	Hortobágy, H	Agrosti- Glycerietum poiformis
002802	10			Agrosti - Beckmannietum
002803	15			Agrosti -Alopecuretum pratensis
002804	17			Pholiuro - Plantaginetum
002805	18			Puccinellietum limosae hungaricum
002806	12			Camphorosmetum annuae
002807	17			Artemisio - Festucetum pseudovinae
002808	15			Achilleo - Festucetum pseudovinae
002901	8	Bodrogközy(1966)	Orosháza, H	Suaedetum maritimae hungaricum
002902	10			Lepidio - Puccinellietum limosae crisicum
002903	8			Bolboschoenetum maritimi continentale
002904	8			Agrosti - Caricetum distantis
002905	5			Agrosti - Caricetum distantis asteretosum
003001	20	Bodrogközy(1965)	Orosháza, H	Pholiuro - Plantaginetum tenuiflorae
003002	25			Agrosti - Alopecuretum pratensis
003003	15			Agrosti - Beckmannietum
003101	22	Moore(n.p.)	North Bull Island, IRL	Puccinellietum maritimae, Juncetum gerardii (12 resp. 10 rel.)
003102	19			Puccinellietum maritimae, Juncetum gerardii(15 resp. 4 rel.)
003104	22			Juncetum gerardii, Puccinelietum maritimae (10 resp. 12 rel.)
003105	28			Puccinellietum maritimae, Juncetum gerardii (20 resp. 8 rél.)
003201	24	Ivimey-Cook&Proctor (1966)	The Burren,IRL	Puccinellietum maritimae, Juncetum gerardii (20 resp. 4 rel.)
003301	16	Westhoff(n.p.)	N. Ireland, IRL Ameland,NL	various comm.
003401	43	Westhoff(n.p.)	Vendée, Charente maritime, F	various comm.

Publ.nr. and table nr.	number of relevés	Author(s)	Location	Plant community
003501	10	Westhoff(1947)	Texel,Vlieland,	Artemisietum maritimae
003502	10		Terschelling, NL	Juncus maritimus - Oenanthe lachenalii soc.
003503	22			Puccinellietum maritimae typicum
003504	14			Puccinellietum maritimae
003505	18			Armerieto - Festucetum
003506	11			Armerieto - Festucetum
003507	12			Puccinellietum distantis
003508	49			Junceto - Caricetum extensae
003601	19	Westhoff(n.p.)	Halland, Skåne, Blekinge, S.	various comm.
003704	15	Frei(1937)	Sicily, I.	Suaeda maritima - Kochia hirsuta ass.
003705	1			Haloplepis amplexicaulis ass.
003706	10			Salicornia radicans ass.
003707	10			Salicornia fruticosa - Statice bellidifolia ass.
003708	10			Arthrocnemum glaucum - Sphenopus divaricatus ass.
003709	1			Limoniastrum - Statice lychnidifolia ass.
003824	1	Braun-Blanquet &	IRL	Salicornietum europaeae
003825	10	Tüxen(1952)		Puccinellietum maritimae
003826	17			Juncetum gerardii, Plantago coronopus - Cerastium tetrandrum - ass. (8resp. 9 rel.)
003904	8	Oberdorfer(1952)	Vardar delta, GR	Suaeda maritima - Bassia hirsuta - ass.
003905	2			Salicornia radicans - ass.
003906	10			Arthrocnemum glaucum - Halocnemum strobilaceum - ass.
003907	14			Petrosimonia crassifolia - Bupleurum gracile - ass.
003908	2			Camphorosma ruthenica - comm.
003909	5			Salsola soda - Suaeda splendens- ass.
003910	4			Crypsis - Heleochloa - comm.
004001	15	Nordhagen(1954)	Finmark, N	various comm.
004002	8			Puccinellietum phryganodis & Puccinellietum retroflexae
004003	13			Caricetum subspathaceae and two other ass.
004004	13			Festuceto - Caricetum glareosae
004101	3	Nègre(1956)	Sedd-el-Messjoun, MA	Scirpus maritimus ass.
004102	15			Aeluropus litoralis ass.
004106	16			Suaeda fruticosa - Lepturus incurvatus ass.
004110	8			Frankenia corymbosa var. ballsiana - Koelpinia linearis subass. Diplotaxis ollivieri
004113	7			Frankenieto - Koelpinietum subass. Mesembryanthemum nodiflorum
004202	24	Wolff(1968)	Lagoons of Mesolonghi, GR	various comm.
004203	13			Salicornietum europaeae
004204	30			Salicornietum radicantis
004205	28			Salicornietum fruticosae
004206	37			Salicornio - Halocnemetum strobilacei
004207	32			Arthrocnemetum
004208	11			Romuleo - Saginetum maritimae
004209	4			Juncetum maritimi
004309	22			Artemisietum maritimae typicum, Artemisietum maritimae armerietosum
004310	22	Beeftink(1965)	SW Netherlands, NL	Halimionetum portulacoidis
004311	15			Agrostis stolonifera soc.
004312	12			Artemisietum maritimae agrostidetosum
004313	8			Armerieto - Festucetum, var. Juncus gerardii
004314	8			Armerieto - Festucetum, var. Festuca rubra f. litoralis
004315	8			Puccinellietum maritimae, terminal phase with Halimione pedunculata
004316	12			Glaux maritima soc.
004317	35			Puccinellietum maritimae typicum
004318	22			Puccinellietum distantis
004319	8			Puccinellietum maritimae typicum, var. Limonium vulgare
004320	9			Puccinellietum maritimae typicum, terminal phase with Plantago maritima & Limonium vulgare
004322	14			Festuca arundinacea - Potentilla anserina - ass.
004323	10			Puccinellietum maritimae agrostidetosum
004324	6			Puccinellietum distantis
004325	15			Armerieto - Festucetum
004326	11			Plantagineto - Lolietum
004327	10			Atripliceto - Agropyretum littoralis
004328	30			Phragmites communis consoc.
004401	6	Micevski(1965)	Macedonia, YU	Suaedetum maritimae balcanicum
004402	5			Camphorosmetum annuae balcanicum
004403	5			Puccinellietum convolutae
004406	5			Pholiureto - Plantaginetum balcanicum
004407	7			Crypsidetum aculeatae balcanicum
004505	25	Von Hübschmann(n.p.)	North Sea coast, D	Puccinellietum maritimae
004506	8			Artemisietum maritimae
004507	7			Puccinellietum distantis
004605	5	Guinea(1953)	Santander, ES	Puccinellion maritimae

Publ.nr. and table nr.	Number of relevés	Author(s)	Location	Plant community
004606	5			Salicornion fruticosae
004701	6	Corbetta(1970)	Lesina,Varano, I.	Scirpetum maritimi
004702	24			Salicornietum fruticosae
004802	9	Tallon(1958)	Camargue, F	Arthrocnemetum
004901	6	Šmarda(1953)	Moravia, CS	Puccinellia distans comm.
005004	6	Bochnig(1959)	Vilm, DDR	Atriplex hastata - comm.
005110	7	Krausch(1965)	Danube delta,	Aeluropo - Salicornietum
005208	5	Vicherek(1962)	Bohemia, CS	Salicornietum europeae hungaricum
005209	7			Suaedetum pannonicae
005210	17			Puccinellietum limosae hungaricum
005211	5			Crypsidetum aculeatae spergulariosum marginatae
005212	15			Juncetum gerardi pannonicum
005310	16	Košir(1963) in	Dalmatian islands	Staticeto - Artemisietum caerulescentis
005316	6	Horvatic(1963)	Pag and Rab, YU	Juncetum maritimo - acuti
005322	9			Schoeneto - Plantaginetum maritimae
005401	5	Corillion & Dizerbo (1952)	Brittany , F	veg. Salicornia fruticosa
005501	14	Géhu(1960)	Chausey, F	various comm.
005602	8	Beeftink(1958)	Camargue, F	various comm.
				Arthrocnemetum glauci
				Sagino maritimae - Tortelletum flavovirentis
005706	10	Jeschke(1962)	Sylt, D	various comm.
				Puccinellietum maritimae
				transition
				Juncetum gerardii
005802	8	Den Hartog(1958)	Texel,Terschelling,NL	Zosteretum nano-stenophyllae
005901	9	Gaertner(1961)	Lake of Barsbek, D	Scirpus maritimus comm.
005908	9			Juncetum gerardii
005911	13			Juncetum gerardii,
				Armerion maritimae (resp. 9 rel.)
006001	46	Westhoff,Van Leeuwen & Adriani(1961)	Goeree, NL	Saginetum maritimae
006102	2	Jage & Jage(19)	Altmark, D	Juncetum gerardii
				Triglochin maritima soc.
006201	11	Schwabe(1972), n.p.	Trischen, D	Zosteretum nanae
006202	7			Spartinetum townsendii
006203	11			Salicornietum strictae
006204	13			Salicornietum patulae
006205	14			Puccinellietum distantis
006206	28			Puccinellietum maritimae
006207	9			Halimionetum portulacoidis
006208	30			Juncetum gerardii
006209	11			Artemisietum maritimae
006210	6			Agropyretum littoralis
006303	4	Runge(1968 n.p.)	Eastern Frisian Islands, D.	Blysmetum rufi
006305	4			Halimionetum portulacoidis
006306	5			Juncus maritimus - Oenanthelachenalii -ass.
006311	5			Artemisietum maritimae
006401	8	Christiansen(1955)	Eastern Frisian Islands, D	Salicornietum strictae (4 rel.)
				trans. Salicornietum strictae - Salicornietum patulae (1 rel.)
				Salicornietum patulae (3 rel.)
006501	58	Woodell (1968 n.p.)	Scolt Head Island,GB	Puccinellietum maritimae (33 rel.)
				Halimionetum portulacoidis (21 rel.)
				trans. salicornietum strictae - Halimionetum portulacoidis (4 rel.)
006601	4	Beeftink&van der Maarel(1972 n.p.)	Praia de Bordeira,P	Juncetum maritimi (1 rel.)
				Juncus acutus veg. (3 rel.)
006602	7		Lagos, P	various comm.
006603	8		Lagos, P	various comm.
006604	5		Rio Guadiana, P	various comm.
006605	4		Figueira da Foz ,P	various comm.
006606	10			various comm.
006607	5		Torreira, P	various comm.
006608	13		Viana do Castelo,P	various comm.
006609	8		Caminha,P	Scirpetum maritimi (4 rel.)
				Juncetum maritimi (4 rel.)
006610	6			Scirpetum maritimi (1 rel.)
				Juncetum maritimi (5 rel.)
006611	10		Lagos,P	various comm.
006703	9	Libbert(1940)	Darss, DDR	Atriplex - litorale ass.
006708	5			Spergularia salina - ass.
006709	9			Junco - Caricetum extensae
006710	13			Armerietum maritimae
006802	8	Horst(1966-1968,n.p.)	Wangerooge, D	Armerion maritimae
006915	20	Rivas-Goday & Galiano (1958)		Puccinellieto - Plantaginetum crassifoliae

Publ.nr. and table nr.	Number of relevés	Author(s)	Location	Plant community
007001	7	Den Hartog(1955)	Terschelling, NL	Puccinellietum maritimae (5 rel.)
				Halimionetum portulacoidis (2 rel.)
007104	4	Braun-Blanquet &	Ameland, NL	Puccinellietum maritimae
007105	5	De Leeuw(1936)		Armerieto - Festucetum arenariae
007201	10	Dahl & Hadač(1941)	Ostøy, N	Halo - Scirpetum maritimi
007202	5			Caricetum paleaceae
007203	10			Juncetum gerardii
007204	10			Astereto - Puccinellietum maritimae
007205	5			Glaucetum maritimae
007206	5			Atriplicetum littoralis
007301	3	Pignatti(1953)	Ebrodelta, ES	Limonium densissimum - Limoniastrum monopetalum ass.
007409	9	Altehage & Rossman	DDR	Puccinellia distans - Obione pedunculata ass.
007410	23	(1939)		Triglochin maritima - Scorzonera parviflora ass.
007501	3	Hadač(1970)	SW IS	Plantagini - Puccinellietum maritimae
007601	5	Simonetti & Musi	Venice, I	Spartina juncea veg.
007602	12	(1969/1970)		Spartina juncea veg.
007702	9	Kloss(1969)	Wismar, DDR	Spergularia marina - Atriplex litoralis comm.
007704	16			Trifolium fragiferum - Carex distans comm.
007705	16			Suaeda maritima - Spergularia marina comm.
007707	9			Centaurium pulchellum - Juncus gerardii comm.
007708	21			Puccinellia distans - maritima comm.
007709	21			Limonium vulgare - Juncus gerardii comm.
007801	18	Valsecchi(1967)	Sardinia, I	Spartina juncea veg.
007904	2	Hadač (1946)	W. Spitzbergen, N	Caricetum ursinae
007905	3			Caricetum subspathaceae
008001	14	Şerbănescu(1965)	R	Bassia hirsuta - ass.
008002	64			Salicornia herbacea ass.
008004	50			Obione pedunculata ass.
008005	30			Obione verrucifera ass.
008006	23			Aeluropus littoralis ass.
008007	27			Agropyron elongatum ass.
008008	32			Taraxacum bessarabicum ass.
008009	46			Aster tripolium ass.
008010	16			Cyperus pannonicus ass.
008011	42			Spergularia marginata ass.
008012	68			Crypsis aculeata ass.
008013	29			Petrosimonia triandra ass.
008018	23			Pholiurus pannonicus ass.
008020	45			Bassia sedoides ass.
008021	40			Atriplex littoralis ass.
008023	110			Carex divisa ass.
008024	120			Camphorosma annua ass.
008025	28			Plantago maritima ass.
008026	16			Lepidium crassifolium ass.
008027	34			Statice gmelini ass.
008101	14	Beeftink(1973,n.p.)	Rother estuary, GB	various comm.
008102	4			various comm.
008103	1			various comm.
008104	13			various comm.
008105	7		Poole Harbour, GB	various comm.
008106	7			various comm.
008107	8			various comm.
008108	14			various comm.
008109	19		West-Wittering, GB	various comm.
008110	2			various comm.
008111	2			various comm.
008112	5		Rother estuary, GB	various comm.
008113	6			various comm.
008202	51	Spekman & Groten (1963)	Terschelling, NL	Saginetum maritimae
008301	6	Gillner(1955)	N	Juncus gerardii - Gentiana detonsa veg.
008403	22	Tyler(1969)	Södermanland, Östergötland, S	Eleocharetum uniglumis, caricetosum mackenziei
008404	33			Juncetum gerardii, juncetosum gerardii
008405	35			Juncetum gerardii, festuceto - caricetosum nigrae
008406	12			Caricetum pulchellae, caricetosum extensae
008407	13			Caricetosum pulchellae, Blysmetosum rufi
008501	3	Tyler(1967, n.p.)	Roslagen, S	Caricetum pulchellae
008502	23			Eleocharetum uniglumis, eleocharetosum (13 rel.)
				Eleocharetum uniglumis, caricetosum mackenziei (7 rel.)
				Agrostis stolonifera - Triglochin palustre comm. (3 rel.)
008503	12			Juncetum gerardii, juncetosum
008504	14			Juncetum gerardii, Festuco -caricetosum nigrae
008601	10	De Molenaar(1974)	Angmagssalik district, SE Greenland (DK)	Puccinellietum phryganodis
008602	7			Caricetum subspathaceae
008603	4			Sagino - Phippsietum algidae
008604	30			Caricetum glareosae (25 rel.)
				Potentillo - Caricetum rariflorae (5 rel.)

010121	2			Salicornietum europaeae
010122	14			Puccinellietum maritimae
010123	8			Juncetum gerardii
010124	9			Saginetum maritimae
012010	6	Birse & Robertson (1976)	Scotland, GB	Puccinellietum maritimae
012011	8			Juncetum gerardii
009407	12	Rivas-Martinez & Costa (1975)	La Mancha, ES	Parapholi - Frankenietum pulverulentae
009510	12	Hesjedal (1969)	Horaland, N	Carex recta veg.

Spartina communities

Publ.nr. and table nr.	Number of relevés	Author(s)	Location	Plant community
000522	3	Corillion (1953)	Brittany, F	various comm.
001880	10	Molinier & Tallon (1965)	Camargue, F	Spartina maritima veg.
003509	9	Westhoff (1947)	Texel, Vlieland, Terschelling, NL	Salicornieto - Spartinetum
004301	21	Beeftink (1965)	SW Netherlands, NL	Spartinetum maritimae
004302	30			Spartinetum townsendii
004303	19			Salicornietum strictae
004304	13			Suaedetum maritimae
004305	20			Scirpetum maritimi compacti
004306	32			Atriplicion littoralis
004307	23			Puccinellietum maritimae typicum with Spartina maritima
004308	19			Spartinetum townsendii, Spartinetum maritimae (13 resp. 6 rel.)
010101	21	Beeftink (1959)	Skallingen, DK	various comm.
010202	27	Bilio (1964)	North and Baltic Sea Coast, D	various comm.
010302	16	Westhoff & Segal (961)	Terschelling, NL	various comm.
010402	20	Roosma (1961)	Oostvoorne, NL	Spartina townsendii comm.
010403	70			Puccinellia maritima comm.
010505	4	Ebbers (1965)	Goeree, NL	Spartinetum townsendii
010601	19	Bos & Simons (1964)	Bergen op Zoom, NL	Salicornietum strictae
010602	20			Spartinetum townsendii with Salicornia europaea & Suaeda maritima
010603	8			Salicornietum strictae with Aster tripolium
010604	17			Spartinetum townsendii
010605	16			Spartinetum townsendii with Salicornia europaeae
010606	11			Spartinetum townsendii with Aster tripolium
010607	8			Aster tripolium soc.
010608	9			Aster tripolium soc. with Spartina townsendii
010609	19			Puccinellietum maritimae typicum with Puccinellia maritima
010610	18			Halimionetum portulacoidis
010611	9			Puccinellietum maritimae typicum with Limonium vulgare
010612	11			Puccinellietum maritimae typicum
010613	5			Scirpus maritimus ver. compactus soc.
010626	23			Phragmites communis soc., consoc.
010701	5	Duvigneaud & Lambinon (1963)	Estuary of the Yser, B	Spartinetum townsendii, Salicornietum strictae (3 resp. 2 rel.)
010703	13			Puccinellia maritima veg.
010828	7	Géhu & Ghestem (1963)	Bay of the Canche, F	various comm.
010956	2	Géhu (1963)	Bay of the Canche, F	various comm.
011116	10	Beeftink (1964)	Brittany, F	various comm.
011117	10			various comm.
011120	4			various comm.
011123	15			various comm.
011136	8			various comm.
011201	28	Vanden Berghen (1965)	Oléron, F	various comm.
011301	3	Beeftink (1969 n.p.)	Lagoon of Grado, I	various comm.
011401	12	Pignatti (1969 n.p.)	Lagoon of Grado, I	Spartinetum strictae
011402	3			Salicornietum europaeae
011601	5	Lemée (1933)	Estuary of the Orne, F	Salicornietum europaea
011602	7			Atropidetum maritimae
011703	3	Anonymus (1959)	Ambleteuse, F	various comm.
011827	2	Weevers (1940)	Goeree, NL	Salicornio - Spartinetum
011901	13	Géhu (1970 n.p.)	Rade de Brest, F	Spartinetum alterniflorae
011902	13	(1970 n.p.)	Bidassoa, F	Spartinetum alterniflorae
011903	27	(1971 n.p.)	Marennes, La Rance, F	Spartinetum strictae
011904	7	(1971 n.p.)	Arcachon, F	Spartinetum strictae
011905	35	(1971 n.p.)	Slack, F	Spartinetum townsendii
011906	15	(1971 n.p.)	Sable d'Or, F	Spartinetum townsendii
011907	10	(1971 n.p.)	Gironde, F	Spartinetum townsendii
011908	10	(1971 n.p.)	Holy Isles, Blakeney, GB	Spartinetum townsendii

Appendix 2

List of references to publications in alphabetical order of the author's name(s). (Manuscripts are not included).

Anon. 1959. Vegetatie monding van de Slack bij Ambleteuse. Kruipnieuws 21.
0117

Adriani, M.J. 1945. Sur la phytosociologie, la synécologie et la bilan d'eau de halophytes. Thèse Amsterdam.
0001 SIGMA Comm. 88. Montpellier.

Altehage, C. & B. Rossmann. 1939. Vegetationskundliche Untersuchungen der Halophytenflora binnenländischer
0074 Salzstellen im Trockengebiet Mitteldeutschlands. Beih. Bot. Centr. Bl. 60B:135-180.

Beeftink, W.G. 1958. De halofiele vegetaties van het Franse Mediterrane Gebied.
0056 Bericht. Excursion Intern. Ver. für Vegetationskunde 28 Mai - 6 Juni 1958. Mimeographed, 22 pp.

Beeftink, W.G. 1959. Some notes on Skallingens salt marsh vegetation and its habitat.
0101 Acta Bot. Neerl. 8:449-472.

Beeftink, W.G. 1964. Rapport inzake de 8e excursie van de Internationale Vereniging voor Vegetatiekunde,
0111 gehouden van 28 mei tot 8 juni 1962 in het kustgebied van NW- en W-Frankrijk. Report Delta Instituut voor Hydrobiologisch Onderzoek, Yerseke, 84 pp.

Beeftink, W.G. 1965. De zoutvegetaties van ZW. Nederland beschouwd in Europees verband. Meded. Landbouwhoge-
0043 school Wageningen 65(1):1-167.

Berghen, C. vanden.1965. Notes sur la végétation du Sud-ouest de la France. III. La Végétation de quelques
0112 prés salés d'Oléron (Charente-Maritime). Bull. Jardin Bot. Etat. 35:363-369.

Bilio, M. 1964. Die aquatische Bodenfauna von Salzwiesen der Nord- und Ostsee. I. Biotop und Okologische
0102 Faunenanalyse: Turbellaria. Int. Revue Ges. Hydrobiol. 49:509-562.

Birse, E.L. & J.S. Robertson. 1976. Plant communities and soils of the lowland and southern upland regions
0120 of Scotland. Soil survey of Scotland, Aberdeen, 226 pp.

Bochnig, E. 1959.Vegetationskundliche Studien im Naturschutzgebiet Insel Vilm bei Rügen. Arch. Freunde
0050 Naturgesch. Mecklenburg. 5:139-183.

Bodrogközy, G. 1959. Ökologische Verhältnisse der Standortstypen der Sandweiden von falschem Schafschwingel
0024 (Potentillo - Festucetum pseudovinae danubiale) in Süd- Kiskunság. Acta Biol. Nova Ser. 5:145-160.

Bodrogközy, G. 1960. Phytozönologische und bodenökologische Untersuchungen an den Sumpfwiesen im süden des
0026 Gebietes Kiskunság (Klein-Kumanien). Acta Bot. Hung. 6:171-207.

Bodrogközy, G. 1962. Die standortökologischen Verhältnisse der halophilen Pflanzengesellschaften des Pannonicum.
0027 I. Untersuchungen an den Solontschak-Szikböden der Südlichen Kiskunság. Acta Bot. Hung. 8:1-37.

Bodrogközy, G. 1965. Ecology of the halophilic vegetation of the Pannonicum. III. Results of the investigation
0025 of the Solonetz of Orosháza. Acta Biol. 11:3-25.

Bodrogközy, G. 1965. Ecology of the halophilic vegetation of the Pannonicum. IV. Investigations on the
0030 Solonetz meadow soils of Orosháza. Acta Bot. Hung. 11:207-227.

Bodrogközy, G. 1965. Ecology of the halophilic vegetation of the Pannonicum. II. Correlation between Alkali
0028 ("Szik") plant communities and genetic soil classification in the Northern Hortobágy. Acta. Bot. Hung. 11:1-51.

Bodrogközy, G. 1966. Ecology of the halophilic vegetation of the Pannonicum. V. Results of the investigation
0029 of the "Fehértó" of Orosháza. Acta Bot. Hung. 12:9-26.

Bolós, O. de. 1962. El paisaje vegetal barcelonés. Fac. Filos. Let. Univ. Barcelona.
0019

Bolós, O. & Molinier, R. 1958. Recherches phytosociologiques dans l'Île de Majorque. Collect. Bot. 5:699-865.
0011

Bos, E.S. & Simons, H.H.J. 1964. Vegetatie-onderzoek van het schorrencomplex ten zuiden van Bergen op Zoom.
0106 Report State University Utrecht, 47 pp.

Braun-Blanquet, J. 1957. Les prés salés du Languedoc méditerranéen. Bull. Mus. Hist. Nat. Marseille 27.
0014

Braun-Blanquet, J. & Bolós, O. de. 1957. Les groupements végétaux du Bassin Moyen de l'Ebre et leur dynamisme.
0020 Consejo superior de investigaciones cientificas. Anal. Estacion Exp. Aula Dei:5(1/4).

Braun-Blanquet, J. & De Leeuw, W.C. 1936. Vegetationskizze von Ameland. Ned. Kruidk. Arch. 46:359-393.
0071

Braun-Blanquet, J. & Tüxen, R. 1952. Irische Pflanzengesellschaften. SIGMA Comm. 117. Veröff. Geobot. Inst. Rübel,
0038 Zürich 25:224-421.

Carvalho Fontes, F. 1945. Algumas caracteristicas fitosociológicas das "Salgados" de Sacavém.
0017 Bol. Soc. Broteriana, Alcobaça 19.

Christiansen, W. 1955. Salicornietum. Mitt. Flor. Soz. Arbeitsgem. N.F. 5:64-65.
0064

Corillion, M.R. 1953. Les Halipèdes du Nord de la Bretagne (Finistère, Côtes-du-Nord, Ille-et-Vilaine).
0005 Revue Gén. Bot. 60.

Corillion, R. & A. Dizerbo. 1952. Sur la présence du Salicornia fruticosa dans le N.W. de la France.
0054 Bull. Soc. Bot. France 99:58-60.

Corbetta, F. 1970. Lineamenti della vegetazione macrofitica dei Laghi di Lesina e di Varano. Giorn. Bot.
0047 Ital. 104:165-191.

Dahl, E. & E. Hadač. 1941. Strandgesellschaften der Insel Ostøy in Oslofjord. Eine pflanzensoziologische
0072 Studie . Nytt. Mag. Naturv. 82:252-312.

Dierschke, H. 1975. Beobachtungen zur Küstenvegetation Korsikas. Anal. Inst. Cavanilles 32:967-991.
0091

Duvigneaud, J . & Lambinon, J. 1963. Flore et végétation halophiles de la rive droite de l'estuaire de
0107 l'Yser entre Lombartzijde et Nieuport. Lejeunia NS 17:1-62.

Ebbers, C.S. 1965. Vegetatiekartering van een gedeelte van de Kwade Hoek. Report State University Utrecht, 38 pp
0105
Frei, M. 1937. Studi fitosociologici su alcune associazioni littorali in Sicilia. Giorn. Bot. Ital. 44(2).
0037
Gaertner, K.H. 1961. Die Vegetationsverhältnisse der Wiesen am Barsbeker See. Mitt. Arbeitsgem. Flor.
0059 Schleswig-Holst. u. Hamburg 9:67-175. Kiel.
Géhu, J.M., 1960. Quelques observations sur la végétation et l'écologie d'une station reputée de
0055 l'Archipel des Chausey: l'île aux Oiseaux. Bull. Lab. Marit. Dinard 46:78-92.
Géhu, J.M. 1963. L'excursion dans le Nord et l'Ouest de la France de la Société Internationale de Phyto-
0109 sociologie. Bull. Soc. Bot. Nord France 16:105-189.
Géhu, J.M., B. Caron & M.Bon. 1975. Données sur la végétation des prés salés de la Baie de Somme.
0098 Coll. Phytos 4, Les vases salées, p. 197-225.
Géhu, J.M. & C. Delzenne. 1975. Apport a la connaissânce phytosociologique des prairies salées de l'Angleterre.
0097 Coll. Phytos.4, Les vases salées, p. 227-247.
Géhu, J.M. & J. Géhu-Franck. 1975. Données nouvelles sur les végétations à Frankenia laevis des hauts
0092 de schorre sablonneux des côtes atlantiques. Phytocoenologia 2:154-168.
Géhu, J.M. & Ghestem, A. 1963. Zonation végétale en Baie de Canche. Bull. Soc. Bot. Nord France 16:27-33.
0108
Gillner, V. 1955. Strandängsvegetation i Nord-Norge. Svensk Bot. Tidskr. 49:217-228.
0083
Gillner, V. 1960. Vegetations- und standortsuntersuchungen in den strandwiesen der schwedischen westküste.
0003 Acta Phytogeogr. Suecica, 43.
Guinea, E. 1953. Ammophiletea, Crithmo - Staticetea, Salicornieteaque santanderiensis. An. Inst. Bot.
0046 A.J. Cavanilles 11:545-568.
Hadač, E. 1946. The plant-communities of Sassen Quarter, Vestspitsbergen. Studia Botanica Čechoslovaca 7(2/4).
0079
Hadač, E, 1970. Sea-shore communities of Reykjanes Peninsula, S.W. Iceland. (Plant communities of
0075 Reykjanes Peninsula, Part 2). Folia Geob. Phytotax. 5:133-144.
Hartog, C. den. 1955. Roodwieren van de kwelders van Terschelling. Levende Natuur 58:169-173.
0070
Hartog, C. den. 1958. De vegetatie van het Balgzand en de oeverterreinen van het Balgkanaal. Wetensch.
0058 Meded. K.N.N.V. 27:1-28.
Hesjedal, O.H. 1969. Strandvegetasjon i indre, midtre OG ytre Fjordomrade, Hordaland. Hovedfagsoppgave i
0095 botanikk. Bergen.
Hocquette, M., J.M. Géhu & M. Fauquet, 1965. Contributions à l'étude phytosociologique de l'estuaire de
0023 l'Authie. Bull. Soc. Bot. Nord France, 18:114-143.
Horvatic, S. 1963. Vegetacijska karta otoka Paga s opcim pregledom vegetacijskih jedinica hrvatshog primorja
0053 Zagreb. 187 pp. izdaje Jugoslavenska akademija znanosti i umjetnosti.
Ivimey-Cook, R.B. & Proctor, M.C.F. 1966. The application of association analysis to phytosociology.
0032 J. Ecol. 54:179-192.
Jage, H. & Ingrid Jage, 19 . Zur Flora der Altmark. Verh. Bot. Ver. Prov. Brandenburg 104:54-62.
0061
Jeschke, L. 1962. Vegetationskundliche Beobachtungen in Listland (Insel Sylt). Beitr. Naturkundemus. Stral-
0057 sund 1:67-84.
, K. 1969. Salzvegetation an der Boddenküste Westmecklenburgs (Wismar-Bucht). Natur und Naturschutz in
0077 Mecklenburg 7:77-114.
Korczagin, A.A. 1935. Die vegetation (der Wiesen und Moore) des Meeralluviums der Mesener Bucht und Tsche-
0096 schkaja Guba. Acta Inst. Bot. Acad. Scient. USSR 3(2),344pp.
Krausch, H.D. 1965. Vegetationskundliche Beobachtungen im Donaudelta. Limnologica 3:271-313.
0051
Kristiansen, J.N. 1977. A phytosociological and synchorological contribution to the Caricetum subspathaceae
0088 and the Festuco - Caricetum glareosae on salt marshes in Northern Norway. Astarte 10(2):107-121.
Lemée, G. 1933. Etude sur la végétation halophile de l'estuaire de l'Orne. Bull. Soc. Linn. Normandie
0116 Sér. 8(6):25-47.
Libbert, W. 1940. Die Pflanzengesellschaften der Halbinsel Darss. Rep. Spec. Nov. Regni Veg. Beih. 114:
0067 1-95.
Menke, B. 1969. Vegetationskundliche und vegetationsgeschichtliche Untersuchungen an Strandwällen. Mitt.
0100 Flor. -Soz. Arbeitsgem. N.F. 14:95-120.
Micevski, K. 1965. Halofitska vegetacije Ovčeg Polja. Acta Mus. Maced. Scient. Natur 10:67-90. Skopje.
0044
Molenaar, J.G. de. 1974. Vegetation of the Angmagssalik district Southeast Greenland. I. Littoral vegetation.
0086 Meddelelser om Grønland. 198(1) København.
Molinier, R. & Tallon, G. 1965. Etudes botaniques en Camargue: II. vers la foret en Camargue. Rev. Hist.
0018 Nat. Paris.
Nègre, R. 1956. Recherches Phytosociologiques sur le Sedd-El-Messjoun. Travaux de l'Inst. Scient. Sér. Bot. 10.
0041 Chérifien. Rabat.
Nordhagen, R. 1954. Studies on the vegetation of salt and brackish marshes in Finmark (Norway). Vegetatio
0040 5/6:381-394.
Oberdorfer, E. 1952. Beitrag zur Kenntnis der Nordägäischen Küsten-vegetation. Vegetatio 3:329-349.

0039
Pignatti, S. 1953. Si alcune Plumbaginacee interessanti raccolte alla joce dell'Ebro. Collect. Bot. 3:377-383.
0073
Pignatti, S. 1966. La vegetazione alofila della laguna veneta. Ist. Veneto Sci. Let. Arti, Venezia 33
0002
Rebassoo, H.E. 1975. Sea-shore plant communities of the Estonian Islands. Acad. of Sciences of the Estonian
0099 SSR. Tartu 1975, 175 pp.
Rivas-Goday, S. 1945. Fàcies subhàlofitas del Schoenetum nigricantis: origen y sucesion. Bol. Soc. Broteriana,
0016 Alcobaça. 19.
Rivas-Goday, S. 1955. Aportaciones a la Fitosociologîa hispanica en Anales. Ist. Bot. Cavanilles. Madrid 13.
0015
Rivas-Goday, S. & E.F. Galiano. 1958. Glaux maritima, L. en las praderas salinas de la alta Meseta Ibérica.
0069 Anal. Inst. Bot. A.J. Cavanilles 16:505-509.
Rivas-Martinez, S. & M. Costa. 1975. Datos sobre la vegetacion halofila de la Mancha (España). Coll. Phytos. 4,
0094 Les vases salées, p. 89-97.
Roosma, M. 1961. De halophiele vegetatie van het schor te Oost-Voorne. Report State University Utrecht, 36 pp.
0104
Şerbănescu, J. 1965. Asociatiile Halofite din Cîmpia Romănă. Studii Techn. Econ. Ser. C. Pedologie
0080 15:1-131.
Simonetti, G. & F. Musi. 1969/1970. Segnalazione di Spartina juncea Willd., nuova per il litorale
0076 Veneto - Atti. Ist. Veneto Sci. Lett. Arti Classe Sci. Mat. Nat. 128:87-95.
Šmarda, I. 1953. Halofytni Květana jéžni Moravy Práce Moravskoslezské Akad. Věd. přirod.
0049 25(13/4).
Soó, R. de. 1947. Conspectus des groupement végétaux dans les Bassins Carpathiques. Inst. Bot.
0012 Univ. Debrecen. 1.
Spekman, J. & J. Groten. 1963. Enkele aspekten van het Saginetum maritimae op de Beer en de Boschplaat.
0082 Dokt. Skriptie Utrecht, 32 pp.
Steindórsson, S. 1976. Some notes on the shore vegetation of Iceland. Acta Bot. Isl. 4:19-35.
0087
Tallon, M.G. 1958. Les sols alcalins en Camargue et leur végétation. Colloque sur la Camargue.
0048 C.R. Congr. Soc. Savantes, p. 503-518, Paris.
Thannheiser, D. 1974. Beobachtungen zur Küstenvegetation der Varanger - Halbinsel (Nord-Norwegen).
0090 Z. Deutsch. Ges. Polarforschung 44:148-159.
Thannheiser, D. & W. Hofmann. 1977. Pflanzengesellschaften am Meeresstrand im Kongs- und Krossfjord
0089 (West-Spitzbergen). Doc. Phytos. N.S. 1:297-303.
Tüxen, R. & Westhoff, V. 1963. Saginetea maritimae, eine Gesellschaftsgruppe im wechselhalinen
0007 Grenzbereich der europäischen Meeresküsten. Mitt. Flor. Soziol. Arbeitsgem. N.F. 10.
Tyler, G. 1969. Studies in the ecology of Baltic sea-shore meadows II. Flora and vegetation.
0084 Opera Botanica 25, 103 pp.
Valsecchi, F. 1967. Contributo alla conoscenza dell'areale e dell' ecologia della Spartina juncea
0078 Willd. in Sardegna. Bull. Ist. Bot. Univ. Sassari 7:10-12.
Vicherek , J. 1962. Rostlinná - spolecenstva Jihomoravské halofytnî vegetace. Publ. Fac. Sci. Univ. J.E.
0052 Purkyňe 430:65-96.
Weevers, Th. 1940. De flora van Goeree en Overflakkee dynamisch beschouwd. Ned. Kruidk. Arch. 50:285-354.
0118
Wendelberger, G. 1950. Zur Soziologie der kontinentalen halophyten Vegetation Mitteleuropas. Osterr.
0004 Akademie der Wissenschaften, 108,5. Wien.
Westhoff, V. 1947. The vegetation of dunes and salt marshes on the Dutch islands of Terschelling,
0035 Vlieland and Texel. Den Haag 1947, 131 pp. (tables not published in full).
Westhoff, V., C.G. van Leeuwen & M.J. Adriani. 1961. Enkele aspekten van vegetatie en bodem der duinen
0060 van Goeree, in het bijzonder de contactgordels tussen zout en zoet milieu. Jrb. 1961. Wetensch. Genootsch. Goeree - Overflakkee, p. 47-92.
Westhoff, V. & Segal, S. 1961. Cursus vegetatiekunde 12-17 juni 1961 op Terschelling. Report Hugo de
0103 Vries Lab. Amsterdam.
Wilkon-Michalska, J. 1963. Halofity Kujaw (the Halophytes from Kujawy). Studia Soc. Torunensis Torun-
0022 Polonia, Sectio D (Botanica), 27(1):3-121.
Wolff, W.J. 1968. The halophilous vegetation of the lagoons of Mesolonghi Greece. Vegetatio 16:195-234.
0042

Appendix 3

List of plant species occurring in the relevés of the salt marsh data set with author's name(s) and code number according to the Trieste System.

Synonyms are indicated with S and have the same code as belonging to the correct name. Some species names could not be checked, to them an ad hoc code is attached.

4720 001		Acer platanoides L.
9332 027		Achillea asplenifolia Vent.
9332 029		Achillea collina J.Becker
9332 024		Achillea millefolium L.
9332 025		Achillea pannonica Scheele
9332 026		Achillea setacea Waldst. et Kit.
0459 011	S	Acorellus pannonicus (Jacq.)Palla
9830 720		Acrocladium cuspidatum Lindb.
0370 001		Aeluropus litoralis (Gouan) Parl.
0370 002		Aeluropus repens Parl.
9605 901		Aetheorhiza bulbosa (L.)Cass.
0405 010	S	Agropyron acutum (DC.) Roem. et Schult.
0405 021		Agropyron curvifolium Lge
0405 006		Agropyron elongatum (Host)Beauv.
0505 101	S	Agropyron glaucum
0405 008		Agropyron junceum (L.)Beauv.
0405 066		Agropyron junceum x pungens
0405 010	S	Agropyron litorale Auct.
0405 010	S	Agropyron littoreum Schumach.
0405 010		Agropyron pungens (Pers.) R. et S.
0405 010		Agropyron pycnantum Gren. et Godron
0405 014		Agropyron repens (L.)Beauv.
0242 012	S	Agrostis alba L.
0242 022		Agrostis densior Hackel
0242 011		Agrostis gigantea Roth
0242 012	S	Agrostis maritima Lam.
0242 024		Agrostis nebulosa Bss. et Reut.
0242 024		Agrostis stolonifera L.
0242 021		Agrostis tenuis Sibth.
0242 023		Agrostis verticillata Vill.
0242 021	S	Agrostis vulgaris With.
2401 001		Aizoon hispanicum L.
7211 002		Ajuga genevensis L.
0070 002		Alisma lanceolatum With.
0070 003		Alisma plantago - aquatica L.
7097 007		Alkanna tinctoria (L.)Tausch
1049 025		Allium sphaerocephalon L.
1049 032		Allium vineale L.
1882 002		Alnus glutinosa (L.)Vill.
0225 001		Alopecurus aequalis Sobol.
0225 013		Alopecurus arundinaceus Poir.
0225 002		Alopecurus bulbosus Gouan
0225 004		Alopecurus geniculatus L.
0225 008		Alopecurus pratensis L.
0225 013	S	Alopecurus ventricosus Pers.
4991 004		Althaea officinalis L.
2299 009		Amaranthus albus L.
2299 004		Amaranthus retroflexus L.
9830 814		Amblystegium riparium (Hedw.) B.S.G.
9830 810		Amblystegium serpens ssp. juratzkanum (Schimper)Dixon
9146 001		Ambrosia maritima L.
6016 001		Ammi visnaga (L.) Lam.
0249 001		Ammophila arenaria (L.) Link
0249 003		Ammophila arundinacea Host
3707 001		Amorpha fruticosa L.
2275 002		Anabasis aphylla L.
9331 005		Anacyclus marrocanus Ball
6338 004		Anagallis arvensis L.
6338 003		Anagallis tenella (L.)L.
0134 001		Andropogon ischaemum L.
6082 003		Angelica archangelica L. var. litoralis (Fries)
6082 001		Angelica palustris (Besser)Hoffm.
6082 002		Angelica sylvestris L.
0792 001		Anthelia juratzkana
0205 002		Anthoxanthum odoratum L.
5938 001		Anthriscus sylvestris (L.) Hoffm.
7482 901	S	Antirrhinum orontium L.
0251 002		Apera spica - venti (L.)Beauv.
6004 001		Apium graveolens L.
9581 023		Arachnospermum canum C.A. Mayer
2443 033		Arenaria serpyllifolia L.
6350 017		Armeria maritima (Mill.)Willd.
0275 001		Arrhenatherum elatius (L.)Presl.
9358 008		Artemisia alba Turra
9358 031		Artemisia austriaca Jacq.
9358 010		Artemisia caerulescens L.
9358 061		Artemisia caerulescens L. ssp. gallica (Willd.)K.Persson
9358 056		Artemisia campestris L.
9358 061		Artemisia gallica Willd.
9358 045		Artemisia herba - alba Asso
9358 008	S	Artemisia incanescens Jordan
9358 009		Artemisia maritima L.
9358 015	S	Artemisia monogyna W. et K.
9358 047		Artemisia rupestris L.
9358 009	S	Artemisia salina Willd.
9358 015		Artemisia santonicum L. ssp. santonicum
9358 001		Artemisia vulgaris L.
2255 002		Arthrocnemum fruticosum (L.)Moq.
2255 003		Arthrocnemum glaucum (Delile)Ung.-Sttg.
2255 001		Arthrocnemum perenne (Miller)Moss
0333 001	S	Arundo phragmites L.
1113 003		Asparagus maritimus (L.) Milles
1113 004		Asparagus officinalis L.
8485 031		Asperula cynanchica L.
8486 063	S	Asperula glauca (L.) Besser
0980 003		Asphodelus microcarpus Viv.
8900 025		Aster linosyris (L.) Bernh.
8900 020	S	Aster punctatus W. et K.
8900 012		Aster squamatus (Sprengel)Hieron.
8900 019		Aster tripolium L.
8900 019	S	Aster tripolium L. var. longicaulis
8900 019	S	Aster tripolium L. ssp. pannonicus (Jacq.)Soo
8900 999		Aster spec.
9093 001		Asteriscus aquaticus (L.)Less
3766 084		Astragalus asper Jacq.
3766 086		Astragalus austriacus Jacq.
3766 024		Astragalus hamosus L.
9450 004		Atractylis cancellata L.
2229 001	S	Atriplex arenaria Woods
2229 016		Atriplex calotheca (Rafn)Fries
2229 001		Atriplex halimus L.
2229 017		Atriplex hastata L.
2229 017	S	Atriplex hastata L. var. Salina Wahl.
2229 017	S	Atriplex hastata L. var. triangulare (Willd.)Moq.
2229 011		Atriplex laciniata L.
2229 017	S	Atriplex latifolia Wirtgen
2229 013		Atriplex littoralis L.
2229 019		Atriplex longipes Drejer
2229 011	S	Atriplex maritima L.
2229 015		Atriplex patula L.
2229 103	S	Atriplex pedunculata L.
2229 101	S	Atriplex portulacoides L.
2229 019		Atriplex praecox Hülpers
2229 010		Atriplex rosea L.
2229 012		Atriplex tatarica L.
2229 012	S	Atriplex tornabenii tineo
2229 017	S	Atriplex triangulare Willd.
0384 008	S	Atropis maritima Griseb.
0273 999		Avena spec.
0204 001		Baldingera arundinacea (L.)Dum.
9830 074		Barbula spec.
9830 252	S	Barbula spadicea (Mitt.)Braithw.
2239 002		Bassia hirsuta (L.) Asch.
2239 003		Bassia sedoides (Pallas)Asch.
2546 127	S	Batrachium aquatile (L.)Dumort
0303 001		Beckmannia eruciformis (L.)Host
8879 001		Bellis annua L.
8879 002		Bellis perennis L.
6038 901		Berula erecta (Hudson)Coville
2221 002		Beta macrocarpa Guss.
2221 001	S	Beta maritima L.
2221 001		Beta vulgaris L. ssp. maritima (L.) Arcangeli
9237 001		Bidens tripartita L.
2201 801		Bilderdykia convolvulus (L.)Dumort.
6497 001	S	Blackstonia imperfoliata (L.fil.)Samp.
6497 001		Blackstonia perfoliata (L.) Hudson
6497 002		Blackstonia perfoliata (L.)Huds.ssp. imperfoliata L.fil.
6497 005		Blackstonia perfoliata (L.)Huds.ssp. serotina (Koch)Vollm.
6497 005	S	Blackstonia serotina (Koch)Beck
0468 701	S	Blysmus compressus (L.) Panzer
0468 702	S	Blysmus rufus (Huds.)Link
0468 016	S	Bolboschoenus maritimus (L.)Palla
0393 005		Brachypodium phoenicoides (L.)R. et S.
0393 002		Brachypodium pinnatum (L.) Beauv.
9830 867		Brachythecium albicans (Hedw.) B.S.G.
9830 862		Brachythecium mildeanum Schimper
9830 868		Brachythecium rutabulum (Hedw.) B.S.G.
0367 001		Briza maxima L.
0367 002		Briza media L.
0367 003		Briza minor L.
0389 005	S	Bromus arenarius Thom.
0389 025		Bromus commutatus Schrader
0389 004		Bromus erectus Hudson
0389 005		Bromus hordeaceus L.
0389 023		Bromus macrostachys Desf.
0389 012		Bromus molliformis Lloyd
0389 007		Bromus mollis L.
0389 013		Bromus racemosus L.
0389 022		Bromus rubens L.
0389 018		Bromus squarrosus L.
0389 020		Bromus tectorum L.
9830 536		Bryum argenteum Hedwig
9830 532		Bryum caespiticium Hedwig
9830 534		Bryum marratii Wils.
9830 520		Bryum pallens Rohling
9830 566		Bryum pendulum (Hornsch.)Schimper
9830 535		Bryum pseudotriquetrum (Hedw.)Schwaegr.
9830 533		Bryum salinum Hag.
5994 023	S	Bupleurum gracile (Bieb.)DC.
5994 024		Bupleurum semicompositum L.
5994 023		Bupleurum tenuissimum L.
5994 023	S	Bupleurum tenuissimum L. ssp. gracile (Bieb.)Wolff
0081 001		Butomus umbellatus L.
2920 001		Cakile maritima Scop.
0247 017		Calamagrostis deschampsioides Trin.
0247 005		Calamagrostis epigejos (L.)Roth

0247 016 Calamagrostis holmii Lange
0247 008 Calamagrostis neglecta (Ehrh.)G., M. et Sch.
0247 011 Calamagrostis pseudophragmites (Halier)Koeler
9423 001 S Calendula microphylla Lange
9423 001 Calendula suffruticosa Vahl ssp. algarbiensis (Boiss.)Nyman
9830 821 Calliergon stramineum (Brid.)Kindb.
9830 720 Calliergonella cuspidata (Hedw.)Laeske)
4530 005 Callitriche stagnalis Scop.
4530 002 Callitriche truncata Guss.
9700 273 Caloplaca fulgens Körb.
2524 001 Caltha palustris L.
6994 002 Calystegia sepium (L.)R.Br.
6994 001 Calystegia soldanella (L.)R.Br.
2235 003 Camphorosma annua Pallas
2235 001 Camphorosma monspeliaca L.
2235 003 S Camphorosma ovata W. et K.
2235 001 S Camphorosma ruthenica Bieb.
9830 852 Camptothecium lutescens (Hedw.) B.S.G.
9830 802 Campylium polygamum (B.S.G.) C. Jens
9830 803 Campylium stellatum (Hedw.)Lang & Jensen
2986 001 Capsella bursa - pastoris (L.)Medicus
2966 020 *Cardamine pratensis L.*
2883 901 Cardaria draba (L.)Desv.
9461 012 Carduus acanthoides L.
9461 005 Carduus nutans L.
0525 001 Carex acutiformis Ehrh.
0525 005 Carex arenaria L.
0525 016 Carex bigelowii Torrey
0525 043 Carex distans L.
0525 044 Carex disticha Hudson
0525 045 Carex divisa Hudson
0525 178 Carex elata All.
0525 050 Carex extensa Good.
0525 052 Carex flacca Schreb.
0525 112 *Carex fusca All.*
0525 120 S Carex glareosa Wahlenb.
0525 047 Carex glauca Scop.
0525 065 Carex gracilis Curtis
0525 071 Carex hirta L.
0525 072 Carex hispida Willd.
0525 074 Carex hordeistichos Vill.
0525 206 Carex hostiana DC
0525 115 Carex lamprophysa Sam.
0525 089 Carex leporina L.
0525 092 Carex liparicarpos Gaudin
0525 180 Carex lyngbyei Hornem.
0525 121 Carex mackenzie V. Krecz.
0525 122 Carex maritima Gunn.
0525 103 Carex melanostachya Willd.
0525 115 S Carex nemorosa (Rebent.)Schinz et Keller
0525 077 Carex nigra (L.) Reichard
0525 113 Carex norvegica Retz.
0525 068 Carex nutans Host
0525 118 Carex otrubae Podp.
0525 119 Carex paleacea Wg.
0525 123 Carex panicea L.
0525 124 Carex paniculata L.
0525 137 Carex praecox Schreb.
0525 140 Carex punctata Gaudin
0525 123 Carex rariflora (Wahlenb.)Sm.
0525 124 Carex recta Boott.
0525 148 Carex riparia Curtis
0525 125 Carex salina Wahlenb.
0525 127 Carex scandinavica Davies
0525 049 Carex serotina Merat.
0525 165 Carex stenophylla Wahlenb.
0525 182 Carex stygia Fr.
0525 126 Carex subspathacea Wormsk.
0525 130 Carex ursina Dewey
0525 176 Carex vesicaria L.
0525 177 Carex vulpina L.
0525 115 S Carex vulpina L. ssp. nemorosa (Rebent.)Schinz et Keller
9449 013 Carlina racemosa L.
2405 004 Carpobrotus edulis (L.)N.E.Br.
2936 001 Carrichtera annua (L.)DC.
6020 001 Carum carvi L.
0347 001 Catabrosa aquatica (L.)Beauv.
0387 001 Catapodium marinum (L.) Hubbard
0174 002 Cenchrus longispinus (Hackel)Fernald
6073 001 Cenolophium fischeri (Sprengel)Koch
6073 001 S Cenolophium denudatum (Hornem.)Tutin
9476 096 S Centaurea arenaria Bieb. ssp. Tauscheri (Kerner)Soo
9476 148 Centaurea calcitrapa L.
9476 186 Centaurea debeauxii Gren. et Godron ssp. thuillieri Dostal
9476 177 Centaurea dracunculifolia Dufour
9476 178 Centaurea jacea L.
9476 177 S Centaurea jacea L. ssp. dracunculifolia
9476 175 Centaurea pannonica (Heuffel)Simonkai
9476 182 S Centaurea pratensis Thuill.
9476 052 Centaurea sadlerana Janka
9476 052 S Centaurea scabiosa L. ssp. sadlerana (Janka)Asch. et Gr.
9476 152 Centaurea sphaerocephala L.
9476 096 Centaurea tauscheri Kerner
6496 015 Centaurium barellieri Dufour
6496 002 Centaurium erythraea Rafn
6496 002 Centaurium erythraea Rafn ssp. erythraea
6496 002 S Centaurium glomeratum (Wittrock)Ronniger
6496 005 Centaurium littorale (Turner)Gilmour)
6496 002 S Centaurium minus Auct.
6496 002 Centaurium minus Moench
6496 011 Centaurium pulchellum (Swartz)Druce
6496 013 Centaurium spicatum (L.)Fritsch
6496 012 Centaurium tenuiflorum (Hoffmans. & Link) Fritsch
6496 002 Centaurium triphyllum (Schmidt)Melderis
6496 002 S Centaurium umbellatum Auct.
6496 005 S Centaurium vulgare Rafn
9810 010 Cephaloziella divaricata (Sm.)Schiffner
9819 999 Cephaloziella spec.
2430 002 S Cerastium anomalum W. et K.
2430 038 S Cerastium caespitosum Gilib.
2430 001 Cerastium cerastioides (L.)Britton
2430 049 Cerastium diffusum Pers.
2430 002 Cerastium dubium (Bast.)Schwartz
2430 038 Cerastium fontanum Baumg.
2430 038 S Cerastium fontanum Baumg. ssp. triviale (Lk.)Jalas
2430 044 Cerastium glomeratum Thuill.
2430 038 S Cerastium holosteoides Fr.
2430 038 S Cerastium holosteoides Fr. var. vulgare (Hartm.)Hyl.
2430 048 Cerastium pumilum curtis
2430 047 *Cerastium semidecandrum L.*
2430 004 S Cerastium viscosum Auct.Mult.
2430 038 S Cerastium vulgatum L.1762, Non L.1755
9830 076 Ceratodon purpureus (Hedw.) Brid.
9700 601 Cetraria delisei (Bory)Th.Fr.
9339 901 Chamomilla recutita (L.)Rauschert
2223 021 Chenopodium album L.
2223 001 Chenopodium botrys L.
2223 009 Chenopodium glaucum L.
2223 010 Chenopodium rubrum L.
6497 003 S Chlora imperfoliata L. Fil.
6497 003 S Chlora perfoliata (L.) L.
6497 005 S Chlora serotina Koch
9341 901 S Chrysanthemum leucanthmum L. ssp. leucanthemum Auct.
0134 801 Chrysopogon gryllus (L.)Trin.
9553 001 Cichorium intybus L.
9462 060 *Cirsium arvense (L.)Scop.*
9462 055 Cirsium brachycephalum Juratzka
9462 050 Cirsium canum (L.) All.
9462 052 Cirsium monspessulanum (L.)Hill
9462 057 Cirsium palustre (L.)Scop.
9462 028 Cirsium vulgare (Savi)Ten.
7787 001 S Cistanche lutea Hoffgg. et Lk.
7787 001 Cistanche phelypaea (L.)Coutinho
0489 001 Cladium mariscus (L.)Pohl
9700 058 Cladonia foliacea (Huds.)Schaer
9700 077 Cladonia furcata (Huds.)Schrad.
9700 066 Cladonia magyarica Vain.
9700 077 Cladonia rangiformis Hoffm.
6069 001 Cnidium dubium (Schkuhr)Thell.
2907 005 Cochlearia aestuaria (Lloyd)Heywood
2907 007 Cochlearia anglica L.
2907 001 Cochlearia danica L.
2907 002 Cochlearia officinalis L.
0978 003 Colchicum arenarium W. et K.
0978 005 Colchicum kochii Parl.
3356 003 S Comarum palustre L.
5970 001 Conium maculatum L.
2539 109 Consolida regalis S.F.Gray
2539 109 Consolida regalis S.F.Gray ssp. regalis
6993 019 Convolvulus arvensis L.
6993 010 Convolvulus lineatus L.
6994 002 S Convolvulus sepium L.
6994 001 S Convolvulus soldanella L.
8926 001 Conyza canadensis (L.)Cronq.
3774 007 Coronilla varia L.
2884 002 S Coronopus procumbens Gilib.
2884 002 Coronopus squamatus (Forsskal) Asch.
6159 005 Cornus suecica L.
0269 001 Corynephorus articulatus (Desf.)Beauv.
0269 002 Corynephorus canascens (L.)
9351 001 Cotula coronopifolia L.
9605 022 Crepis biennis L.
9605 901 S Crepis bulbosa (L.)Tausch
9605 060 Crepis capillaris (L.)Wallr.
9605 051 Crepis foetida L.
9605 051 Crepis foetida L. ssp. rhoeadifolia Bieb.
9605 072 Crepis rhoeadifolia Bieb.
9605 053 Crepis sancta (L.) Babc.
6976 001 Cressa cretica L.
6039 001 Crithmum maritimum L.
0221 001 Crypsis aculeata Aiton
0221 002 Crypsis alopecuroides Schrad.
0221 003 Crypsis schoenoides (L.)Lam.
6968 012 Cuscuta eipthymum (L.)L.
6834 001 Cynanchum acutum L
0282 001 Cynodon dactylon (L.)Pers.
5838 001 Cynomorium coccineum L.
0373 001 Cynodurus cristatus L.
0459 004 Cyperus fuscus L.
0459 009 Cyperus longus L.
0459 011 Cyperus pannonicus L.

0372 001 Dactylis glomerata L.
0372 002 Dactylis hispanica Roth
1368 002 S Dactylorchis incarnata (L.)Verm.
6142 008 Daucus carota L.
6142 008 Daucus catota L. ssp. gummifer Hooker Fil.
6142 008 Daucus carota L. ssp. maritimus Lam.
6142 008 S Daucus gummifer Lam.
6142 008 S Daucus maritimus Lam.

7415 001 Datura stramonium L.
2539 109 S Delphinium consolida L.
0270 003 Deschampsia caespitosa (L.)Beauv.
2997 001 Descurainia sophia (L.)Webb
2502 063 Dianthus deltoides L.
2502 061 Dianthus gallicus Pers.
2502 104 Dianthus pontederae Kerner
9830 252 Didymodon spadiceus (Mitt.)Limpr.
0166 801 Digitaria sanguinalis (L.)Scop.
9333 001 S Diotis candididissima Desf.
1084 001 Dipcadi serotinum (L.)Medicus
9700 531 Diploschistes scruposus (Schreb.)Norm.
2946 010 Diplotaxia ollivieri Maire
2946 004 Diplotaxis tenuifolia (L.)Dc.
8540 006 Dipsacus ferox Loisel.
9830 033 Distichium capillaceum (Hedw.)B.S.G.
9061 901 Dittrichia viscosa (L.)Greuter
3697 004 S Dorycnium gracile Jordan
3697 004 Dorycnium pentaphyllum Scop.
3697 002 Dorycnium rectum (L.)Ser.
2989 001 S Draba verna L.
9830 846 Drepanocladus aduncus (Hedw.)Sensu Moenkm.
9830 847 Drepanocladus uncinatus (Hedw.)Warnst.
0380 001 Duponti fisheri R.Br.

0166 902 S Echinochloa crus-galli (L.)Beauv.
5231 001 Elatine alsinastrum L.
0469 008 Eleocharis multicauslis (Sm.)Sm.
0469 013 Eleocharis palustris (L.)R. et S.
0469 014 Eleocharis parvula (R. et S.)Lk.
0469 015 Eleocharis pauciflora (Lightf.)Lk.
0469 015 S Eleocharis quinqueflora (Hartm.)Schwarz
0469 016 Eleocharis uniglumis (Lk)Schultes
0411 001 Elymus arenarius L.
0405 014 S Elytrigia repens (L.)Nevski
4540 002 S Empetrum hermaphroditum Hagerup
4540 001 Empetrum nigrum L.
4540 002 Empetrum nigrum L. ssp. hermaphroditum (Hagerup)Bocher
5795 005 Epilobium hirsutum L.
5795 015 Epilobium palustre L.
1482 004 Epipactis palustris (Mill.)Crantz
9906 009 Equisetum arvense L.
9906 005 Equisetum fluviatile L.
9906 006 Equisetum palustre L.
9906 002 Equisetum ramoisissimum Desf.
0341 002 Eragrostis pilosa (L.)Beauv.
0341 003 Eragrostis poaeoides Beauv.
0112 002 Erianthus ravennae (L.)Beauv.
8926 001 Erigeron canadensis L.
0466 001 Eriophorum angustifolium Honck.
0466 004 Eriophorum scheuchzeri Hoppe
3927 025 Erodium cicutarium (L.)L. Her.
2989 943 Erophila verna (L.)Chevall.
5923 024 Eryngium campestre L.
5923 025 Eryngium corniculatum Lam.
5923 012 Eryngium creticum Lam.
5923 008 Eryngium maritimum L.
5923 010 Eryngium planum L.
3004 029 Erysimum hieracifolium L.
3004 036 Erysimum repandum L.
6496 011 S Erythraea pulchella (Sweet)Fr.
6496 012 S Erythraea tenuiflora Hoffm. et Lk.
8816 001 Eupatorium cannabinum L.
4498 005 Euphorbia chamaesyce L.
4498 101 Euphorbia cyparissias L.
4498 018 Euphorbia palustris L.
4498 094 Euphorbia paralias L.
4498 003 Euphorbia polygonifolia L.
4498 072 Euphorbia portlandica L.
4498 092 Euphorbia seguierana Necker
7638 013 S Euphrasia curta (Fr.)Wettst.
7638 025 Euphrasia frigida Pugsley
7644 015 S Euphrasia litoralis Fr.
7638 013 Euphrasia nemorosa (Pers.)Wallr.
7644 015 S Euphrasia odontites L.
7644 015 S Euphrasia serotina Lam.
7638 999 Euphrasia spec.
9830 891 Eurynchium praelongum (Hedw.)Hobkirk Synops.
9830 898 Eurhynchium pulchellum (Hedw.)Dixon
9830 891 S Eurhynchium stokesii (Turner) B.S.G.
9830 898 S Eurhynchium strigosum
8966 001 Evax pygmaea (L.)Brot.

6018 001 Falcaria vulgaris Bernh.
0385 044 S Festuca arenaria Kit.
0385 006 Festuca arundinacea Schreb.
0385 018 Festuca duriuscula L.
0385 006 S Festuca elatior L. ssp. arundinacea (Schreb.)Hackel
0385 021 Festuca fenas Lag.
0385 044 Festuca ovina L.
0385 079 Festuca ovina ssp. vulgaris (Hackel) Brid.
0385 051 Festuca pratensis Huds.
0385 055 Festuca pseudovina Hackel
0385 060 Festuca rubra L.
0385 062 Festuca rupicola Heuff.
0385 062 S Festuca sulcata (Hackel)Nym.
0385 072 Festuca vaginata W. et K.
0385 073 Festuca valesiaca Schleicher
0385 079 S Festuca vulgaris Hackel
8969 007 S Filago germanica Huds.
8969 007 Filago pyramidata L.
3374 001 S Filipendula hexapetala Gilib.
3374 002 Filipendula ulmaria (L.)Maxim.
3374 001 Filipendula vulgaris Moench.
3354 001 Fragaria vesca L.
3354 003 Fragaria viridis Duchesne
5233 003 Frankenia corymbosa Desf.
5233 006 S Frankenia hirsuta L.
5233 005 Frankenia laevis L.
5233 001 Frankenia pulverulenta L.
5233 004 S Frankenia reuteri Boiss.
5233 004 Frankenia thymifolia Desf.

1048 008 Gagea pusilla (Schmidt) R. et S.
1048 011 Gagea reticulata A. et Schult.
7270 009 Galeopsis bifida Boenn.
8486 132 Galium aparine L.
8486 034 Galium arenarium Loisel.
8486 038 Galium mollugo L.
8486 016 Galium palustre L.
8486 026 Galium verum L.
0276 001 Gaudinia fragilis (L.)Beauv.
3675 001 Genista tinctoria L.
6509 022 S Gentiana aurea L.
6509 022 Gentiana aurea (L.)H.Sm.
6509 004 S Gentiana detonsa Rottb.
6509 004 Gentiana detonsa (Rottb.)G.Don fil.
3924 007 Geranium pratense L.
3924 033 Geranium pusillum L.
2848 001 Glaucium flavum Crantz.
2848 001 S Glaucium luteum Scop.
6337 001 Glaux maritima L.
7249 001 Glechoma hederacea L.
2388 001 Glinus lotoides L.
0383 001 Glyceria fluitans (L.)R.Br.
0383 003 Glyceria maxima (Hartm.)Holmb.
2497 026 Gypsophila muralis L.
2497 021 Gypsophila perfoliata L.

2229 103 Halimione pedunculata (L.)Aellen
2229 101 Halimione portulacoides (L.)Aellen
2253 001 Halocnemum strobilaceum (Pallas)Bieb.
2249 001 Halopeplis amplexicaulis (Vahl)Ung.-Stbg
6071 007 S Haloscyas scoticum L.
9569 001 Hedypnois cretica (L.)Dum.-Courset
9569 001 S Hedypnois rhagadioloides (L.) Willd
0469 013 S Heleocharis palustris R. et S.
0469 016 S Heleocharis uniglumis (Lk.)Schult.
0222 001 Heleochloa alopecuroides Host
0222 002 Heleochloa schoenoides Host
5245 019 Helianthemum salicifolium (L.)Miller
5245 002 Helianthemum squamatum (L.)Pers.
9006 013 Helichrysum arenarium (L.)Moench
9006 005 Helichrysum stoechas (L.)Moench
7052 012 Helitropium supinum L.
6122 005 Heracleum sphondylium L.
6122 005 S Heracleum sphondylium L. ssp. sibiricum (L.) Simonkai
2476 015 Herniaria fruticosa L.
2476 016 Herniaria lenticulata L.
5013 003 Hibiscus trionum L.
9607 029 Hieracium praealtum Vill. ssp. Bauhinii (Besser)Petunnikov
9607 010 Hieraium pilosella L.
5470 001 Hippophae rhamnoides L.
5837 001 Hippuris vulgaris L.
0257 001 Holcus lanatus L.
0468 029 S Holoschoenus vulgaris Lk. var. romanus L.
2431 001 Holosteum umbellatum L.
2441 001 Honkenya peploides (L.)Ehrh.
0410 007 Hordeum hystrix Roth
0410 010 Hordeum maritimum With.
0410 011 Hordeum murinum L.
0410 012 Hordeum nodosum L.
2985 801 Hornungia petraea (L.)Reichenb.
2985 901 S Hornungia procumbens (L.)Hayek
2985 001 Hutchinsia alpina (L.)R.Br.
2985 901 Hutchinsia procumbens (L.)Desv.
2985 902 Hymenolobus pauciflorus (Koch)Schinz et Thell.
2985 901 Hymenolobus procumbens (L.)Nutt.
5839 003 Hydrocotyle vulgaris L.
9830 981 Hylocomium squarrosum (Hedw.) B.S.G.
5168 054 Hypericum perforatum L.
9830 956 Hypnum cupressiforme Hedw.
9572 003 Hypochoeris achyrophorus L.
9572 011 S Hypochoeris aethnensis (L.)Benth. et Hook.
9572 008 Hypochoeris glabra L.
9572 009 Hypochoeris radicata L.
9572 009 Hypochoeris radicata L. var. salina

0109 001 Imperata cylindrica (L.)Beauv.
9061 008 Inula britannica L.
9061 019 Inula crithmoides L.
9061 006 Inula hirta L.
9061 004 Inula salicina L.
9061 021 Inula viscosa (L.)Aiton
7003 002 Ipomoea sagittata Poiret
1264 010 Iris pseudoacorus L.
1264 018 Iris sisyrhynchium L.
1264 015 Iris spurium L.
1264 019 Iris xiphium L.

9476 178 S Jacea vulgaris
0936 002 Juncus acutus L.
0936 005 Juncus ambiguus Guss.
0936 006 Juncus anceps De la Harpe
0936 007 Juncus arcticus (Willd.)
0936 008 Juncus articulatus L.
0936 009 Juncus atratus Krock.
0936 010 S Juncus balticus Willd.
0936 013 Juncus bufonius L.
0936 017 Juncus compressus Jacq.
0936 018 Juncus conglomeratus L.
0936 020 Juncus effusus L.
0936 011 S Juncus fasciculatus Schousboe
0936 024 Juncus gerardii Loisel.
0936 008 S Juncus lamprocarpus Ehrh.
0936 029 Juncus maritimus Lam.
0936 002 S Juncus megalocarpus Asch. et Gr.
0936 032 Juncus pygmaeus Rich.
0936 005 S Juncus ranarius Song. et Perr.
0936 038 Juncus subnodulosus Schrank
0936 039 Juncus subulatus Forsk.

8543 029 Knautia arvensis (L.) Coulter
2239 002 S Kochia hirsuta (L.)Nolte
2240 003 Kochia laniflora (Gmel.) Borbas
2240 001 Kochia prostrata (L.)Schrader
0346 005 Koeleria albescens Dc.
0346 006 Koeleria gracilis Pers.
0346 012 Koeleria hispida Dc.
0346 001 Koeleria phleoides (Vill.)Pers.
0346 010 Koeleria splendens Presl.
0346 013 Koeleria villosa Pers.
9562 001 Koelpinia linearis Pallas

9596 011 Lactuca saligna L.
9596 009 S Lactuca scariola L.
9596 009 Lactuca serriola L.
9596 004 Lactuca tatarica (L.)C.A.Meyer
9605 053 S Lagoseris sancta (L.)K. Maly
0256 001 Lagurus ovatus L.
7073 005 S Lappula myosotis Moench.
7073 005 Lappula squarrosa (Retz.)Dumort
3854 023 Lathyrus palustris L.
3854 020 Lathyrus pratensis L.
3854 027 Lathyrus tuberosus L.
9593 002 Launaea nudicaulis (L.)Hook.Fil.
4990 001 Lavatera cretica L.
9700 310 Lecidea decipiens (Ehrh.)Ach.
0795 002 Lemna minor L.
9574 004 Leontodon autumnalis L.
9574 018 Leontodon hirtus L.
9574 012 Leontodon hispidulus (Delile)Boiss.
9574 025 S Leontodon nudicaulis Auct.
9574 018 S Leontodon nudicaulis (L.)Banks
9574 025 S Leontodon saxatilis Lam.
9574 025 Leontodon taraxocoides (Vill.)Merat
9574 025 Leontodon taraxacoides (Vill.)Merat ssp. taraxacoides
2883 008 Lepidium cardamines L.
2883 018 Lepidium cartilagineum (Mayer)Thell.
2883 018 Lepidium cartilagineum (Mayer)Thell. ssp. crassifolium
2883 018 S Lepidium crassifolium W. et K.
2883 901 S Lepidium draba L.
2883 019 Lepidium latifolium L.
2883 017 Lepidium perfoliatum (L.)L.
2883 013 Lepidium ruderale L.
9830 490 Leptobryum piriforme (L.)Schimper
0127 001 S Lepturus filiformis Trin.
0127 002 S Lepturus incurvatus (L.)Trin.
9341 901 Leucanthemum vulgare Lam.
6071 007 Ligusticum scoticum L.
6352 001 Limoniastrum monopetalum (L.)Boiss.
6351 017 S Limonium angustifolium (Tausch.)Fourn.
6351 035 Limonium aragonense (Debeaux)Pignatti
6351 029 Limonium articulatum (Loisel.)Kuntze
6351 072 Limonium auriculae-ursifolium (Pourret)Druce
6351 019 Limonium bellidifolium (Gouan)Dumort
6351 067 Limonium binervosum (Sm.)Salmon
6351 072 S Limonium binervosum (Sm.)Salmon ssp. lychnidifolium Fourn.
6351 019 S Limonium caspium (Willd.)Gams
6351 028 Limonium catalaunicum (Willk. et Costa)Pignatti
6351 079 Limonium costae (Willk.)Pignatti
6351 077 Limonium delicatulum (De Girard)O.Kuntze
6351 026 Limonium dichotomum (Cav.) O. Kuntze
6351 067 S Limonium dodartii (Girard)O.Kuntze
6351 087 Limonium echioides (L.)Miller
6351 005 Limonium ferulaceum (L.)O.Kuntze
6351 064 Limonium girardianum (Guss.)Fourr.
6351 060 S Limonium globulariifolium O.Kuntze
6351 013 Limonium gmelinii (Willd.)O.Kuntze
6351 018 Limonium humile Miller
6351 072 S Limonium lychnidifolium Syme.
6351 026 S Limonium occidentale Lloyd
6351 056 Limonium oleifolium Miller
6351 056 Limonium oleifolium Miller ssp. oleifolium
6351 074 Limonium ovalifolium (Poiret)O.Kuntze
6351 060 Limonium ramosissimum (Poiret)Maire
6351 003 Limonium thouinii (Viv.)O.Kuntze
6351 069 Limonium transwallianum (Pugsley)Pugsley
6351 056 S Limonium virgatum (Willd.)Fourr.
6351 017 Limonium vulgare Miller
6351 017 Limonium vulgare Miller ssp. angustifolium
6351 017 S Limonium vulgare Miller var. macrocladum
7480 034 Linaria vulgaris Miller ssp. serotinum (Reichenb.) Gams
3945 023 S Linum angustifolium Huds.
3945 023 Linum bienne Miller
3945 036 Linum catharticum L.
3945 029 Linum maritimum L.
3945 018 Linum perenne L.
3945 034 Linum strictum L.
0395 001 Lolium multiflorum Lam.
0395 088 Lolium parabolicum Sennen
0395 002 Lolium perenne L.
0395 003 Lolium rigidum Gaudin
0395 007 Lolium subulatum Vis.
8523 017 Lonicera periclymenum L.
3698 007 Lotus corniculatus L.
3698 001 S Lotus corniculatus L. ssp. tenuifolius (L.) Hartman
3698 022 Lotus creticus L.
3698 012 S Lotus decumbens Poiret
3698 012 Lotus palustris Willd.
3698 001 S Lotus siliquosus L.
3698 001 Lotus tenuis W. et K.
3698 009 Lotus uliginosus Schkuhr
0937 005 Luzula campestris (L.)Dc.
2491 004 Lychnis flos-cuculi L.
7326 001 Lycopus europaeus L.
0199 001 Lygeum spartum L.
6330 006 Lysimachia nummularia L.
6330 013 Lysimachia thyrsiflora L.
6330 003 Lysimachia vulgaris L.
5476 011 Lythrum borysthenicum (Schrank)Litv.
5476 007 Lythrum hyssopifolia L.
5476 001 Lythrum salicaria L.
5476 002 Lythrum virgatum L.

4992 003 Malva stipulacea Cav.
4992 003 S Malva trifida Cav.
7238 009 Marrubium peregrinum L.
9339 901 S Matricaria bajeri Kanitz
9339 901 S Matricaria chamomilla L.
9339 901 S Matricaria chamomilla L. ssp. bajeri Neumay
9339 005 S Matricaria inodora L.
9339 004 Matricaria maritima L.
9339 005 Matricaria perforata Merat
9339 009 Matricaria pubescens ssp. maroccana
3042 002 Matthiola sinuata (L.) R.Br.
3042 010 Matthiola tricuspidata (L.) R.Br.
3688 031 S Medicago denticulata Willd.
3688 005 S Medicago falcata L.
3688 031 S Medicago hispida Gaertner
3688 025 Medicago littoralis Rhode
3688 021 Medicago marina L.
3688 037 Medicago minima (L.)Bartal.
3688 031 Medicago polymorpha L.
3688 005 Medicago sativa L.
3688 005 Medicago sativa L. ssp. falcata (L.) Arcangeli
7635 024 Melampyrum pratense L.
2490 105 S Melandrium album (Mill.)Garcke
2490 028 S Melandrium viscosum (L.)Čelak.
3689 003 Melilotus alba Medicus
3689 001 Melilotus dentata (W. et K.)Pers.
3698 010 Melilotus indica (L.)All.
3698 015 Melilotus messanensis (L.)All.
3698 005 Melilotus officinalis (L.)Pallas
7328 009 Mentha aquatica L.
7328 014 S Mentha longifolia Auct.
7328 012 Mentha longifolia (L.) Huds.
7328 002 Mentha pulegium L.
7328 011 S Mentha rotundifolia Auct.
7328 014 Mentha spicata L.
7328 011 Mentha suaveolens Ehrh.
2405 001 Mesembryanthemum nodiflorum L.
2258 001 Microcnemum coralloides (Loscos et Pardo)Font Quer
2441 042 Minuartia verna (L.) Hiern.
7482 901 Misopates orontium (L.)Rafin.
9830 579 Mnium hornum Hedw.
9830 593 Mnium rostratum Web. et Mohr.
2444 001 Moehringia trinervia (L.)Clairv.
0340 002 Molinia caerulea (l.)Moench.
0400 001 Monerma cylindricum (Willd.)Coss.
2416 001 Montia fontana L.
2416 001 S Montia lamprosperma Cham.
1095 002 Muscari comosum (L.)Mill.
1095 003 Muscari racemosum (l.)Mill.
7100 035 S Myosotis caespitosa Schultz.
7100 007 S Myosotis hispida Schldl.
7100 035 Myosotis laxa Lehm
7100 013 S Myosotis micrantha Auct.
7100 036 S Myosotis palustris (L.)Hill
7100 007 Myosotis ramosissima Rochel
7100 036 Myosotis scorpioides L.
7100 013 Myosotis stricta Link
2543 001 Myosurus minimus L.
1874 001 Myrica gale L.

0064 004 Najas marina L.
0394 001 Nardus stricta L.
2965 001 Nasturtium officinale R.Br.
2985 901 S Noccaea procumbens (L.)Reich

2229 103 S Obione pedunculata (L.)Moq.
2229 101 S Obione portulacoides (L.)Moq.
7644 015 S Odontites rubra Besser
7644 015 Odontites verna (Bell.)Dum.
7644 015 Odontites verna (Bell.)Dum. ssp. litoralis (Fr.)Nyman
7644 015 Odontites verna (Bell.)Dum. ssp. serotona (Dum.)Corb.
6046 003 Oenanthe fistulosa L.
6046 009 Oenanthe lachenalii Gmelin
6046 009 S Oenanthe lachenalii Gmel. var. multiradiata
6046 008 Oenanthe peucedanifolia Pollich
6046 007 Oenanthe silaifolia Bieb.
9830 051 Oncophorus wahlenbergii Brid.
3685 034 Ononis arvensis L.
3685 034 S Ononis hircina Jacq.
3685 006 Ononis natrix L.
3685 032 Ononis spinosa L.
9467 002 Onopordum acanthium L.
9907 003 Ophioglossum vulgatum L.
1396 042 Orchis coriophora L.
1396 802 Orchis incarnata L.
1396 025 Orchis laxiflora Lam.
1396 026 Orchis palustris Jacq.
1039 010 Ornithogalum umbellatum L.
7791 031 Orobanche caryophyllacea Sm.
7791 033 Orobanche lutea Baumg.
6082 001 S Ostericum palustre (Besser)Besser
9333 001 Otanthus maritimus (L.)Hoffm. et Link
9830 891 Oxyrhynchium praelongum var. stokesii (Turn.)Dixon et Jameson

1202 001 Pancratium maritimum L.
0166 003 Panicum crus-galli L.
2853 003 Papaver dubium L.
0127 001 Parapholis incurva (L.)Hubb.
0127 002 S Parapholis strigosa (Dum.)Hubb.
7642 001 Parentucellia viscosa (L.)Caruel
3203 001 Parnassia palustris L.
0161 001 Paspalum distichum L.
5453 017 S Passerina annua Wikstr.
6120 001 Pastinaca sativa L.
7648 018 Pedicularis palustris L.
7648 015 Pedicularis verticillata L.
3958 001 Peganum harmala L.
9700 099 Peltigera canina (L.)Willd.
5476 012 Peplis portula L.
2277 005 S Petrosimonia crassifolia Auct.
2277 005 Petrosimonia oppositifolia (Pallas)Litv.
2277 002 Petrosimonia triandra (Pallas)Simonkai
6116 018 Peucdeanum latifolium (Bieb.)Dc.
0204 001 Phalaris arundinacea L.
9830 621 Philonotis fontana (Hedw.)Brid.
9830 621 S Philonotis tomentella Mol.
0229 001 Phippsia algida (Soland.)R.Br.
0223 002 Phleum arenarium L.
0223 012 Phleum pratense L.
0127 001 Pholiurus filiformis Schinz. et Thell.
0127 002 Pholiurus incurvus Schinz. et Thell.
0127 003 Pholiurus pannonicus Trin.
0333 001 Phragmites communis Trin.
9575 013 Picris coronopifolia Dc.
9575 002 Picris echioides L.
9575 009 Picris hieracioides L.
6033 016 Pimpinella saxifraga L.
0022 006 Pinus nigra Arnold
8116 024 Plantago albicans L.
8116 003 Plantago cornuti Gouan
8116 004 Plantago coronopus L.
8116 004 S Plantago coronopus L. var. minor Auct.
8116 007 Plantago crassifolia Forsk.
8116 023 Plantago lagopus L.
8116 020 Plantago lanceolata L.
8116 001 Plantago major L.
8116 001 Plantago major L. ssp. intermedia (Dc.)Arcangeli
8116 036 Plantago major L. ssp. winteri (Wirtg.)Ludw.
8116 008 Plantago maritima L.
8116 012 Plantago media L.
8116 025 Plantago ovata Forsk.
8116 001 S Plantago pauciflora Domin.
8116 011 Plantago schwarzenbergiana Schur
8116 061 Plantago subpolaris
8116 002 Plantago tenuiflora W. et K.
8116 036 Plantago winteri
0378 002 Poa alpina L.
0378 004 Poa annua L.
0378 040 Poa arctica R.Br.
0378 011 Poa bulbosa L.
0378 011 Poa bulbosa L. var. vivipara
0378 048 S Poa cenisia Auct.
0378 014 Poa compressa L.
0378 022 Poa irrigata Lindm.
0378 024 Poa jubata Kern.
0378 033 Poa pratensis L.
0378 038 Poa silvicola Guss.
0378 022 S Poa subcoerulea Sm.
0378 044 Poa trivialis L.
9581 001 S Podospermum canum Meyer
9830 478 Pohlia nutans (Hedw.)Lindb.
2214 002 Polycnemum arvense L.
4273 023 Polygala comosa Schkuhr
4273 024 Polygala vulgaris L.
2201 018 S Polygonum aequale Lindm.
2201 018 Polygonum arenastrum Boreau
2201 015 Polygonum aviculare L.
2201 901 Polygonum convolvulus L.
2201 002 Polygonum equisetiforme Sibth. et Sm.
2201 026 Polygonum lapathifolium L.
2201 005 Polygonum maritimum L.
2201 011 Polygonum patulum Bieb.
0233 002 Polypogon maritimus Willd.
0233 004 S Polypogon monspeliensis L.
9830 023 Polytrichum alpinum Hedw.
9830 033 Polytrichum norvegicum Hedw.
9830 032 Polytrichum piliferum Bridel
2421 001 Portulaca oleracea L.
2421 001 Portulaca oleracea L. ssp. sativa (Haw.)Celak
2421 001 S Portulaca sativa Haw.
0058 006 Potamogeton crispus L.
0058 011 Potamogeton natans L.
0058 020 Potamogeton pusillus L.
3356 004 Potentilla anserina L.
3356 004 Potentilla anserina L. ssp. egedii
3356 051 S Potentilla arenaria Borkh.
3356 018 Potentilla argentea L.
3356 051 Potentilla cinerea Chaix
3356 004 S Potentilla egedii Wormsk.
3356 052 Potentilla erecta (L.)Rausch.
3356 003 Potentilla palustris (L.)Scop.
3356 054 Potentilla reptans L.
9830 304 Pottia heimii (Hedw.)Furnr.
6315 011 S Primula finmarchica Jacq.
6315 011 Primula nutans Georgi
6315 010 Primula stricta Hornem.
6315 003 Primula veris L.
7254 003 Prunella vulgaris L.
0384 015 Puccinellia capillaris
0384 014 S Puccinellia coarctata Fern. et Weatherby
0384 002 Puccinellia convoluta (Hornem.)Hay
0384 003 Puccinellia distans (L.)Parl.
0384 009 S Puccinellia faesticiformis Heynh.
0384 004 Puccinellia fasciculata Bickn.
0384 007 Puccinellia limosa Holmb.
0384 008 Puccinellia maritima (Huds.)Parl.
0384 009 Puccinellia palustris Hayek
0384 011 Puccinellia phryganodes Scribn. et Merr.
0384 013 Puccinellia retroflexa (Curt.)Holmb.
0384 013 S Puccinellia retroflexa ssp. borealis var. virescens Holmb.
0384 009 S Puccinellia salinaria Auct.
9078 004 Pulicaria arabica (L.)Cass.
9078 002 Pulicaria dysenterica (L.)Bernh.
9078 006 Pulicaria graeca Sch.-Bip.
9078 002 S Pulicaria uliginosa Steven
9078 003 Pulicaria vulgaris Gaertn.

2546 008 Ranunculus acris L.
2546 127 Ranunculus aquaticus L.
2546 123 Ranunculus aquaticus L. ssp. baudotii Godron
2546 031 Ranunculus bulbosus L.
2546 101 Ranunculus flammula L.
2546 077 Ranunculus hyperboreus Rottb.
2546 109 Ranunculus lateriflorus Dc.
2546 049 Ranunculus pedatus W. et K.
2546 006 Ranunculus repens L.
2546 102 Ranunculus reptans L.
2546 032 Ranunculus sardous Crantz
2546 078 Ranunculus sceleratus L.
2546 128 Ranunculus trichophyllus Chaix
9599 001 Reichardia tingitana (L.)Roth.
3125 021 Reseda battandieri Pitard
3125 016 Reseda stricta Pers.
9830 408 Rhacomitrium canescens (Hedw.)Bridel
7647 023 Rhinathus angustifolius C.C. Gmelin
7647 023 Rhinanthus angustifolius Gmel. ssp. angustifolius
7647 023 S Rhinanthus glaber Lam.
7647 002 Rhinanthus minor L.
7647 023 S Rhinanthus serotinus (Schönheit) Oborny
3161 901 Rhodiola rosea L.
9830 981 Rhytidiadelphus sqaurrosus(Hedw.)Warnst.
1261 001 Romulea bulbocodium (L.)Seb. et Mauri
1261 002 Romulea rollii Parl.
2965 807 Rorippa amphibia (L.)Besser
2965 804 Rorippa sylvestris (L.)Besser
3353 114 Rubus ulmifolius Schott.
2195 014 Rumex acetosa L.
2195 050 Rumex bucephalophorus L.
2195 036 Rumex crispus L.
2195 032 Rumex hydrolapathum Huds.
2195 030 Rumex longifolius D.C.
2195 046 Rumex maritimus L.
2195 043 Rumex obtusifolius L.
2195 042 Rumex pulcher L.
2195 037 Rumex stenophyllus Ledeb.
2195 015 Rumex thyrsiflorus Fingerh.
0059 001 Ruppia maritima L.

2433 002 Sagina intermedia Fenzl.
2433 013 Sagina maritima Don Fil.
2433 001 Sagina nodosa (L.)Fenzl.
2433 011 Sagina procumbens L.
2433 006 Sagina subulata (Swartz)Presl.
2255 002 S Salicornia arabica Auct.
2257 007 Salicornia dolichostachya Moss.

2257	001		Salicornia europaea L.
2255	002	S	Salicornia fruticosa (L.)L.
2257	001	S	Salicornia herbacea (L.)L.
2257	001	S	Salicornia herbacea (L.)L. ssp. cf. emerici
2255	001	S	Salicornia perennis Mill.
2255	001	S	Salicornia radicans Sm.
2257	002		Salicornia ramosissima J. Woods.
2257	008	S	Salicornia veneta Pignatti et Lausi
1873	040		Salix atrocinerea Brot.
1873	039		Salix cinerea L.
1873	007		Salix herbacea L.
1873	049		Salix repens L.
2269	008		Salsola kali L.
2269	005		Salsola soda L.
2269	022		Salsola vermiculata L.
7290	031		Salvia austriaca Jacq.
7290	025		Salvia nemorosa L.
7290	021		Salvia pratensis L.
8515	002		Sambucus nigra L.
6328	001		Samolus valerandi L.
3381	001		Sanguisorba officinalis L.
3189	027		Saxifraga tridactylites L.
8546	014		Scabiosa atropurpurea L.
8546	023		Scabiosa maritima L.
8546	040		Scabiosa ochroleuca L.
9810	001		Scapania irrigua (Nees)Dum.
0376	001		Schismus calycinus Koch
0478	603		Schoenoplectus lacustris (L.)Palla
0468	606		Schoenoplectus tabernaemontani (Gmel.)Palla
0477	001		Schoenus nigricans L.
1086	002		Scilla autumnalis L.
0468	001		Scirpus acicularis L.
0468	027	S	Scirpus americanus Pers.
0468	005		Scirpus australis L.
0468	008		Scirpus cernuus Vahl.
0468	011		Scirpus holoschoenus L.
0468	029		Scirpus holoschoenus L. var. romanus L.
0468	013		Scirpus lacustris L. ssp. lacustris
0468	036		Scirpus lacustris L. ssp. glaucus (Sm.)Hartm.
0468	014		Scirpus litoralis Schrader
0468	016		Scirpus maritimus L.
0468	021		Scirpus palustris L.
0468	001		Scirpus planifolius Grimm
0468	027		Scirpus pungens Vahl.
0468	030		Scirpus rufus Schrad.
0468	031		Scirpus savii Seb. et Mauri
0468	036		Scirpus tabernaemontani C.C. Gmelin
0468	038		Scirpus uniglumis Lam.
2483	002		Scleranthus annuus L.
0388	001		Scleropoa rigida (L.)Griseb.
9581	001		Scorzonera cana (Meyer) O. Hoffm.
9581	002		Scorzonera laciniata L.
9581	010		Scorzonera parviflora Jacq.
9581	777		Scorzonera pumila
7234	012		Scutellaria hastifolia L.
3161	014		Sedum acre L.
3161	016	S	Sedum boloniense Loisel.
3161	901		Sedum rosea (L.)Scop.
3161	016		Sedum sexangulare L.
6070	001		Selinum carvifolia (L.)L.
9411	045		Senecio aquaticus Hill.
9411	024		Senecio auricula Bourgeau
9411	040		Senecio congestus (R.Br.)D.C.
9411	019		Senecio doria L.
9411	080		Senecio erraticus Bertol.
9411	040	S	Senecio palustris (L.)Hook.
9411	040	S	Senecio tubicaulis Mansfeld
9411	064		Senecio viscosus L.
9411	065		Senecio vulgaris L.
9474	001		Serratula tinctoria L.
0171	004		Setaria viridis (L.)Beauv.
0280	001		Sieglingia decumbens (L.)Berhn.
6065	001		Silaus pratensis Besser
2490	105		Silene alba (Miller) E.H.L. Krause
2490	164		Silene conica L.
2490	056	S	Silene cucubalus Wibel
2490	056	S	Silene maritima With.
2490	027		Silene multiflora (W. et K.)Pers.
2490	044		Silene otites (L.)Wibel
2490	028		Silene viscosa (L.) Pers.
2490	056		Silene vulgaris (Moench)Garcke
2490	056		Silene vulgaris (Moench)Garcke ssp. angustifolia (Miller)Hayek
2490	056		Silene vulgaris (Moench)Garcke ssp. maritima (With.) A. et D. Ove
2917	018		Sisymbrium officinale (L.)Scop.
2997	001	S	Sisymbrium sophia L.
6038	901	S	Sium erectum Huds.
6038	001		Sium latifolium L.
7407	007		Solanum dulcamara L.
7407	001		Solanum nigrum L.
9595	007		Sonchus arvensis L.
9595	001		Sonchus asper (L.)Hill var. pungens Bisch.
9595	004	S	Sonchus aquatilis Pourret
9595	006		Sonchus crassifolius Pourret
9595	004		Sonchus maritimus L.
9595	004		Sonchus maritimus L. ssp. aquatilis (Pourret)Nyman
9595	003		Sonchus oleraceus L.
9595	005		Socnhus palustris L.
0054	003		Sparganium erectum L.
0054	003	S	Sparganium ramosum Huds.
0283	001		Spartina alterniflora Loisel.
0283	003	S	Spartina maritima Fern.
0283	005		Spartina patens Muhl.
0283	003		Spartina stricta Roth.
0283	004		Spartina townsendii Groves
2450	006	S	Spergula marginata (Dc.)Murb.
2450	007	S	Spergula salina (J. et C. Presl.) Dietr.
2450	010		Spergularia diandra (Guss.)Bioss.
2450	011	S	Spergularia longipes Rouy
2450	006	S	Spergularia marginata (Dc.)Kittel
2450	007		Spergularia marina (L.)Griseb.
2450	006		Spergularia media (L.)Presl.
2450	011		Spergularia purpurea (Pers.)Don Fil.
2450	012		Spergularia rubra (L.) Presl.
2450	003		Spergularia rupicola Lebel
2450	007	S	Spergularia salina Presl.
0384	001		Sphenopus divaricatus (Gouan)Rchb.
0230	002		Sporobolus arenarius (Gouan)Duv.-J.
6350	017	S	Statice armeria L.
6351	018	S	Statice bahustensis Fr.
6351	019	S	Statice bellidifolia (Gouan)Dc.
6351	016	S	Statice bubanii Gird
6351	016	S	Statice dichotoma Mut.
6351	067	S	Statice dodartii Girard
6351	005	S	Statice ferulacea L.
6351	064	S	Statice girardiana Guss.
6351	013	S	Statice gmelinii Willd.
6351	016	S	Statice lanceolata Rchb.
6351	017	S	Statice limonium L.
6351	017	S	Statice limonium L. ssp. serotina (Rchb.)Nyman
6351	072	S	Statice lychnidifolia Girard
6351	047	S	Statice minutiflora Guss.
6351	026	S	Statice occidentalis Lloyd
6351	018	S	Statice rariflora Dreyer
6351	017	S	Statice serotina Rchb.
6351	026	S	Statice spathulata Begingt.
6351	056	S	Statice virgata Willd.
2429	016		Stellaria crassifolia Ehrh.
2429	010		Stellaria graminea L.
2429	017		Stellaria humifusa Rottb.
2429	003		Stellaria media (L.)Vill.
2261	010		Suaeda altissima (L.)Pallas
2261	001	S	Suaeda fruticosa Auct.
2261	002	S	Suaeda fruticosa Auct. var. brevifolia Moq.
2261	007		Suaeda maritima (L.)Dumort
2261	007	S	Suaeda maritima L. var. procumbens
2261	015		Suaeda prostata Pallas
2261	002		Suaeda pruinosa Lange
2261	009		Suaeda splendens (Pourr.)Gren. et Godron
2261	001		Suaeda vera J.F. Gmelin
2881	001		Subularia aquatica L.
8542	001		Succisa pratensis Moench.
9830	351		Synthrichia ruralis Bridel
5239	001		Tamarix africana Poiret
5239	003		Tamarix gallica L.
5239	012		Tamarix laxa Willd.
5239	012	S	Tamarix pallasii Desv.
5239	008		Tamarix ramosissima Ledeb.
9592	002		Taraxacum bessarabicum (Hornem.)Land.-Mazz.
9592	070	S	Taraxacum lissocarpum (Dahlst.)Dahlst.
9592	070		Taraxacum lividum (WetK.)Peterm.
9592	068		Taraxacum palustre (Lyons)Simons
9592	069		Taraxacum suecicum Hagl.
3699	001		Tetragonolobus maritimus (L.)Roth.
3699	001	S	Tetragonolobus siliquosus (L.)Roth
7212	015		Teucrium scordium L.
2548	013		Thalictrum flavum L.
2548	013	S	Thalictrum glaucum Desf.
2548	009		Thalictrum minus L.
2548	010		Thalictrum simplex L.
2118	007		Thesium arvense Horvatovszky
2118	007	S	Thesium ramosum Hayne
9574	025	S	Thrincia hirta Roth
9574	025	S	Thrincia hirta Roth var. arenaria
9830	790		Thuidium abietinum Br.Eur.
5453	003		Thymelaea hirsuta (L.)Endl.
5453	017		Thymelaea passerina (L.)Cosson et Germ.
5945	003		Torilis japonica (Houtt)Dc.
9830	214		Tortella turtuosa (L.)Limpr.
9830	351		Tortula ruraliformis Dixon
0143	001		Tragus racemosus (L.)All.
3690	019		Trifolium angulatum W. et K.
3690	049		Trifolium arvense L.
3690	044		Trifolium campestre Schreber
3690	046		Trifolium dubium Sibth.
3690	029		Trifolium fragiferum L.
3690	014		Trifolium hybridum L.
3690	016		Trifolium isthmocarpum Brot.
3690	093	S	Trifolium maritimum Huds.
3690	047		Trifolium micranthum Viv.
3690	046	S	Trifolium minus Sm.
3690	063		Trifolium pratense L.
3690	010		Trifolium repens L.
3690	030		Trifolium resupinatum L.
3690	048		Trifolium striatum L.
3690	093		Trifolium squamosum L.
3690	031		Trifolium tomentosum L.
0066	001	S	Triglochin barrelieri Loisel.
0066	001		Triglochin bulbosum L.
0066	003		Triglochin maritimum L.

0066 004   Triglochin palustre L.
3687 017   Trigonella monspeliaca L.
0271 005   Trisetum flavescens (L.)P.B.
0271 017   Trisetum loeflingianum P.B.
0271 010   Trisetum pumilum Kunth
0405 008 S Triticum junceum L.
0405 010 S Triticum litorale Host
0405 010 S Triticum maritimum L.
9380 001   Tussilago farfara L.
0049 007   Typha angustata Bor.
0049 001   Typha angustifolia L.
0049 002   Typha latifolia L.

3681 001   Ulex europaeus L.
1974 002   Urtica dioica L.

8532 001   Valeriana officinalis L.
8532 001 S Valeriana salina Pleije!
8532 999   Valeriana spec.
8529 016   Valerianella dentata (L.)Pollich
8529 010   Valerianella locusta (L.)Laterrade
2936 001 S Vella annua L.
7460 082   Verbascum lychnitis L.
7460 020   Verbascum phoeniceum L.
7138 003   Verbena officinalis L.
7138 004   Verbena supina L.
7579 046   Veronica arvensis L.
7579 030   Veronica chamaedrys L.
7579 062 S Veronica orchidea Crantz
7579 042   Veronica praecox All.
7579 017   Veronica prostrata L.
7579 035   Veronica scutellata L.
7579 003   Veronica serpyllifolia L.
7579 062   Veronica spicata L.
7579 043   Veronica triphyllos L.
7579 047   Veronica verna L.
3852 010   Vicia cracca L.
3852 031   Vicia hirsuta (L.)S.F. Gray
3852 047   Vicia lathyroides L.
3852 051   Vicia lutea L.
3852 036   Vicia tetrasperma (L.)Schreber
5274 020   Viola canina L.
5274 027   Viola palustris L.
0385 004 S Vulpia fasciculata (Forsk.)Fritsch
0385 004 S Vulpia longiseta (Brot.)Hackel
0385 904   Vulpia membranacea (L.)Link

9148 004 S Xanthium italicum Moretti
9148 002   Xanthium spinosum L.
9148 001   Xanthium strumarium L.

0062 001   Zannichellia palustris L
0055 001   Zostera marina L.
0055 003   Zostera nana Roth.
0055 002   Zostera noltii Hornem.
3965 004   Zygophyllum album L.Fil.

REFERENCES TO ORIGINAL PUBLICATIONS IN VEGETATIO

E. van der Maarel et al.	1976. Vegetatio 32: 65–72
L. Orlóci	1976. Vegetatio 32: 117–120
J.L. Guillerm et al.	1976. Vegetatio 33: 1–11
S. Pignatti	1976. Vegetatio 33: 23–32
W.M. Kortekaas et al.	1976. Vegetatio 33: 51–60
F. Romane et al.	1977. Vegetatio 33: 99–106
E. Feoli	1977. Vegetatio 33: 119–125
E. Feoli	1977. Vegetatio 33: 147–152
M. Rejmánek	1977. Vegetatio 35: 55–61
R. Neuhäusl	1977. Vegetatio 35: 115–122
J. Moravec	1978. Vegetatio 37: 33–42
L. Orlóci	1978. Vegetatio 37: 123–125
J.J. Moore et al.	1978. Vegetatio 38: 89–93
E. van der Maarel et al.	1978. Vegetatio 38: 143–156
W. Holzner et al.	1978. Vegetatio 38: 157–164
E. van der Maarel	1979. Vegetatio 39: 97–114
E. Feoli et al.	1979. Vegetatio 39: 123–125
D. Lausi et al.	1979. Vegetatio 39: 171–184
E. Feoli et al.	1979. Vegetatio 40: 49–54
E. Feoli et al.	1979. Vegetatio 40: 55–59
J.M.W. Louppen et al.	1979. Vegetatio 40: 107–114
P. David et al.	1979. Vegetatio 40: 117–123

GPSR Compliance

The European Union's (EU) General Product Safety Regulation (GPSR) is a set of rules that requires consumer products to be safe and our obligations to ensure this.

If you have any concerns about our products, you can contact us on ProductSafety@springernature.com

In case Publisher is established outside the EU, the EU authorized representative is:

Springer Nature Customer Service Center GmbH
Europaplatz 3
69115 Heidelberg, Germany

Batch number: 10593309

Printed by Printforce, the Netherlands